Discrete and
Computational Geometry:
Papers from
the DIMACS Special Year

DIMACS
Series in Discrete Mathematics and Theoretical Computer Science

Volume 6

Discrete and Computational Geometry:
Papers from the DIMACS Special Year

Jacob E. Goodman
Richard Pollack
William Steiger
Editors

NSF Science and Technology Center
in Discrete Mathematics and Theoretical Computer Science
A consortium of Rutgers University, Princeton University,
AT&T Bell Labs, Bellcore

American Mathematical Society
Association for Computing Machinery

This DIMACS volume is a collection of invited research and survey papers growing out of the Special Year in Discrete and Computational Geometry, which took place at DIMACS from September 1989 through August 1990.

1991 *Mathematics Subject Classification.* Primary 03, 05, 12, 13, 14, 15, 32, 51, 52, 57, 68.

Library of Congress Cataloging-in-Publication Data

Discrete and computational geometry: papers from the DIMACS special year/Jacob E. Goodman, Richard Pollack, William Steiger, editors.
    p. cm.—(DIMACS series in discrete mathematics and theoretical computer science; ISSN 1052-1798; v. 6)
    Includes bibliographical references.
    AMS: ISBN 0-8218-6595-1 (acid-free paper)
    ACM: ISBN 0-89791-464-3
    1. Geometry–Data processing–Congresses. I. Goodman, Jacob E. II. Pollack, Richard D. III. Steiger, William L., 1939– . IV. DIMACS (Group). V. American Mathematical Society. VI. Association for Computing Machinery. VII. Series.
QA448.D38D57   1991   91–36806
516′.00285–dc20   CIP

To order through AMS contact the AMS Customer Services Department, P.O. Box 6248, Providence, Rhode Island 02940-6248 USA. For VISA or MASTERCARD orders call 1-800-321-4AMS. Order code DIMACS/6.

To order through ACM contact ACM Order Department, P.O. Box 64145, Baltimore, Maryland 21264. Phone 301-528-4261. Order number 222915.

10 9 8 7 6 5 4 3 2 1   95 94 93 92 91

Foreword

This DIMACS volume is a collection of invited research and survey papers growing out of the Special Year in Discrete and Computational Geometry, which took place at DIMACS from September 1989 through August 1990. We especially thank the members of the Special Year Organizing Committee and the organizers of the workshops, Bernard Chazelle (Chair), Louis J. Billera, Kenneth Clarkson, Jacob E. Goodman, Richard Pollack, Peter Shor, William Steiger, Bernd Sturmfels, and Subhash Suri, for planning the various activities of the year and in particular for organizing the workshops that brought together many outstanding speakers and visitors representing the many different communities that contribute to the field.

Daniel Gorenstein, Director
Robert Tarjan, Co-Director
Fred S. Roberts, Associate Director

Preface

The Center for Discrete Mathematics and Theoretical Computer Science (DIMACS) was established by the National Science Foundation early in 1989. During 1989–1990 more than 200 scientists visited DIMACS to participate in the Special Year in Discrete and Computational Geometry, the first DIMACS "special year." This activity brought together both long-term and short-term visitors whose interests represent the many different communities that contribute to the field. The present volume is a reflection of some of the work that took place during the year.

The year was highlighted by six workshops that defined the focus for much of activity of the Special Year. The topics were: Geometric Complexity (October 16–20, 1989), Probabilistic Methods in Discrete and Computational Geometry (November 27–December 1, 1989), Polytopes and Convex Sets (January 8–12, 1990), Arrangements and Their Realizations (March 19–23, 1990), Practical Issues in Geometric Computation (April 16–20, 1990), and Algebraic Issues in Geometric Computation (May 21–25, 1990).

The volume at hand consists of a number of refereed papers invited from participants in the Special Year. Some were presented at the workshops, including several survey talks, while others were written after their authors' DIMACS visits. All relate in one way or another to the main geometric themes that occupied the attention of the participants during the year. The diversity of the twenty-three papers appearing in this volume attests to the fact that geometry continues to be both a vital source of ideas in theoretical computer science and discrete mathematics as well as a fertile arena in which the two disciplines stimulate each other.

Jacob E. Goodman
Richard Pollack
William Steiger

Contents

ix

DIMACS Series in Discrete Mathematics
and Theoretical Computer Science
Volume **6**, 1991

Geometric Partitioning and its Applications

PANKAJ K. AGARWAL

ABSTRACT. In this survey paper we review some results related to *geometric partitioning*, i.e., given a set of objects in $\mathbb{R}^d$, partition the space into a few regions so that each region intersects a small number of objects. We first describe the known bounds on the size of such a partitioning and present some of the algorithms for computing a geometric partitioning. We then discuss several applications of geometric partitioning.

1. Introduction

Divide-and-conquer is one of the most commonly used paradigm in designing efficient algorithms. The general idea of this approach is to decompose the problem into two or more subproblems, solve each of them recursively, and then combine the results of the subproblems to obtain the overall result. The efficiency of the algorithm depends on the size of each subproblem, the number of subproblems, and the complexity of the divide and merge steps. In the standard divide-and-conquer approach, the objects are partitioned into subsets of roughly equal size either arbitrarily or by using some very simple criterion, which ensures that the size of each subproblem is small and that the divide step is straightforward. In this case, the efficiency of the overall algorithm is mainly governed by the complexity of the merge step. Such an approach has been successfully applied to obtain fast algorithms for computing convex hull, Voronoi diagram, closest-pair, etc. See [51, 95, 86]. But this approach fails to give efficient algorithms for many other geometric problems.

An alternative divide-and-conquer approach to solve a geometric problem involving a set of objects, Γ, in $\mathbb{R}^d$ is to decompose the space into a few regions, each of which intersects a small number of objects. Generally, one would like each region τ to be of simple shape, so that one can easily compute Γ_τ, the set of objects that intersect τ. Each region now induces a subproblem

1991 *Mathematics Subject Classification.* Primary 52B30, 68P05, 68Q20, 68Q25, 68V05.

Work on this paper has been supported by National Science Foundation Grant CCR-91-06514.

1

involving the objects of Γ_τ, which can be solved either recursively or using some direct method. Finally, the results of the subproblems are combined to obtain the overall result. The advantage of this scheme is that the merge step becomes considerably simpler. To illustrate this point, consider the following problem: *Given a set Γ of n segments in $\mathbb{R}^2$, count the number of intersection points between the segments of Γ.* If we followed the standard divide-and-conquer approach, we would partition the set Γ into two subsets Γ_1, Γ_2 of size $\lceil n/2 \rceil$ and $\lfloor n/2 \rfloor$, respectively, and would count the number of intersections within Γ_1 and Γ_2. The merge step would then consist of counting the number of intersection points between the segments of Γ_1 and Γ_2. But this subproblem does not seem to be any easier than the original problem. Therefore, it is not clear how an efficient algorithm can be obtained following this strategy. On the other hand, if we partition the plane into few regions, each of which intersects a small number of segments, and count the number of intersection points within each region, then the merge step becomes trivial—add the intersection points computed by each subproblem.

Of course, the divide step in the second approach is not trivial. It is not obvious how to obtain a "good partition" of the space, i.e., partition the space into a few regions so that each region intersects a small number of objects. To make the notion of "good partition" more precise, we need to introduce a few definitions.

DEFINITION 1.1. A region $\tau \subseteq \mathbb{R}^d$ is called a *cell* if it is a real semialgebraic[1] (closed) set and is homeomorphic to a d-dimensional disk (see [35] for a more formal definition). Given a set Γ of n objects in $\mathbb{R}^d$ and a parameter $r \le n$, a set of (possibly unbounded) cells Ξ with pairwise disjoint interiors is called a $\frac{1}{r}$-*cutting* if Ξ covers $\mathbb{R}^d$ and the interior of each cell of Ξ intersects at most n/r objects of Γ. For a set of hyperplanes, we will assume Ξ to be a collection of simplices. The size of Ξ, denoted $|\Xi|$, is the number of cells in Ξ.

In order to obtain a fast divide-and-conquer algorithm, using the second approach, one needs a fast algorithm for computing a $\frac{1}{r}$-cutting of small size. This raises several very interesting questions—What is the smallest size of a $\frac{1}{r}$-cutting for a given set of objects? How fast can one compute a $\frac{1}{r}$-cutting? What are the problems that can be solved using geometric partitioning? In the last few years, these questions have received considerable attention. The aim of this survey paper is to summarize the known results in this area. The paper is organized as follows.

(1) The first part of the paper discusses the known upper bounds on the size of a smallest $\frac{1}{r}$-cutting (§2) and mentions some of the known algorithms to compute $\frac{1}{r}$-cuttings of small size §3.

---

[1]A subset of $\mathbb{R}^d$ is called *real semialgebraic* if it is obtained as a union, intersection, or complementation of real algebraic varieties.

(2) The second part of the paper is devoted to applications of geometric partitioning. We consider three types of problems:

(i) *Combinatorial problems*: Geometric partitioning has been applied to obtain optimal or almost optimal upper bounds for several combinatorial problems. Some of these problems are discussed in §4.

(ii) *Algorithmic problems*: Geometric partitioning has also been used to obtain efficient divide-and-conquer algorithms for a variety of geometric problems; we sketch some of these algorithms in §5.

(iii) *Data structures*: Finally, we turn our attention to some of the data structures based on geometric partitioning §6.

2. Bounds on cuttings and randomized constructions

In this section we discuss the bounds on the minimal size of $\frac{1}{r}$-cuttings for a given set of objects. Most of the time, unless otherwise mentioned, we will assume the objects to be hyperplanes in $\mathbb{R}^d$.

Before giving an upper bound on the size of $\frac{1}{r}$-cuttings, let us first try to obtain a lower bound on its size. Let Γ be a collection of n hyperplanes in $\mathbb{R}^d$ in general position. Let $\mathscr{A}(\Gamma)$ denote the arrangement[2] of Γ, and let Ξ be a $\frac{1}{r}$-cutting for Γ. For the sake of simplicity, assume that no vertex of $\mathscr{A}(\Gamma)$ lies on the boundary of a simplex of Ξ. Since Ξ is a $\frac{1}{r}$-cutting, every cell of Ξ intersects at most n/r hyperplanes of Γ, which implies that τ contains at most $\binom{n/r}{d}$ vertices of $\mathscr{A}(\Gamma)$. On the other hand, $\mathscr{A}(\Gamma)$ has $\binom{n}{d}$ vertices, and since the simplices of Ξ cover the entire space, we get

$$\sum_{\tau \in \Xi} \binom{n/r}{d} \geq \binom{n}{d} \quad \text{or} \quad |\Xi| \geq \frac{\binom{n}{d}}{\binom{n/r}{d}} = \Omega(r^d).$$

A similar lower bound can be attained for other kinds of objects, provided that the number of vertices in their arrangement is $\Omega(n^d)$.

We will now show that there exists a $\frac{1}{r}$-cutting of size $O(r^d)$ for a set of hyperplanes in d dimensions. An upper bound of $O(r^d \log^d r)$ on the size of the smallest cutting follows immediately from a fairly general probabilistic argument of Haussler and Welzl [69], and of Clarkson and Shor [48]. In order to state the results of [69], we need to define some terms.

DEFINITION 2.1. A *range space* is a pair $\Sigma = (X, \mathscr{R})$, where X is a set of *objects*, and $\mathscr{R} \subseteq 2^X$ is a set of *ranges*. For a subset $Y \subseteq X$, the *subspace*

---

[2]The *arrangement* of Γ is the decomposition of the space into connected pieces of various dimensions induced by the hyperplanes of Γ. A connected piece of dimension $k \leq d$ is called a k-face of $\mathscr{A}(\Gamma)$; the zero-dimensional faces are called *vertices*, the one-dimensional faces are called *edges*, and the $(d-1)$-dimensional faces are called *facets*. See the book of Edelsbrunner [51] for details on arrangements.

Σ_Y is defined as

$$\Sigma_Y = (Y, \{R \cap Y \mid R \in \mathscr{R}\}).$$

A subset $A \subset X$ is *shattered* (with respect to $\mathscr{R}$) if

$$\{A \cap R \mid R \in \mathscr{R}\} = 2^A.$$

The *Vapnik-Chervonenkis dimension* of Σ, denoted as VC-dim(Σ), is defined as the cardinality of the maximum size shattered subset of X, i.e., VC-dim$(\Sigma) = d$ if there is no subset of X of size $d + 1$ that can be shattered by $\mathscr{R}$. Finally, a subset $S \subset X$ is called an *ε-net* ($0 < \varepsilon < 1$) for Σ if $S \cap R \neq \emptyset$ for every range $R \in \mathscr{R}$ with $|R| \geq \varepsilon|X|$ (i.e., S intersects every range that contains more than $\varepsilon|X|$ items of X).

The notion of ε-net was introduced by Haussler-Welzl [69] and, using a technique similar to that of Vapnik and Chervonenkis [106], they proved

THEOREM 2.2 (Haussler-Welzl [69]). *A range space of VC-dimension d admits an ε-net of size at most $\frac{8d}{\varepsilon} \log \frac{8d}{\varepsilon}$.*

This bound has been improved to $(1 + o(1))\frac{d}{\varepsilon} \log \frac{d}{\varepsilon}$ by Komlós et al. [70] (see also [26]). A result of Pach and Woeginger [93] shows that, for $d \geq 2$, the above bound is optimal up to within a constant factor.

An immediate consequence of the above theorem is

COROLLARY 2.3. *A set Γ of n hyperplanes in $\mathbb{R}^d$ admits a $\frac{1}{r}$-cutting of size $O(r^d \log^d r)$.*

Sketch of Proof. Given an open segment $s \subseteq \mathbb{R}^d$, let R_s denote the set of hyperplanes of X intersecting s. Set

$$\mathscr{R} = \{R_s \mid \text{for all open segments } s \subset \mathbb{R}^d\}.$$

Consider the range space $\overline{\Sigma} = (\Gamma, \mathscr{R})$. It is an easy exercise to show that VC-dim$(\overline{\Sigma})$ is finite. Therefore, by Theorem 2.2, it admits a $\frac{1}{dr}$-net H of size $O(r \log r)$ (for any fixed d). Triangulate the arrangement of H, $\mathscr{A}(H)$, and let Ξ be the resulting set of simplices. Any open segment lying in the interior of a simplex of Ξ does not intersect any hyperplane of H, therefore it intersects at most $\frac{n}{dr}$ hyperplanes of Γ. For a simplex $\triangle \in \Xi$, let $e_1, \dots, e_d$ be the edges incident to one of its vertices. A hyperplane intersecting the interior of $\triangle$ has to intersect one of e_i ($i \leq d$), therefore the interior of τ intersects at most $d\frac{n}{dr} = \frac{n}{r}$ hyperplanes of Γ.

As for the triangulation of Ξ, the following recursive procedure (called *bottom-vertex triangulation* or *bv-triangulation* for short) triangulates $\mathscr{A}(H)$ into $O(r^d \log^d r)$ simplices [45, 6]: Suppose we have already triangulated all faces of $\mathscr{A}(H)$ of dimension less than k. Let f be a k-face of $\mathscr{A}(H)$ and let v be the lowest vertex of f (in the x_d direction). We triangulate f by connecting v to every $(k-1)$-simplex of f's $(k-1)$-faces. Note that $(k-1)$-faces of f have already been triangulated.

Hence, Ξ is a $\frac{1}{r}$-cutting of size $O(r^d \log^d r)$ for Γ. $\square$

Following a different approach, Clarkson-Shor [48] proved a somewhat stronger result. They showed that if we randomly choose r hyperplanes of Γ and triangulate their arrangement, then for the triangulated arrangement $\mathscr{A}^*(H)$

$$
E\left[\sum_{\tau \in \mathscr{A}^*(H)} n_\tau\right] = O(nr^{d-1}) \,,
$$

where n_τ is the number of hyperplanes of Γ intersecting the interior of the simplex τ. This result shows that $\mathbb{R}^d$ can be decomposed into $O(r^d)$ simplices so that, on average, every simplex intersects only n/r hyperplanes of Γ. But their argument fails to prove that every simplex intersects n/r hyperplanes.

Chazelle and Friedman [36] refined the above arguments and showed that there exists a $\frac{1}{r}$-cutting of size $O(r^d)$ for hyperplanes in $\mathbb{R}^d$. Their idea is as follows:

(1) Pick a random subset H of Γ of size r, triangulate the arrangement $\mathscr{A}(H)$, and, for each simplex τ of the triangulated arrangement, compute Γ_τ, the set of hyperplanes intersecting its interior.

(2) For every simplex τ intersecting more than n/r hyperplanes, do the following:

   (i) Compute a $\frac{1}{t_\tau}$-cutting Ξ_τ of size $O(t_\tau^d \log^d t_\tau)$, where $t_\tau = |\Gamma_\tau| \frac{r}{n}$. Corollary 2.3 guarantees the existence of such a cutting.

   (ii) If a simplex $\triangle$ of Ξ_τ intersects τ, triangulate $\triangle \cap \tau$ into $O(1)$ simplices.

The resulting set of simplices is obviously a $\frac{1}{r}$-cutting for Γ. Chazelle-Friedman showed that the expected number of simplices in the above construction is $O(r^d)$. Roughly speaking, they prove that the expected number of simplices in $\mathscr{A}^*(H)$, for which $t_\tau > t$, decreases exponentially with t. We thus have

THEOREM 2.4 (Chazelle-Friedman [36]). *Given a collection Γ of n hyperplanes in $\mathbb{R}^d$ and a parameter $r \le n$, there exists a $\frac{1}{r}$-cutting of size $O(r^d)$ for Γ.*

In some applications (e.g., see [39, 77]), one needs to compute a *weighted* $\frac{1}{r}$-*cutting*. Let $w : \Gamma \to \mathbb{R}^+$ be a weight function on Γ. For a subset $H \subseteq \Gamma$, let $w(H) = \sum_{h \in H} w(h)$. Ξ is called a *weighted* $\frac{1}{r}$-*cutting* for (Γ, w) if $w(\Gamma_\tau) \le \frac{w(\Gamma)}{r}$ for every simplex $\tau \in \Xi$, where Γ_τ is as usual the set of hyperplanes intersecting the interior of τ. One can easily extend Theorem 2.4 to show that

COROLLARY 2.5 (Matoušek [77]). *Given a set Γ of n hyperplanes and a weight function w, there exists a weighted $\frac{1}{r}$-cutting of size $O(r^d)$ for (Γ, w).*

It is easily seen that Corollary 2.3 and Theorem 2.4 can be converted into randomized algorithms for constructing a $\frac{1}{r}$-cutting. For example, in order to construct a $\frac{1}{r}$-cutting of size $O(r^d \log^d r)$, we choose a random subset of Γ of size $cr \log r$ (for some suitable constant c), triangulate their arrangement as described in Corollary 2.3, and for each simplex count the number of hyperplanes intersecting its interior. If any simplex intersects more than $\frac{n}{r}$ hyperplanes, we restart. By being somewhat careful, the number of hyperplanes intersecting the interior of each simplex can be counted in time $O(nr^{d-1})$, which gives an $O(nr^{d-1})$ randomized expected time algorithm for computing a $\frac{1}{r}$-cutting of size $O(r^d \log^d r)$. Similarly, one can convert Theorem 2.4 into an $O(nr^{d-1})$ randomized expected time algorithm for computing a $\frac{1}{r}$-cutting of size $O(r^d)$. Hence, we have

THEOREM 2.6 (Chazelle-Friedman [36]). *Given a set Γ of n hyperplanes in $\mathbb{R}^d$ and a parameter $r < n$, a $\frac{1}{r}$-cutting of size $O(r^d)$ for Γ can be computed in $O(nr^{d-1})$ randomized expected time.*

So far we have assumed that objects were hyperplanes. If Γ is a collection of nonlinear objects, say, a set of real algebraic varieties $\bigvee f_i \equiv \{f_i = 0\}$, where f_i is a polynomial of fixed maximum degree, then the situation is quite different.

Let c be a constant size d-dimensional *cell*. For a cell c, let $R_c \subseteq \Gamma$ be the set of varieties intersecting the interior of c; then one can show that the range space

$$\widetilde{\Sigma} = (\Gamma, \{R_c \mid c \text{ is a constant size cell}\})$$

has finite VC-dimension. Therefore, one can still apply Theorem 2.2 to prove that $\widetilde{\Sigma}$ has a $\frac{1}{r}$-net H of size $O(r \log r)$. In two dimensions, one can decompose $\mathscr{A}(H)$ into $O(r^2 \log^2 r)$ cells (see §3 for details), but it is not obvious how to obtain such a decomposition in higher dimensions. The currently best known result partitions $\mathscr{A}(H)$ into $O((r \log r)^{2d-3} 2^{\alpha^c(r)})$ cells [35], where $\alpha(r)$ is the inverse Ackermann function and c is some constant.

THEOREM 2.7 (Chazelle et al. [35]). *Given a set Γ of real algebraic varieties of fixed maximum degree in $\mathbb{R}^d$ and a parameter r, there exists a $\frac{1}{r}$-cutting of size $O((r \log r)^{2d-3} 2^{\alpha^e(r)})$ for Γ.*

It is an open question whether there exists a $\frac{1}{r}$-cutting of size $O(r^d)$ (or close to this bound) for a set of real algebraic varieties in $\mathbb{R}^d$.

3. Deterministic algorithms

In the previous section, we presented randomized algorithms for constructing an optimal size $\frac{1}{r}$-cutting for hyperplanes. But often randomized algorithms are not satisfactory, and one wants deterministic algorithms. Chazelle and Friedman [36] showed that Theorem 2.4 can be converted into a deterministic algorithm using a weighted version of Lovász's greedy hypergraph covering algorithm [73] and the general method of derandomizing probabilistic algorithms, called the method of *conditional probabilities* [96, 102]. The method of conditional probabilities converts probabilistic proofs of existence of some objects into polynomial algorithms constructing such objects, provided that certain probabilities can be computed efficiently. But the disadvantage of this method is that the algorithms have a large exponent in the complexity bound. The time complexity of the deterministic algorithm proposed by Chazelle-Friedman has running time of $O(n^{d(d+3)/2+1}r)$, which is impractical in most cases. For example, in $3d$, the running time of the Chazelle-Friedman algorithm is $O(n^{10}r)$ while Theorem 2.6 gives a randomized algorithm with $O(nr^2)$ expected running time.

Berger and Rompell [24, 25] succeeded in parallelizing the method of conditional probabilities in certain cases, which in turn gives an NC algorithm[3] for computing a $\frac{1}{r}$-cutting of size $O(r^d)$ for a set of n hyperplanes in $\mathbb{R}^d$ (see also [87]).

If one wants to compute the set of hyperplanes intersected by each simplex of the cutting, then the lower bound argument described in the beginning of the previous section gives an $\Omega(nr^{d-1})$ lower bound on the time complexity of the cutting algorithms [2]. This raises the question of whether a $\frac{1}{r}$-cutting of size $O(r^d)$ can be computed by a deterministic algorithm, whose running time is close to the lower bound.

It turns out that computing a $\frac{1}{r}$-cutting in two dimensions is significantly easier than in higher dimensions, therefore we will first consider the two-dimensional case, and then study some of the higher-dimensional algorithms.

3.1. Computing a cutting in the plane. For $d = 2$, the following very simple algorithm computes a $\frac{1}{r}$-cutting of size $O(r^2)$ in $O(n^2)$ time for a set of lines in the plane. But we first need to mention some simple observations related to arrangements of lines in the plane.

DEFINITION 3.1. Let Γ be a set of n lines in the plane in general position, and let p be an arbitrary point. The *level* of p is defined to be the number of lines in Γ that lie above p, not including the line(s) passing through p itself. The *k-level* ($0 \le k \le n$) of $\mathscr{A}(\Gamma)$ is the closure of the union of all edges whose level is k. The k-level of $\mathscr{A}(\Gamma)$ is an x-monotone polygonal chain starting and ending with unbounded rays. The number of edges in a

---

[3]An algorithm is in NC if the running time is $O(\log^{O(1)} n)$ using a polynomial number of processors.

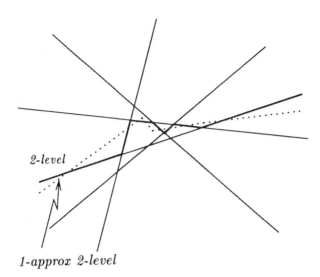

2-level

1-approx 2-level

FIGURE 1. Levels and approximate levels in arrangement of lines.

level λ, denoted $|\lambda|$, is called the *size* of λ. An x-monotone polygonal chain is called the *ε-approximate k-level* if it lies in the strip bounded by the $k - \varepsilon$ and $k + \varepsilon$ levels of $\mathscr{A}(\Gamma)$ (see Figure 1).

LEMMA 3.2. *Let* $\Lambda_1, \ldots, \Lambda_u$ *be disjoint collections of levels in an arrangement of* n *lines* $\mathscr{A}(\Gamma)$. *If each* Λ_i *contains at least* v *levels, then we can pick a level* $\lambda_i \in \Lambda_i$ $(1 \leq i \leq u)$ *such that*

$$\sum_{i=1}^{u} |\lambda_i| \leq \frac{n^2}{v}.$$

LEMMA 3.3 (Edelsbrunner-Welzl [58]). *Let* $q_0, q_1, \ldots, q_t$ *be the vertices of the k-level in* $\mathscr{A}(\Gamma)$, *and let* u *be a natural number. Let* $\Pi(k, u)$ *denote the polygonal path* $q_0 q_u q_{2u} \cdots q_{\lfloor t/u \rfloor u} q_t$ *augmented with the initial and final rays of the k-level. Then*

(i) *each edge of* $\Pi(k, u)$ *intersects fewer than* u *lines of* Γ.
(ii) $\Pi(k, u)$ *is a* $\lfloor u/2 \rfloor$*-approximate k-level of* $\mathscr{A}(\Gamma)$.

We can now describe the algorithm. It consists of the following three steps:

Step I: Construct the arrangement $\mathscr{A}(\Gamma)$ and divide the levels of $\mathscr{A}(\Gamma)$ into groups of size $m = cn/r$, where $c < 1$ is some suitable constant. That is, for every i $(1 \leq i \leq n/m)$, let Λ_i denote the collection of all k-levels with $(i - 1)m < k \leq im$. (For the sake of simplicity assume that r is divisible by cn.) Assume that the k_i-level is the shortest member of Λ_i. By Lemma 3.2, the total sum of the sizes of the k_i-levels $(1 \leq i \leq n/m)$ is at most n^2/m.

Step II: For every *even* i $(1 \leq i \leq n/m)$, "short-cut" the k_i-level to obtain an infinite polygonal path $\Pi(k_i, m)$, as described in Lemma 3.3, with

$u = \frac{n}{\gamma}$. It follows from Lemma 3.3(ii) that these paths are pairwise disjoint. Furthermore, the total number of vertices of $\Pi(k_i, m)$, for all even i, $1 \le i \le n/m$, is clearly at most $\frac{n^2}{m^2} + \frac{n}{m}$.

Step III: Draw vertical segments (or rays) in both directions through every vertex of $\Pi(k_i, m)$ until they hit $\Pi(k_{i-2}, m)$ or $\Pi(k_{i+2}, m)$ ($1 \le i \le n/m$, i is even). The resulting partition of the plane consists of $s = O(r^2)$ (generalized) trapezoids $\triangle_1, \ldots, \triangle_s$, each of which can be partitioned into two triangles.

By Lemma 3.3(i), every edge of $\Pi(k_i, m)$ intersects at most $m \le c\frac{n}{r}$ lines of Γ. On the other hand, by the construction and by Lemma 3.3(ii), any vertical edge of a trapezoid can intersect at most $O(m) = O(n/r)$ lines. Thus, every triangle intersects $O(n/r)$ lines of Γ. By choosing the value of c appropriately, we can guarantee that each triangle intersects at most n/r lines of Γ. Since $\mathscr{A}(\Gamma)$ can be constructed in $O(n^2)$ time, the overall running time of the algorithm is also $O(n^2)$.

Matoušek [74] gave a faster deterministic algorithm, whose running time was $O(nr^2 \log^2 r)$, by constructing $\lfloor cn/r \rfloor$-approximate levels directly without computing the entire arrangement. The running time was subsequently improved by Agarwal [2] to $O(nr \log n \log^{3.33} r)$, and then by Matoušek [77] and Chazelle [33] to $O(nr)$. A nice property of these algorithms is that, as a side product, they also compute the set of lines intersected by each triangle, which is required in most of the applications.

THEOREM 3.4 (Matoušek [77]; Chazelle [33]). *Given a collection Γ of n lines in $\mathbb{R}^2$ and a parameter r, a $\frac{1}{r}$-cutting of size $O(r^2)$ for Γ can be constructed deterministically in time $O(nr)$.*

Note that Theorem 3.4 is optimal if one wants to return the set of lines intersected by each triangle, but the lower bound does not hold if one is interested only in computing the cutting, not in computing the lines intersected by each triangle of the cutting. Matoušek has recently shown that a $\frac{1}{r}$-cutting ($r \le n^{1-\delta}$ for arbitrarily small but fixed $\delta > 0$) of size $O(r^2)$ for a set of n lines in the plane can be computed in time $O(n^{1/2} r^{3/2} + n \log r)$ [79].

3.2. Computing a cutting in higher dimensions. In higher dimensions, the first efficient algorithm for computing a $\frac{1}{r}$-cutting was proposed by Matoušek [76] and improved by him in [77], where he showed that given a set Γ of n hyperplanes in $\mathbb{R}^d$ and a parameter $r \le n^{1-\delta}$ ($\delta > 0$ is an arbitrarily small but fixed constant), a $\frac{1}{r}$-cutting of size $O(r^d)$ for Γ can be computed in time $O(nr^{d-1})$, which, as mentioned in the beginning of the section, is optimal if one explicitly wants to compute the hyperplanes intersected by each simplex of the cutting. Recently Chazelle settled the question for larger values of r by presenting an optimal algorithm for all values of r [33].

THEOREM 3.5 (Chazelle [33]). *Given a collection Γ of n hyperplanes in $\mathbb{R}^d$ and a parameter $r \leq n$, one can compute a $\frac{1}{r}$-cutting of size $O(r^d)$ for Γ in time $O(nr^{d-1})$.*

In order to sketch the proof of this theorem we need some auxiliary results.

DEFINITION 3.6. For a range space $\Sigma = (X, R)$ and a parameter r, a subset $A \subset X$ is called a $\frac{1}{r}$-*approximation* for Σ if

$$\left| \frac{|A \cap R|}{|A|} - \frac{|R|}{|X|} \right| \leq \frac{1}{r} \quad \text{for every } R \in \mathscr{R}.$$

The notion of $\frac{1}{r}$-approximation was originally introduced by Vapnik-Chervonenkis [106]. They proved that range spaces with VC-dimension d admit a $\frac{1}{r}$-approximation of size $O(r^2 \log r)$. The bound has been improved by Matoušek et al. [83] to $O(r^{2(1-1/d)} \log^{1+1/2d} r)$. Matoušek has also shown that under certain fairly general assumptions, which are satisfied by most of the geometric range spaces, a $\frac{1}{r}$-approximation of size $O(r^2 \log r)$ for a range space of VC-dimension d can be computed in time $O(nr^{2d} \log^d r)$ [78].

The following simple observation leads to an efficient algorithm for computing a $\frac{1}{r}$-cutting for Γ. Let $\overline{\Sigma}$ be the same as defined in Corollary 2.3.

LEMMA 3.7. *Let A be a $\frac{1}{r}$-approximation for the range space $\overline{\Sigma}$ and let Ξ be a $\frac{1}{t}$-cutting for A. Then Ξ is a $\left(\frac{d}{r} + \frac{d}{t} \right)$-cutting for Γ.*

COROLLARY 3.8 (Matoušek [78]). *Let Γ be a set of n hyperplanes in $\mathbb{R}^d$ and r some fixed constant. Then a $\frac{1}{r}$-cutting of size $O(r^d)$ for Γ can be computed in time $O(n)$.*

PROOF. First compute a $\frac{1}{2dr}$-approximation A in time $O(n)$ of constant size for the range space $\overline{\Sigma}$, and then compute, in constant time, a $\frac{1}{2dr}$-cutting Ξ of size $O(r^d)$ for A using the Chazelle-Friedman algorithm. By Lemma 3.7, Ξ is a $\frac{1}{r}$-cutting for Γ. □

The above corollary gives an optimal algorithm if r is bounded by some constant, say α, but it is very inefficient if r is large. For large values of r, say $r \geq \alpha$, an obvious approach is to apply Corollary 3.8 repeatedly. That is, for $i = 1, 2, \ldots, k = \log_\alpha r$, compute a $\frac{1}{\alpha^i}$-cutting Ξ_i by refining Ξ_{i-1}. In particular, Ξ_0 is just a simplex containing all vertices of $\mathscr{A}(\Gamma)$. Suppose we have already computed Ξ_{i-1}. For each simplex τ of Ξ_{i-1}, let $\Gamma_\tau \subseteq \Gamma$ denote the subset of n_τ hyperplanes intersected by the interior of τ. If $n_\tau > \frac{n}{\alpha^i}$, we do the following:

(i) Compute a $\frac{1}{2d\alpha}$-approximation A of size $O(\alpha^2 \log \alpha)$ for the subspace $\overline{\Sigma}_{\Gamma_\tau}$.

(ii) Compute a $\frac{1}{2d\alpha}$-cutting Ξ_τ for A of size $O(\alpha^d)$.

(iii) For every simplex $\triangle \in \Xi_\tau$ that intersects τ triangulate $\triangle \cap \tau$.

It is easy to see that Ξ_k is a $\frac{1}{\alpha^i}$-cutting for Γ, but the size of Ξ_k is not $O(r^d)$, because the constant hidden in big-O notation in step (ii) accumulates at every stage. Chazelle however observed that, by modifying the last two steps of the above algorithm, one can obtain an optimal algorithm:

(ii′) Compute a subset $H \subseteq A$ of $O(\alpha \log \alpha)$ hyperplanes such that H is a $\frac{1}{2d\alpha}$-net for the subspace $\overline{\Sigma}_A$ and that the number of vertices of $\mathscr{A}(H)$ lying inside τ is at most $4\left(\frac{|H|}{n_\tau}\right)^d \nu_\tau$, where ν_τ is the number of vertices of $\mathscr{A}(A)$ that lie inside τ.

(iii′) Compute the arrangement $\mathscr{A}(H)$ and triangulate every face of $\tau \cap \mathscr{A}(H)$ as described in the proof of Corollary 2.3.

A standard probabilistic argument shows that a random subset $H \subseteq A$ of size $c\alpha \log \alpha$ (c is some suitable constant) satisfies the properties of (ii′) with probability $> \frac{1}{2}$. Moreover, using the method of conditional probabilities, one can compute the set H in time polynomial in α.

Using a very clever argument, Chazelle has shown that the number of simplices in the triangulation of $\mathscr{A}(H) \cap \tau$ is

$$O\left(\left(\frac{|H_\tau|}{n_\tau}\right)^d \nu_\tau + \alpha^{d-1} \log^d \alpha\right) = O\left(\left(\frac{\alpha^i \log \alpha}{n}\right)^d \nu_\tau + \alpha^{d-1} \log^d \alpha\right).$$

Summing this quantity over all simplices of Ξ_{i-1} and using the fact that $\sum_{\tau \in \Xi_{i-1}} \nu_\tau = O(n^d)$, we obtain the following recurrence for the size of Ξ_i:

$$|\Xi_i| \le C\alpha^{d-1} \log^d \alpha |\Xi_{i-1}| + O(\alpha^{di} \log^d \alpha),$$

where C is some fixed constant. The solution of the above recurrence is $|\Xi_i| = O(\alpha^{(i+1)d})$, provided that α is sufficiently large. Hence, the size of Ξ_k is $O(r^d)$. Finally, a similar analysis shows that the running time of the above procedure is $O(nr^{d-1})$ (see Chazelle [33] for details). This finishes the proof of Theorem 3.5.

3.3. Cutting algorithms for nonlinear objects. So far we have assumed that the objects for which we wanted to compute a $\frac{1}{r}$-cutting were hyperplanes in $\mathbb{R}^d$, but in several problems one has to deal with nonlinear objects. We now survey some deterministic algorithms for computing a $\frac{1}{r}$-cutting for a set of n real algebraic varieties $\Gamma = \{\bigvee f_i, \bigvee f_2, \ldots, \bigvee f_n\}$, where f_i is a d-variate polynomial with rational coefficients. We also assume that the degree of all polynomials is bounded by some constant, say b. Let $\widetilde{\Sigma}$ be the range space defined in §2. By Theorem 2.2, $\widetilde{\Sigma}$ admits a $\frac{1}{r}$-net H of size $O(r \log r)$. Moreover, H can be computed in polynomial time using the method of conditional probabilities [36, 102]. In fact, H can be computed more efficiently, by following the same approach as in Corollary 3.8. That is, we first compute a $\frac{1}{2r}$-approximation A of $\widetilde{\Sigma}$ of size $O(r^2 \log r)$, and then

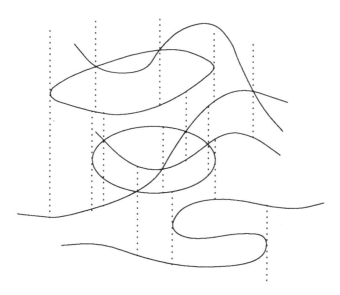

FIGURE 2. Vertical decomposition.

compute a $\frac{1}{2r}$-net of A using the Chazelle-Friedman algorithm. The running time of this procedure is $O(nr^{2k} + r^{O(k)})$, where k is the VC-dimension of $\tilde{\Sigma}$.

In two dimensions one can decompose $\mathscr{A}(H)$ into $O(r^2 \log^2 r)$ constant size cells as follows: Draw a vertical segment in both directions through every vertex of $\mathscr{A}(\Gamma)$ and points of vertical tangencies until it hits an edge of $\mathscr{A}(\Gamma)$ (see Figure 2). It is easy to see that the resulting decomposition has $O(r^2 \log^2 r)$ cells.

In higher dimensions, the problem is much harder. It is not known how to decompose a set Γ of n surfaces into $O(n^d)$ cells. The famous paper of Collins [49] shows that the *cylindrical algebraic decomposition* of Γ can decompose $\mathscr{A}(\Gamma)$ into $O(n^{2^{d}-1})$ cells, but this bound is far from the lower bound of $\Omega(n^d)$. Significant progress in this direction has been made by Chazelle et al. [35], who developed a procedure that can decompose $\mathscr{A}(\Gamma)$ into $O(n^{2d-2})$ cells. Since their method is rather technical, we take a very simple example to illustrate the basic idea of their algorithm—we assume that $\Gamma = \{\gamma_1, \ldots, \gamma_n\}$ is a collection of n spheres in $\mathbb{R}^3$. For a sphere γ_i, let $\mathscr{C}_i = \{\gamma_j \cap \gamma_i \mid j \neq i, 1 \leq i \leq n\}$, and let γ_i^* denote the set of points of vertical tangencies (i.e., equator of γ_i). For every pair $\gamma_l, \gamma_m \in \Gamma$ do the following:

(i) Let $\Gamma_{l,m}$ be the xy-projection of the set $\mathscr{C}_l \cup \mathscr{C}_m \cup \{\gamma_1^*, \ldots, \gamma_n^*\}$; $\Gamma_{l,m}$ is a collection of ellipses and circles in the plane.

(ii) Compute the vertical decomposition $\mathscr{A}^*(\Gamma_{l,m})$ of $\mathscr{A}(\Gamma_{l,m})$; $\mathscr{A}^*(\Gamma_{l,m})$ consists of $O(r^2)$ cells.

(iii) For every cell $c \in \mathscr{A}^*(\Gamma_{l,m})$, let ψ_c denote the cylinder

$$\psi_c = \bigcup \{ (x, y, z) \mid (x, y) \in c, \ z \in \mathbb{R} \}.$$

See Figure 3. It is easy to check that every sphere γ_i either does not intersect ψ_c or cuts it completely (i.e., partitions ψ_c into three regions). Moreover no γ_i intersects γ_l or γ_m inside ψ_c.

(iv) Let $\tau_1, \ldots, \tau_k$ ($k \leq 5$) be the connected components of $\psi_c - (\gamma_l \cup \gamma_m)$, i.e., the layers formed by the spheres γ_l, γ_m inside ψ_c. We choose the closure of a cell τ_i ($i \leq k$) in the final decomposition $\mathscr{A}^*(\Gamma)$ if τ_i does not intersect any sphere of Γ (see Figure 3).

If a cell τ is selected more than once, we keep only one copy of τ. Since each cell $c \in \mathscr{A}^*(\Gamma_{l,m})$ contributes at most five cells to $\mathscr{A}^*(\Gamma)$, the total number of cells in the decomposition, over all pairs of Γ, is $O(n^4)$.

In higher dimensions, $\mathscr{A}^*(\Gamma_{l,m})$ is computed by a recursive call to the decomposition procedure in $d - 1$ dimensions.

In order to compute $\mathscr{A}^*(\Gamma)$ we solve $\binom{n}{2}$ subproblems recursively in $\mathbb{R}^{d-1}$ (each involving $O(n)$ surfaces), and each cell of the $(d - 1)$-dimensional decomposition contributes $O(1)$ cells (the constant depending on the degree, b) to the final decomposition, therefore the total number of cells in the overall decomposition is $O(n^{2d-2})$.

For $\mathbb{R}^3$, using the theory of Davenport-Schinzel sequences [68, 15], one can show that the algorithm of Chazelle et al. [35] decomposes $\mathscr{A}(\Gamma)$ into $O(n^3 2^{\alpha^c(n)})$ cells, where $c = (2b)^{2^{d-1}}$ and $\alpha(n)$ is the inverse Ackermann function. This in turn improves the bound for higher dimensions to $O(n^{2d-3} 2^{\alpha^c(n)})$.

THEOREM 3.9 (Chazelle et al. [35]). *Given a collection Γ of n real algebraic varieties of fixed maximum degree (say, at most b) in $\mathbb{R}^d$, one can decompose $\mathscr{A}(\Gamma)$, in time $O(n^{2d-1} \log n)$, into $O(n^{2d-3} 2^{\alpha^c(n)})$ constant size cells, where $c = (2b)^{2^{d-1}}$.*

As mentioned above, it is an open question whether one can decompose the arrangement of a collection Γ of n real algebraic varieties in $\mathbb{R}^d$ into $O(n^d)$ constant size cells. Actually, even improving the space complexity of Theorem 3.9 to $O(n^d)$ seems to be hard.

A consequence of the above theorem is that if r is bounded by some constant then a $\frac{1}{r}$-cutting of size $O((r \log r)^{2d-3} 2^{\alpha^c(r)})$ for Γ can be computed in time $O(n)$. For larger values of r, we can apply this procedure repeatedly as in the case of hyperplanes. We thus obtain

THEOREM 3.10. *Given a collection Γ of n real algebraic varieties of fixed degree in $\mathbb{R}^d$ and a parameter r, one can compute a $\frac{1}{r}$-cutting of size*

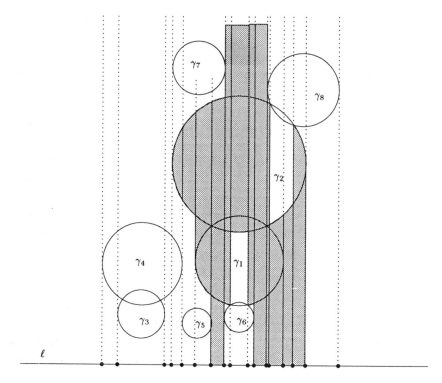

FIGURE 3. Decomposing arrangements of circles into cells. The partitioning of ℓ gives the decomposition $\mathcal{A}^*(\Gamma_{1,2})$ for $\Gamma_{1,2}$. Vertical strips denote the cylinders ψ_c for $c \in \mathcal{A}^*(\Gamma_{1,2})$, and shaded regions denote the cells of $\mathcal{A}^*(\Gamma)$ chosen in step (iv) (for the pair γ_1 and γ_2).

$O(r^{2d-3+\varepsilon})$ *in time* $O(n^{1+\varepsilon}r^{2d-3})$, *or of size* $O((r \log r)^{2d-3}2^{\alpha^c(r)})$ *in time* $O(nr^{O(1)})$.

Although we are not aware of any better result in the general case, one can obtain a $\frac{1}{r}$-cutting of almost optimal size for some special cases. For example, one can compute a $\frac{1}{r}$-cutting of size $O(r^d \log r)$ in time $O(nr^{d+1})$ for a collection of n spheres in $\mathbb{R}^d$ [10].

4. Applications to combinatorial problems

Most of the combinatorial problems related to arrangements of manifolds or to configuration of points can be formulated as an extremal problem in hypergraph theory, i.e., what is the maximum number of hyperedges a hypergraph H with n vertices can have, if it satisfies certain properties. Geometric partitioning has been used to obtain an upper bound for a variety of such problems. In this section, we will discuss some of these problems. For the sake of clarity, we will state the problems in the geometric setting, not as extremal hypergraph problems.

4.1. Incidence problems. In the *incidence problem* we are given a collection Γ of n surfaces of a certain kind and another collection S of m points in $\mathbb{R}^d$. The goal is to count the number of pairs $(\gamma, p) \in \Gamma \times P$ such that $p \in \Gamma$. Let $\mathscr{I}(m, n)$ denote the maximum number of incidence pairs.

If Γ is a set of n lines in $\mathbb{R}^2$, a simple graph-theoretic argument shows that $\mathscr{I}(m, n) = O(m\sqrt{n} + n)$, which was improved to $O(m^{2/3}n^{2/3} + m + n)$ by Szemerédi and Trotter [105] (see also [117]). The following simple and elegant proof is due to Clarkson et al. [47].

THEOREM 4.1 (Szemerédi-Trotter [105]; Clarkson et al. [47]). *The maximum number of incidence pairs between a set of n lines and a set of m points in the plane is*

$$\mathscr{I}(m, n) = O(m^{2/3}n^{2/3} + m + n).$$

SKETCH OF PROOF. Let Γ be a set of n lines in $\mathbb{R}^2$, S be a set of m points, and let Ξ be a $\frac{1}{r}$-cutting of size $s = O(r^2)$ for Γ, where $r \le n$ is some parameter, whose value will be specified later. It can be shown that the points of S that lie on the vertices of Ξ can contribute at most $O(nr)$ incidence pairs. Let $S' \subseteq S$ be the subset of points not lying on the vertices of the triangles of Ξ. We now have to estimate the number of incidence pairs between Γ and S'.

For a triangle $\triangle \in \Xi$, let $\Gamma_\triangle$ be the set of lines intersecting the interior of $\triangle$ and $S_\triangle \subseteq S'$ the set of points lying in $\triangle$. Let $n_\triangle = |\Gamma_\triangle|$ and $m_\triangle = |S_\triangle|$. Obviously, $n_\triangle \le \frac{n}{r}$ and $\sum_{\triangle \in \Xi} m_\triangle \le 2m$ (a point lying on an edge of Ξ is counted twice). We thus get the following recurrence:

$$\mathcal{I}(\Gamma, S) \leq \sum_{\triangle \in \Xi} \mathcal{I}(\Gamma_{\triangle}, S_{\triangle}) + O(nr).$$

Using the fact that $\mathcal{I}(\Gamma_{\triangle}, S_{\triangle}) \leq \mathcal{I}(m_{\triangle}, n_{\triangle}) = O(m_{\triangle}\sqrt{n_{\triangle}} + n_{\triangle})$ [51] and choosing $r = \lceil m^{2/3}/n^{1/3} \rceil$, we obtain the desired bound (see [47] for details). $\square$

A construction of Erdős (see Edelsbrunner and Welzl [59]) shows that the above bound is tight in the worst case. One can obtain the same upper bound on $\mathcal{I}(m, n)$ for pseudo-lines and unit circles. For general circles Clarkson et al. [47] proved that $\mathcal{I}(m, n) = O(m^{3/5}n^{4/5} + m + n)$.

The following corollaries are immediate consequences of Theorem 4.1.

COROLLARY 4.2 (Szemerédi-Trotter [105]). *Given a set of m points, a set of n lines, and a parameter $k \leq m$, the number of lines each containing at least k points is $O(m^2/k^3 + m/k)$.*

COROLLARY 4.3 (Szemerédi-Trotter [105]; Beck [20]). *Given a set of n points in the plane, not all of them on a line, there is a point $p \in S$ that lies on at least $\Omega(n)$ distinct lines connecting p to other elements of S.*

The incidence problem can be formulated in higher dimensions as well, e.g., counting incidences between planes and points in $\mathbb{R}^3$. If both planes and points are arbitrary, then the problem is not interesting, because one can easily obtain an $\Omega(mn)$ lower bound on $\mathcal{I}(m, n)$ by choosing m points on a line ℓ and n planes passing through ℓ. But if we assume that no three points are collinear, one can use Corollary 4.3 and Theorem 4.1 to prove that $\mathcal{I}(m, n) = O(m^{2/3}n + n^2)$ [4]; the bound is tight in the worst case. See [55] for related results.

Clarkson et al. [47] have shown that, for a collection of n spheres in $\mathbb{R}^3$ in general position,

$$(4.1) \qquad \mathcal{I}(m, n) = O(m^{3/4}n^{3/4}\beta(m, n) + m + n),$$

where $\beta(m, n)$ is a function of the inverse Ackermann function, and therefore grows very slowly with m and n. The additional factor $\beta(m, n)$ in the above bound is due to the fact that the best known bound on the size of a $\frac{1}{r}$-cutting for spheres in $\mathbb{R}^3$ is $O(r^3 2^{O((\alpha(n))^2)})$.

In some cases one wants to bound the number of incidence pairs where the point sets satisfy certain properties. For example, consider the following problem: *Given a set Γ of n unit spheres in $\mathbb{R}^3$ and a set S of m points such that, for each sphere $\gamma \in \Gamma$, all points of S lie only on one side of γ, what is the maximum number of incidence pairs?*

Again, one can use the geometric partitioning to obtain an upper bound on $\mathscr{I}(m, n)$. Observe that in this case all points of S lie in a single cell of $\mathscr{A}(\Gamma)$, therefore it is enough to decompose only the cell of $\mathscr{A}(\Gamma)$ that contains all the points of S instead of decomposing the entire space. Using this observation, Edelsbrunner and Sharir proved that

$$(4.2) \qquad \mathscr{I}(m, n) = O(m^{2/3}n^{2/3} + m + n),$$

which is better than the upper bound mentioned in (4.1).

4.2. Distance related problems. About 45 years ago Erdős asked the following question: *What is the maximum number of pairs in a set of n points in two (or three) dimensions that are exactly at distance 1?*

Let $f_d(n)$ denote the maximum number of such pairs in $\mathbb{R}^d$. Erdős [61] showed that

$$n^{1+c/\log\log n} \leq f_2(n) = O(n^{3/2}).$$

The lower bound is achieved by a set of points on an integer lattice. For $d = 3$, Erdős proved that $f_3(n) = \Omega(n^{4/3}\log\log n)$ [62] and Chung showed that $f_3(n) = O(n^{8/5})$ [42]. Using the results on the number incidences for unit circles (or spheres) one can show

THEOREM 4.4 (Spencer et al. [103]; Clarkson et al. [47]). *The maximum number of unit distances that can occur among a set of n points in two and three dimensions is $f_2(n) = O(n^{4/3})$ and $f_3(n) = O(n^{3/2}\beta(n))$, respectively.*

PROOF. Let S be a set of n points in $\mathbb{R}^2$. Draw a circle C_p of radius 1 around each point p of S. The distance between two points $p, q \in S$ is 1 if and only if $p \in C_q$ and $q \in C_p$, which implies $f_2(n) = \mathscr{I}(n, n)/2 = O(n^{4/3})$. The same argument gives the desired upper bound on $f_3(n)$. □

Erdős has conjectured that the true value of $f_2(n)$ is close to his lower bound. He has repeatedly offered \$500 for a proof or disproof of the bounds $f_2(n) = n^{1+c/\log\log n}$. Surprisingly, asymptotically tight bounds are known on $f_d(n)$ for $d \geq 4$ [63, 64].

The dual of the unit distance problem is the *distinct distance problem*: *What is the minimum number of distinct distances in a set of n points in the plane?* Let $g(n)$ denote this quantity. Erdős showed that $g(n) = O(\frac{n}{\sqrt{\log n}})$ [61]. Using the upper bound on the number of incidences for circles and points in the plane one can easily show that $g_2(n) = \Omega(n^{3/4})$ [47] (see also [41]). This bound has been improved by Chung et al. [43] to $O(\frac{n^{4/5}}{\log^c n})$ for some suitable constant $c > 0$.

A variant of the unit distance problem is the *bichromatic minimum distance*: *Given a set P of m red points and another set Q of n blue points in $\mathbb{R}^d$, how many pairs $(p, q) \in P \times Q$ realize the bichromatic minimum distance (i.e., $d(p, q) = \min_{p' \in P, q' \in Q} d(p', q'))$?* Let $\varphi_d(m, n)$ denote the maximum number of red-blue pairs that can realize the bichromatic mini-mum distance in $\mathbb{R}^d$. It is well known that $\varphi_2(m, n) = \Theta(m + n)$ and, for $d \geq 4$, $\varphi_d(m, n) = \Omega(m, n)$. For $d = 3$, Edelsbrunner and Sharir proved that

THEOREM 4.5 (Edelsbrunner-Sharir [57]). *The maximum number of times the bichromatic closest distance can occur among a set of* m *points and another set of* n *points in* $\mathbb{R}^3$ *is*

$$\varphi_3(m, n) = O(m^{2/3}n^{2/3} + m + n).$$

SKETCH OF PROOF. Let P be a set of m points in $\mathbb{R}^3$, Q another set of n points in $\mathbb{R}^3$, and $\delta = \min_{p \in P, q \in Q} d(p, q)$. Draw a sphere B_p around each point $p \in P$. Let $\mathscr{B}$ be the resulting set of m spheres. Then $d(p, q) = \delta$ if and only if q is incident to B_p. Moreover, all points of Q lie in the exterior of every sphere of $\mathscr{B}$, therefore, by (4.2), we obtain $\varphi_3(m, n) = O(m^{2/3}n^{2/3} + m + n)$. □

Proving a superlinear lower bound on $\varphi_3(m, n)$ is an interesting open problem. A tight bound on $\varphi_3(m, n)$ will yield a similar bound for other problems too. For example, using Theorem 4.5, Agarwal and Matoušek showed that the maximum number of edges in the relative neighborhood graph of n points in $\mathbb{R}^3$ is $O(n^{4/3})$ [9].

4.3. Repeated angles and related problems. The results on incidences can also be applied to the following types of problems: *What is the maximum number of triples* p_i, p_j, p_k *in a set of* n *points in the plane such that* $\angle p_i p_j p_k = \theta$, *for some fixed angle* θ.

A straightforward application of Theorem 4.4 implies that at most $O(n^{7/3})$ triples in a set of n points realize the $\pi/2$ angle. However, Corollary 4.2 and a more careful analysis gives an upper bound of $O(n^2 \log n)$, which is tight in the worst case [92]. See also [112, 113, 114].

One can ask various other similar questions involving triples in a set of points in the plane. For example, the number of triples that form unit area (perimeter) triangles, the number of triples that form isosceles triangles, etc. Theorem 2.2 can be applied to all of these problem to obtain an upper bound of $O(n^{7/3})$; see [92] for details.

4.4. Many faces problems. In the *many faces* problem one wants to bound the complexity of m distinct cells in an arrangement of n surfaces of certain kind. Let $\mathscr{C}(m, n)$ denote the maximum complexity in the generic sense (the complexity of a cell τ is the total number of faces of all dimensions bounding τ).

In 1969 Canham showed that $\mathscr{C}(m, n) = O(m\sqrt{n} + n)$ for lines in the plane (see also [59]). On the other hand, the lower bound construction for incidences between points and lines in the plane can be extended to show

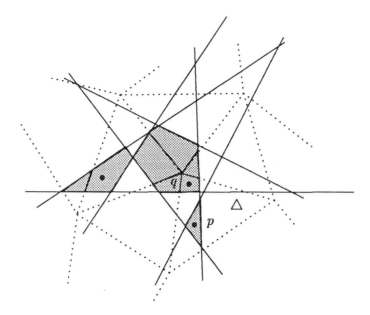

FIGURE 4. $f_p(\Gamma) = f_p(\Gamma_\triangle)$, but a portion of $f_q(\Gamma)$ lies in other triangles.

that $\mathscr{C}(m, n) = \Omega(m^{2/3}n^{2/3} + n)$. A matching upper bound was proved by Clarkson et al. [47].

THEOREM 4.6 (Clarkson et al. [47]). *The maximum complexity of m distinct cells in an arrangement of n lines in the plane is*

$$\mathscr{C}(m, n) = O(m^{2/3}n^{2/3} + n).$$

SKETCH OF PROOF. Let Γ be a set of points in the plane in general position. Mark m cells of $\mathscr{A}(\Gamma)$ by putting a point in each of the cells. Let S denote the resulting set of m points. Without loss of generality, we can assume that no point of S lies on an edge or a vertex of Ξ. Let Ξ be a $\frac{1}{r}$-cutting of size $O(r^2)$ for Γ (the value of r will be specified later). For a triangle $\triangle \in \Xi$, let $\Gamma_\triangle$ be the set of lines intersecting the interior of $\triangle$ and let $S_\triangle = S \cap \triangle$. For a point $p \in S_\triangle$, let $f_p(\Gamma)$ (resp. $f_p(\Gamma_\triangle)$) denote the cell of $\mathscr{A}(\Gamma)$ (resp. $\mathscr{A}(\Gamma_\triangle)$) containing the point p. If $f_p(\Gamma)$ lies completely in the interior of $\triangle$, then $f_p(\Gamma) = f_p(\Gamma_\triangle)$ (see $f_p(\Gamma)$ in Figure 4). Otherwise, some of the edges of $f_p(\Gamma)$ may lie in a cell f' of some $\mathscr{A}(\Gamma_{\triangle'})$, but then f' has to intersect the boundary of $\triangle'$ (see $f_q(\Gamma)$ in Figure 4). Let $\nu_\triangle$ denote the number of edges in the cells of $\mathscr{A}(\Gamma_\triangle)$ that intersect the boundary of $\triangle$.

Then, we have

$$\mathscr{C}(m, n) = \sum_{\triangle \in \Xi} \mathscr{C}(m_\triangle, n_\triangle) + \nu_\triangle \,,$$

where $n_\triangle \le \frac{n}{r}$ for each $\triangle$, and $\sum_{\triangle \in \Xi} m_\triangle = m$. It is well known that $\nu_\triangle = O(n_\triangle)$ [56, 47]. Since $\mathscr{C}(m_\triangle, n_\triangle) = O(m_\triangle \sqrt{n_\triangle} + n_\triangle)$ [27], setting $r = \lceil m^{2/3}/n^{2/3} \rceil$, we obtain $\mathscr{C}(m, n) = O(m^{2/3} n^{2/3} + m + n)$. $\square$

Following a similar approach one can obtain the bounds on $\mathscr{C}(m, n)$ for pseudo-lines, unit circles, circles, segments, etc. [47, 17]. We summarize the known results on $\mathscr{C}(m, n)$ in Table 1.

TABLE 1. Summary of many faces results.

	$\mathscr{C}(m, n)$
Lines	$O(m^{2/3} n^{2/3} + n)$
Pseudo-lines	$O(m^{2/3} n^{2/3} + n)$
Unit circles	$O(m^{2/3} n^{2/3} \beta(m, n) + n)$
Circles	$O(m^{3/5} n^{4/5} \beta(m, n) + n)$
Segments	$O(m^{2/3} n^{2/3} + n\alpha(n) + m \log n)$

5. Applications to algorithmic problems

In the last section we applied the geometric partitioning scheme to solve a number of combinatorial problems. We now discuss a few algorithmic problems for which efficient divide-and-conquer algorithms have been developed based on the geometric partitioning scheme.

5.1. Hopcroft's problem. Hopcroft's problem is a restricted version of the incidence problem. It can be stated as follows: *Given a set Γ of n lines and a set S of m points in the plane, detect if any point of S is incident to some line of Γ.*

In the previous section we established an upper bound on the maximum number of incidence pairs. A nice feature of this proof is that it can be converted into a deterministic algorithm for detecting an incidence (or more generally counting the number of incidence pairs) in time

$$O(m^{2/3} n^{2/3} \log^{1/3} \frac{m}{\sqrt{n}} + (m + n) \log(m + n))$$

[3, 33, 115]. Further, it can be extended to detect incidences between n hyperplanes and m points in $\mathbb{R}^d$ in time

$$O((mn)^{1-1/(d+1)} \log^{O(1)} n + (m + n) \log(m + n))$$

[33].

Recall that the lower bound on $\mathscr{I}(m,n)$ for m lines and n points in the plane is $\Omega(m^{2/3}n^{2/3} + m + n)$, but it does not yield any nontrivial lower bound for the Hopcroft problem. It is an open problem whether one can improve the running time to $o(m^{2/3}n^{2/3})$ or prove a lower bound of $\Omega(m^{2/3}n^{2/3})$. In fact any nontrivial lower bound will be interesting; such a bound will establish a similar lower bound for various other problems.

The same approach can be used to detect (or count) incidences between points and other objects (e.g., circles, unit circles, etc.). Hence, the number of pairs of points in a given set of n points in the plane, which are unit distance apart, can be counted in time $O(n^{4/3+\varepsilon})$ for any $\varepsilon > 0$.[4]

5.2. Computing many faces. Like the incidence problem, the proof of Theorem 4.5 can also be converted into an efficient deterministic algorithm for computing the faces in arrangement of n lines that contain a given set of m points; see [3] for details. The time complexity of this algorithm is $O(m^{2/3}n^{2/3}\log^{5/3}n + (m+n)\log n)$.

If we are given a set $\mathscr{G}$ of n segments instead of lines, Edelsbrunner et al. [54] showed that on can compute the faces of $\mathscr{A}(\mathscr{G})$ containing the points of S in time $O(m^{2/3-\varepsilon}n^{2/3+2\varepsilon} + n\alpha(n)\log m \log^2 n)$. The running time has been slightly improved by Agarwal [3].

5.3. Intersection problems. One of the well-studied problems in computational geometry is reporting the intersection points in a set of segments in the plane. About twelve years ago Bentley-Ottmann [21] discovered an $O((n+k)\log n)$ algorithm to report all k intersection points based on the line sweep technique. Recently Chazelle and Edelsbrunner improved the running time to $O(n\log n + k)$ [31], but their algorithm is quite involved. Using geometric partitioning Clarkson and Shor presented a very simple randomized algorithm whose expected running time was also $O(n\log n + k)$ [48]. (Mulmuley has independently given another randomized algorithm with the same expected running time [88].)

The above-mentioned algorithms are efficient only if we want to report all intersection points explicitly. In some applications we are interested only in knowing the total number of intersections (not in the actual intersection points). In such cases, we would like to have an algorithm whose running time does not depend on k. Chazelle gave the first subquadratic algorithm [28]. A faster algorithm can be obtained using geometric partitioning.

THEOREM 5.1 (Agarwal [3], Chazelle [33]). *The number of intersection points in a set Γ of n segments can be counted in time $O(n^{4/3}\log^{1/3}n)$.*

SKETCH OF PROOF. Compute a $\frac{1}{r}$-cutting Ξ for Γ of size $O(r^2)$, where r is as usual a parameter to be chosen later. For each triangle $\triangle$ of Ξ, we

---

[4]Throughout this paper ε will denote an arbitrarily small positive constant. The constant of proportionality in the time complexity will depend on ε and will tend to ∞ as $\varepsilon \downarrow 0$.

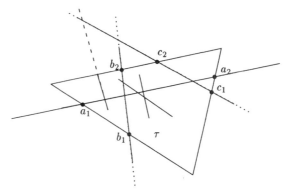

FIGURE 5. Long and short segments; long segments have been extended to full lines and short segments have been clipped at the boundary of $\triangle$.

count the number of intersection points lying inside $\triangle$. Let $\Gamma_\triangle \subseteq \Gamma$ be the set of segments that intersect the interior of $\triangle$. A segment of $\Gamma_\triangle$ is called *short* if at least one of its endpoints lies in the interior of $\triangle$, and it is called *long* otherwise. Let $a_\triangle$ (resp. $b_\triangle$) denote the number of long (resp. short) segments of Γ_τ. The number of intersection points among short segments is counted in time $O(b_\triangle^2)$ by testing all pairs of short segments.

As for long segments, two long segments intersect inside $\triangle$ if and only if their intersection points with $\partial\triangle$ are interlaced (e.g., a_1, c_1, a_2, c_2 in Figure 5). This observation leads to a very simple $O(a_\triangle \log a_\triangle)$ time procedure for counting the number of intersection points among long segments.

Finally, for counting the intersection points between long and short segments, we extend the long segments to full lines and clip short segments within $\triangle$. It is obvious that this step does not change the number of intersection points of Γ lying inside $\triangle$. But a line ℓ intersects a segment e if and only if the point dual to ℓ lies in the double wedge dual to e. The problem thus redues to locating $a_\triangle$ points in an arrangement of $b_\triangle$ double wedges, which can be solved in time $O(b_\triangle^2 + a_\triangle \log b_\triangle)$; see **[1]** for details.

Hence, the intersections lying in $\triangle$ can be counted in $O(b_\triangle^2 + a_\triangle \log b_\triangle)$ time. The running time can be improved to $O(b_\triangle(a_\triangle \log a_\triangle)^{1/2} + a_\triangle \log a_\triangle)$, using a standard batching technique.

Since Ξ can be computed in time $O(nr)$ and $a_\triangle \le n/r$, $\sum_{\triangle \in \Xi} b_\triangle \le 2n$, choosing $r = n^{1/3}/\log^{1/3} n$, gives an upper bound of $O(n^{4/3} \log^{2/3} n)$ on the time complexity. The desired bound on the running time can be attained by using some additional tricks. $\square$

A variant of the segment intersection problem is its bichromatic version, i.e., given a set of m "red" segments and another set of n "blue" segments, report (or count the number of) the red-blue intersection points. The goal is to come up with an algorithm, whose time complexity does not depend on the number of red-red or blue-blue intersection points. Neither the Bentley-Ottmann nor Chazelle-Edelsbrunner algorithm can be modified to report red-blue intersections without paying the cost for red-red or blue-blue intersection points.

Agarwal [3] showed that Theorem 5.1 can be modified to count the number of red-blue intersection points in time

$$O(m^{2/3}n^{2/3}\log^{1/3}\frac{m}{\sqrt{n}} + (m+n)\log(m+n)).$$

It can also report all k red-blue intersections at an additional cost of $O(k)$. Note that a faster algorithm for this problem will yield a similar algorithm for the Hopcroft problem.

Although the Bentley-Ottmann algorithm can be extended to report the intersection points in a set of Jordan arcs, no efficient algorithm is known for counting the number of intersection points in a collection of arcs. Recently, Agarwal et al. [5] presented an $O(n^{4/3+\varepsilon})$ algorithm to count the number of intersections in a collection of unit circles. A more interesting result in this direction is by Agarwal and Sharir [13], who proposed an $O(n^{3/2+\varepsilon})$ (resp. $O(n^{5/3+\varepsilon})$) algorithm to count the number of intersections in a collection of circles (resp. circular arcs) of arbitrary radii.

In three-dimensions, Pellegrini has recently given an $O(n^{8/5+\varepsilon})$ algorithm to detect an intersection between two nonconvex polyhedra, where n is the total number of vertices in the two polyhedron [94].

5.4. Proximity problems. Although it has been known for a long time that the closest pair in a set of n points in $\mathbb{R}^2$ can be computed in $O(n\log n)$ time [22], no such algorithm is known for the bichromatic version, i.e., given a set of m red points and a set of n blue points in $\mathbb{R}^d$, determine the red-blue closest pair. Recently, Agarwal et al. [6] showed that a bichromatic closest pair can be computed in time $O((mn)^{1-1/(\lceil d/2\rceil+1)+\varepsilon})$. Using this algorithm they compute a Euclidean minimum spanning tree of a set S of n points in $\mathbb{R}^d$ in time $O(n^{2(1-1/(\lceil d/2\rceil+1))+\varepsilon})$.

A variant of their algorithm can also compute the diameter of S within the same time bound. Although Clarkson-Shor have given a randomized algorithm for computing the diameter of a set of n points in $\mathbb{R}^3$ whose expected running time is $O(n\log n)$ [48], it remains an open question whether such a bound can be attained by a deterministic algorithm. Furthermore, it is not clear how to modify these algorithms for computing farthest neighbors for each point in S. Using a somewhat different approach, Agarwal et al. [11] presented a randomized algorithm for computing the farthest neighbors. The expected running time of their algorithm is $O(n^{2(1-1/(\lceil d/2\rceil+1))+\varepsilon})$. They apply this algorithm to solve some other proximity problems, including computing a Euclidean maximum spanning tree.

5.5. Parametric search problems. A typical parametric search problem can be formulated as follows: *Given a set* Γ *of objects, a parameter* t*, a function* $f(\Gamma, t)$*, and a condition* χ*, find a value of* $t = t^*$ *such that* $f(\Gamma, t^*)$ *satisfies the condition* χ*.* For instance, consider the problem of computing the kth smallest distance realized by a given n-element point set S in the plane. For the sake of simplicity, assume that all the distances determined by the points of S are distinct. In this case, $\Gamma = S$, $f(\Gamma, t) = |\{(p, q) \mid p, q \in S$ and $d(p, q) \leq t\}|$, and $\chi : f(\Gamma, t^*) = k$.

Megiddo has proposed a very clever general technique to solve parametric search problems [**84, 85**]. Following his general scheme and using geometric partitioning, Agarwal et al. [**5**] presented an $O(n^{4/3} \log^{7/3} n)$ randomized expected time algorithm for selecting the kth smallest distance. See [**16**] for some additional applications of geometric partitioning and parametric search in geometric optimization problems.

It is beyond the scope of this survey paper to describe Megiddo's technique. Instead, we will explain how the geometric partitioning can be used to simplify his technique. A crucial step in Megiddo's parametric search technique is computing the sign of a set of polynomials $\mathscr{F} = \{f_1(t), f_2(t), \ldots, f_k(t)\}$ at $t = t^*$ without knowing the value of t^*. If $t \in \mathbb{R}$ (i.e., $f_1, \ldots, f_k$ are univariate polynomials), then the signs of the above polynomials can be computed as follows: Suppose we have an *oracle* that, given a value $t = \bar{t}$, can determine whether $\bar{t}$ is less than, or equal to, or greater than t^*. Compute the roots $\beta_1 < \cdots < \beta_u$ of all polynomials in $\mathscr{F}$ and perform a binary search, using the oracle, to locate t^* among these roots. If t^* is one of the β_i's, we already know the value of t^*. Otherwise, we know two consecutive roots β_i, β_{i+1} such that $t^* \in (\beta_i, \beta_{i+1})$. Since the sign of every polynomial of $\mathscr{F}$ remains the same for all $t \in (\beta_i, \beta_{i+1})$, it can be easily computed at t^*.

But in some applications t is a d-dimensional vector (i.e., $f_1, \ldots, f_k$ are d-variate polynomials). In this case, one cannot perform the standard binary search. Megiddo had proposed a rather involved approach to handle this case [**85, 51**]. His approach can be simplified using the geometric partitioning. For the sake of simplicity let us assume that $f_1, \ldots, f_k$ are linear functions, i.e., $H = \{\bigvee f_1, \ldots, \bigvee f_k\}$ is a collection of hyperplanes; the general case can be handled similarly.

Compute a $\frac{1}{r}$-cutting Ξ of size $O(r^d)$ for H using the algorithm of Matoušek [**81**] or Chazelle [**33**] (r is some suitable constant). Suppose we know the simplex $\triangle$ of Ξ that contains t^*. Then, for all hyperplanes $\bigvee f_i$ that do not intersect $\triangle$, we can easily determine the sign of f_i. For the remaining n/r hyperplanes that intersect $\triangle$, we recursively compute the signs of the corresponding polynomials. Thus, we can determine the signs of all polynomials in $O(\log n)$ stages. As for determining the simplex that contains t^*, if t^* satisfies certain properties (e.g., the set of feasible solutions of r is a convex set, etc.), then one can compute $\triangle$ by solving a constant number of $(d-1)$-dimensional parametric search problems (i.e., the parameter t is a $(d-1)$-dimensional vector); see [85] for details. Following this approach Matoušek has given an $O(n^{2(1-1/(\lfloor d/2 \rfloor+1))+\varepsilon})$ algorithm to compute the extremal points of a set of n points in $\mathbb{R}^d$ [81]. This general paradigm can also be applied to several other problems.

Geometric partitioning has also been applied to develop an efficient algorithm for computing ham-sandwich cuts in two and three dimensions [72, 71]. (Recall that the ham-sandwich cut of d sets of points in $\mathbb{R}^d$ is a hyperplane that bisects all of the sets simultaneously.)

6. Data structures

In the previous section we surveyed various divide-and-conquer algorithms based on the geometric partitioning scheme. We now turn our attention to query type problems, where one wants to build a data structure on a given set of objects so that a query of certain type can be answered quickly. Actually, the geometric partitioning was originally introduced to construct efficient data structures for various query type problems including closest-point queries [44], point location [45], and simplex range searching [69].

6.1. Point location in arrangements. Consider the problem of locating a point in arrangements of hyperplanes: *Preprocess a set Γ of n hyperplanes in $\mathbb{R}^d$ into a data structure so that, for a query point p, one can quickly determine whether p lies on a hyperplane of Γ.* If the answer is 'no,' some additional information may be required, e.g., the hyperplane of Γ lying immediately above (or below) p. The following very simple data structure, based on the geometric partitioning, was proposed by Clarkson [44, 45].

If $|\Gamma| \leq c$ for some fixed constant c, then no data structure is constructed. Otherwise, we proceed as follows. Compute a $\frac{1}{r}$-cutting Ξ of size $O(r^d)$ for Γ. For each simplex $\triangle$ of Ξ, recursively construct the data structure for the hyperplanes that intersect the interior of $\triangle$. Moreover, if a k-face f of $\triangle$ is not a portion of a hyperplane of Γ, then build a k-dimensional point location structure for the set $\{\text{aff}(f) \cap \gamma \mid \gamma \in \Gamma\}$ and associate it with $\triangle$. (If f is shared by two or more simplices, we create only one copy of the structure.)

The resulting structure is a tree T of height $O(\log n)$ and $O(r^d)$ degree. The space required by T is $O(n^{d+\varepsilon})$ and it can be constructed within the same time. Actually, the space and preprocessing time can be improved to $O(n^d)$ using Chazelle's hyperplane cutting algorithm [33].

To locate a point p in $\mathscr{A}(\Gamma)$, we visit T in a top-down fashion starting from the root. At the root v, we locate p in Ξ. If p lies on a k-face f of a simplex $\triangle$, then we either search recursively in the k-dimensional structure associated with f, or return the hyperplane of Γ containing p, depending on whether a hyperplane of Γ supports f. On the other hand, if p lies in the interior of a simplex $\triangle$, then we recursively locate p in the data structure associated with $\triangle$. Since the height of the structure is $O(\log n)$ and we spend $O(r^d)$ time at each node, the query time is $O(\log n)$ (recall that r is constant). Hence, we have

THEOREM 6.1 (Chazelle [33]; Clarkson [45]). *A set of* n *hyperplanes in* $\mathbb{R}^d$ *can be preprocessed in time* $O(n^d)$ *into a data structure of size* $O(n^d)$ *that can answer point location queries in* $O(\log n)$ *time.*

An additional advantage of this scheme is that it can be maintained dynamically. One can dynamically update the data structure in $O(n^{d-1+\varepsilon})$ amortized time per insert/delete operation. (Some better results in this direction have been obtained by Mulmuley and Sen [89].)

In some applications, instead of processing the entire arrangement one needs to preprocess only a certain kind of cells of $\mathscr{A}(\Gamma)$ for point location, e.g., the upper envelope of $\mathscr{A}(\Gamma)$ (the cell lying above all hyperplanes of Γ), or the cells intersected by some other hyperplane g. Let $\varphi(n)$ denote the maximum complexity of the cells (where the maximum is taken over all arrangements of n hyperplanes) that we want to preprocess. Then, one can modify the above scheme to build a data structure of size $O(\varphi(n)n^\varepsilon)$ so that a point location query can be answered in $O(\log n)$ time. A consequence of this result is that a set S of n points in $\mathbb{R}^d$ can be preprocessed into a data structure of size $O(n^{\lceil d/2 \rceil + \varepsilon})$ so that the closest (or farthest) neighbor of a query point in S can be computed in $O(\log n)$ time [44]. Some other related results can be found in [34, 94, 23].

If Γ is a collection of n algebraic varieties instead of hyperplanes, we can still follow the same approach as above to answer point location queries. The only difference is that we now have to use the algorithm of Chazelle et al. [35] for computing Ξ; the size of Ξ is $O((r \log r)^{2d-3} 2^{\alpha^c(r)})$. This gives a data structure of size $O(n^{2d-3+\varepsilon})$ that supports $O(\log n)$ time point location queries. As shown in [37, 35, 16], a number of geometric problems can be reduced to point location in an arrangement of surfaces. Hence, the above point location algorithm yields efficient algorithms for these problems as well.

6.2. Simplex range searching. The simplex range searching problem can be defined as: *Preprocess a set* S *of* n *points in* $\mathbb{R}^d$ *into a data structure so that the number of points lying inside a query simplex* $\triangle$ *can be counted efficiently.* In a more general setting, each point $p \in S$ is associated with a value $\phi(p) \in \Phi$, where Φ is some commutative semigroup (whose operation is denoted by $+$), and the goal is to compute the sum $\sum_{p \in S \cap \triangle} \phi(p)$. For the sake of simplicity, we will restrict ourselves to the simpler version of the problem.

A very deep result of Chazelle [30] gives a nontrivial lower bound on the simplex range searching under the arithmetic model involving only semigroup operations.[5] In particular, he has shown that, in d dimensions, if only $O(m)$ storage is allowed ($n \le m \le n^d$), then a simplex range query requires $\Omega(\frac{n/\log n}{m^{1/d}})$ time (for $d = 2$ the lower bound can be improved to $\Omega(\frac{n}{\sqrt{m}})$). For example, if we use only $O(n)$ space, the query time will be at least $\Omega(n^{1-1/d}/\log n)$.

Partition trees. There has been a lot of work on obtaining efficient algorithms for answering the simplex range queries. The earlier solutions (see [108, 60, 110]) for this problem used the notion of *partition trees*. In this section, we mention some of the recent results based on geometric partitioning. Haussler-Welzl [69] were the first to apply geometric partitioning for constructing a partition tree. They showed that

THEOREM 6.2 (Haussler-Welzl [69]). *A set S of n points in $\mathbb{R}^d$ can be preprocessed in time $O(n \log n)$ into a data structure of linear size, so that the points of S lying in a query simplex can be counted in time $O(n^{\beta+\varepsilon})$, where $\beta = \frac{d(d-1)}{d(d-1)+1}$.*

SKETCH OF PROOF. For the sake of simplicity, we will sketch the proof only in two dimensions. We construct a partition tree T on the points of S, each of whose node is associated with a region and a subset of points of S. The root v is associated with the entire plane and the set S, and it is processed as follows: Dualize S into a set Γ of n lines in the plane, and compute a $\frac{1}{r}$-cutting Ξ of size $O(r^2)$ for Γ, where r is some suitable constant. Let $\mathscr{R}$ be the set of $O(r^2)$ lines dual to the vertices of Ξ. Compute the arrangement $\mathscr{A}(\mathscr{R})$; it consists of $s = O(r^4)$ cells. We store $\mathscr{A}(\mathscr{R})$ and the value of $|S|$ at v. For each cell τ of $\mathscr{A}(\mathscr{R})$, we create a child w_τ of v, associate the set $S_\tau = S \cap \tau$ with w_τ, and recursively preprocess S_τ. It can be shown that this structure satisfies the following two crucial properties:

(P1) Every line ℓ intersects $O(r^2)$ cells of $\mathscr{A}(\Gamma)$.

(P2) For every line ℓ, there are at most n/r points in the *zone* of $\mathscr{A}(\mathscr{R})$ with respect to ℓ, i.e., in the cells of $\mathscr{A}(\mathscr{R})$ intersected by ℓ.

The first property follows immediately from the well-known 'zone-theorem' [56]. The second property is somewhat tricky and the reader can refer to either the original paper [69] or [3]. Property (P2) also implies that the height of T is $O(\log n)$. Since the time spent at each node is proportional to the number of points associated with it (recall that r is constant, so a $\frac{1}{r}$-cutting

---

[5]The arithmetic model was originally proposed by Fredman [65]. In this model, one stores precomputed partial sums of the values assigned to points, and a query is answered using these precomputed sums. The complexity of the query time is measured in terms of the number of semigroup operations required to compute the overall sum, and the storage is measured in terms of the number of precomputed partial sums. This model ignores various algorithmic issues, e.g., how does one find the set of precomputed sums to answer a query, etc.

can be computed in linear time), the overall data structure can be constructed in time $O(n \log n)$.

To count the number of points of S lying in a query triangle $\triangle$, we visit T in a top-down fashion starting from the root. For each cell τ of $\mathscr{A}(\mathscr{R})$ we determine whether it intersects any of the lines bounding the edges of $\triangle$. If τ intersects any of these lines, then we compute $|S_\tau \cap \triangle|$ recursively by visiting the subtree rooted at w_τ. Otherwise, either all points of S_τ lie in $\triangle$, or none of them lie in $\triangle$. Since we store the value of $|S_\tau|$ at w_τ, $|S_\tau \cap \triangle|$ can be computed in constant time if τ does not intersect any of the lines bounding $\triangle$.

Hence, if $Q(n)$ denotes the maximum query time for a set of n points, then we obtain the recurrence

$$Q(n) \le \sum_{i=1}^{O(r^2)} Q(n_i) + O(r^4),$$

where $\sum_{i=1}^{O(r^2)} n_i \le \frac{n}{r}$. The solution of this recurrence is $O(n^{2/3+\varepsilon})$, as required.

The same approach also works for higher dimensions. $\square$

A nice property of the above structure is that it supports multi-level structures without affecting the asymptotic query time, i.e., one can store a secondary data structure at each node of T. The secondary structure may either be another partition tree or an entirely different structure. The multi-level property of the structure allows it to be used for several other geometric problems; see [50, 66, 7].

Spanning paths with low stabbing number. The first simplex range searching algorithm in two dimensions, whose performance came close to Chazelle's lower bound, was given by Welzl [107] and then refined by Chazelle and Welzl [39]. Their algorithm is based on a data structure called *spanning paths with low stabbing number*. The *stabbing number* of a tree embedded in $\mathbb{R}^d$, whose edges are straight line segments, is the maximum number of its edges intersected by a hyperplane. Chazelle and Welzl [39] proved

THEOREM 6.3 (Chazelle-Welzl [39]). *For a set S of n points in $\mathbb{R}^d$ in general position, there is always a spanning path $\Pi(S)$ whose stabbing number is $O(n^{1-1/d})$.*

Observe that a query simplex $\triangle$ can intersect only $O(n^{1-1/d})$ edges of $\Pi(S)$, therefore $S \cap \triangle$ can be represented as a union of $O(n^{1-1/d})$ subsets, each corresponding to the points lying on $\Pi(S)$ between two consecutive intersection points of $\Pi(S)$ and $\triangle$. Hence, one can precompute $O(n \log n)$ canonical subsets in a straightforward manner, so that $S \cap \triangle$ can be represented by $O(n^{1-1/d} \log n)$ of these canonical subsets. A more sophisticated argument shows that the number of canonical subsets can be reduced to $O(n^{1-1/d}\alpha(n))$, where $\alpha(n)$ is as usual the inverse Ackermann function [109].

For $d = 2, 3$, one can compute the set of canonical subsets in time $O(n^{1-1/d}\log n)$, which leads to almost optimal simplex range searching algorithms [39]. But it is not clear how to extend this approach to higher dimensions.

The proof of Theorem 6.3 is based on the following lemma.

LEMMA 6.4 (Chazelle-Welzl [39]). *Let S be a set of n points in general position and Γ be a multiset of m hyperplanes in $\mathbb{R}^d$. Then there exists a pair $p, q \in S$ such that the relative interior of the segment pq intersects only $O(\frac{m}{n^{1/d}})$ hyperplanes of Γ.*

PROOF. Compute a $\frac{c}{n^{1/d}}$-cutting Ξ of size less than n for Γ, where c is some suitable constant (Corollary 2.5 guarantees the existence of such a cutting). Since $|\Xi| < n$, there are two points $p, q \in S$ that lie in the same simplex $\triangle$. But then the relative interior of the segment pq intersects at most $c\frac{m}{n^{1/d}}$ hyperplanes of Γ as the interior of $\triangle$ itself intersects only $c\frac{m}{n^{1/d}}$ hyperplanes of Γ. $\square$

PROOF OF THEOREM 6.3. Let $S_0 = S$ be a set of $n_0 = n$ points in $\mathbb{R}^d$ in general position and let Γ_0 be a set of $m = \binom{n}{d} + 1$ hyperplanes avoiding these points such that they represent every possible bisection of S by a hyperplane.

By Lemma 6.4, we can find two points $x_0, y_0 \in S$ such that the segment $x_0 y_0$ intersects at most $cm_0/n_0^{1/d}$ elements of Γ_0. Let $S_1 = S_0 - \{x_0\}$, $n_1 = |S_1| = n - 1$, and let Γ_1 be the multiset of hyperplanes obtained from Γ_0 by duplicating every hyperplane crossed by $x_0 y_0$. Let $m_1 = |\Gamma_1|$. We can again find a pair $x_1, y_1 \in S_1$ such that the segment $x_1 y_1$ intersects $\leq cm_1/n_1^{1/d}$ hyperplanes of Γ_1. Let $S_2 = S_1 - \{x_1\}$ and Γ_2 be the multiset of hyperplanes obtained from Γ_1 by duplicating every hyperplane of Γ_1 that crosses $x_1 y_1$.

Repeating this step $n - 1$ times, the segments $x_0 y_0, x_1 y_1, \ldots, x_{n-2} y_{n-2}$ form a spanning tree T of S. By construction, for every $i \leq n - 1$,

$$m_{i+1} \leq m_i \left(1 + \frac{c}{n_i^{1/d}} \right).$$

Hence,

$$m_{n-1} \leq m_0 \prod_{i=0}^{n-2} \left(1 + \frac{c}{n_i^{1/d}} \right).$$

An easy calculation shows that $m_{n-1} \leq \exp(O(n^{1-1/d}))$.

If a hyperplane $\gamma \in \Gamma_0$ intersects t edges of T, then Γ_{n-1} has 2^t copies of γ, therefore $t = O(n^{1-1/d})$. Since the hyperplanes of Γ_0 exhaust all possible ways of how S can be bisected by a hyperplane, the theorem follows from the fact that T can be converted into a spanning path whose stabbing number is twice the stabbing number of T. □

Although spanning paths with low stabbing number were originally used for efficient simplex range searching problems, they have been used for designing efficient data structures for a variety of other query type problems (including ray shooting [1] and computing the face in line arrangements containing a query point [52]). See [1] for an extensive list of such problems, and [83, 18] for some combinatorial applications.

Because of its importance in several geometric problems, a lot of work has been done in the last few years on developing efficient algorithms for computing spanning paths with low stabbing number; see [52, 75, 3, 77, 40, 14]. Most of these papers basically convert the proof of Theorem 6.2 into an efficient algorithm.

Higher-dimensional range searching. As mentioned above, spanning paths with low stabbing number fail to give efficient range searching algorithms in higher dimensions. For $d \geq 4$, Chazelle et al. [38] succeeded in developing a quasi-optimal algorithm for simplex range searching by modifying the Haussler-Welzl algorithm. They observed that property (P2) in the proof of Theorem 6.2 can be strengthened as follows: Although the zone of $\mathscr{A}(\mathscr{R})$ with respect to some query lines can contain $\Omega(n/r)$ points (for $d = 2$), it contains only $O(n/r^2)$ points with respect to a typical line. This observation led them to prove the following surprising result.

THEOREM 6.5 (Chazelle et al. [38]). *Given a set S of n points in $\mathbb{R}^d$ and a parameter r, one can construct a collection $\{\mathscr{A}^*(H_1), \ldots, \mathscr{A}^*(H_k)\}$ of $k = O(\log r)$ triangulated arrangements of hyperplanes, consisting of $O(r \log r)$ hyperplanes each, such that for every hyperplane h there exists an i for which the zone of $\mathscr{A}^*(H_i)$ with respect to h contains at most n/r points of S.*

Based on this theorem, Chazelle et al. construct a data structure in time $O(n^{1+\varepsilon})$ that can answer a simplex range query in time $O(n^{1-1/d+\varepsilon})$. Recently, Matoušek improved the bounds by combining the algorithms of Chazelle-Welzl [39] and Chazelle et al. [38]. He showed

THEOREM 6.6 (Matoušek [79]). *A set S of n points in $\mathbb{R}^d$ can be preprocessed in time $O(n \log n)$ into a data structure of linear size, so that the number of points lying in a query simplex can be counted in time $O(n^{1-1/d} \log^{O(1)} n)$.*

In order to prove this theorem, we need some auxiliary results.

DEFINITION 6.7. Let S be a set of points in $\mathbb{R}^d$ in general position. A *simplicial partition* Π of S is a collection $\Pi = \{(S_0, \triangle_0), \ldots, (S_m, \triangle_m)\}$, where $S_0, \ldots, S_m$ form a partition of S, and each $\triangle_i$ is a simplex containing the set S_i. The simplices of Π are not necessarily pairwise disjoint, and a simplex $\triangle_i$ may contain some points of another set S_j. The *stabbing number* of Π is the maximum number of its simplices intersected by a hyperplane.

LEMMA 6.8 (Matoušek [79]). *Let S be a set of n points in $\mathbb{R}^d$ in general position and $r \leq n$ a parameter. There exists a simplicial partition $\Pi = \{(S_0, \triangle_0), \ldots, (S_m, \triangle_m)\}$ of a subset $S' \subseteq S$ of at least $n/2$ points with $|S_i| = \lfloor n/r \rfloor$ $(0 \leq i \leq m)$, whose stabbing number is $O(r^{1-1/d})$.*

SKETCH OF PROOF. The proof is very similar to that of Theorem 6.3. Let $P_0 = S$ be the set of $n_0 = n$ points in $\mathbb{R}^d$ in general position and let Γ_0 be the same as defined in the proof of Theorem 6.3. Find a $\frac{1}{t}$-cutting Ξ_0 of size at most $r/2$ for Γ_0, where $t = \Theta(r^{1/d})$. For the sake of simplicity, assume that no point of P_0 lies on the boundary of any simplex of Ξ_0. Since $|S_0| \geq n/2$, there is a simplex $\triangle \in \Xi_0$ that contains at least n/r points of P_0. Shrink $\triangle$ until it contains n/r points of P_0. Let $\triangle_0$ be the resulting simplex and $S_0 = \triangle \cap P_0$.

Let $P_1 = P_0 - S_0$ and let Γ_1 be the set of hyperplanes obtained by duplicating every hyperplane of Γ_0 that intersects the interior of $\triangle_0$. Again, we can find a simplex $\triangle_1$ that contains $\lfloor n/r \rfloor$ points of P_1 and intersects at most $|\Gamma_1|/t$ hyperplanes of Γ_1. Set $S_1 = P_1 \cap \triangle_1$. We continue this process until $P_{i+1} < n/2$. The set $\Pi = \{(S_0, \triangle_0), \ldots, (S_i, \triangle_i)\}$ gives the desired simplicial partition of $S' = S - P_{i+1}$ $(|S'| \geq n/2)$. By construction,

$$|S_i| = \left\lfloor \frac{n}{r} \right\rfloor \quad \text{and} \quad |\Gamma_{i+1}| = |\Gamma_i| \left(1 + \frac{c}{r^{1/d}} \right).$$

By the same argument as in Theorem 6.3, we can show that every hyperplane of Γ_0 intersects $O(r^{1-1/d})$ simplices of Π. Choosing a more clever set of hyperplanes instead of Γ_0 and repeating the same procedure, one can show that the stabbing number of Π is $O(r^{1-1/d})$ (see [79] for details). □

An immediate corollary of this lemma is

COROLLARY 6.9 (Matoušek [79]). *Given a set S of n points in $\mathbb{R}^d$ and a parameter r, one can compute a simplicial partition $\Pi = \{(S_0, \triangle_0), \ldots, (S_m, \triangle_m)\}$ of S with $|S_i| = \lfloor n/r \rfloor$ $(1 \leq i \leq m)$, whose stabbing number is $O(r^{1-1/d})$.*

In view of this corollary, by choosing an appropriate value of r, one can construct a recursive data structure of linear size so that a simplex range query can be answered in time $O(n^{1-1/d} \log^{O(1)} n)$. In order to bound the preprocessing time by $O(n \log n)$, one has to introduce several additional tricks; see [79]. This finishes the proof of Theorem 6.6.

If one allows $O(n^{1+\varepsilon})$ preprocessing (space is still linear), the query time of the simplex range searching can be improved to $O(n^{1-1/d} (\log \log n)^{O(1)})$.

Furthermore, the query time of the algorithms of [38] and [79] can be reduced by allowing more space. In particular, if we allow $O(m^{1+\varepsilon})$ space, the query time of Matoušek's algorithm can be improved to $O(\frac{n}{m^{1/d}}(\log\log n)^{O(1)})$.

In some applications [8, 9], the query object is a half-space g and one either wants to detect whether $G \cap S = \emptyset$ or report all the points of $g \cap S$. Matoušek has recently proposed a data structure that can answer the *empty half-space queries* in time $O(n^{1-1/(\lfloor d/2 \rfloor)}2^{\log^* n})$ using $O(n)$ space and $O(n^{1+\varepsilon})$ preprocessing [80]. Moreover, it can report all points of $S \cap g$ in time $O(n^{1-1/(\lfloor d/2 \rfloor)}\log^{O(1)} n + k)$. Using this procedure as a subroutine, Agarwal and Matoušek have presented efficient algorithms for answering ray shooting and nearest (or farthest) neighbor queries and related problems [9, 116]. For example, given a set of n points in $\mathbb{R}^d$ and a parameter $n \le m \le n^{\lceil d/2 \rceil}$, a nearest (or farthest) neighbor query can be answered in time $O(\frac{n}{m^{1/\lceil d/2 \rceil}}\log^3 n)$ using $O(m^{1+\varepsilon})$ space and preprocessing. It can also compute k nearest (or farthest) neighbors at an additional cost of $O(k\log^2 n)$ in the query time.

Acknowledgments

The author would like to thank Jiří Matoušek and an anonymous referee for several useful comments.

REFERENCES

1. P. Agarwal, *Ray shooting and other applications of spanning trees with low stabbing number*, Proc. Fifth Ann. ACM Sympos. on Computational Geometry, 1990, pp. 315–325. (Also to appear in SIAM J. Comput.)

2. _____, *Partitioning arrangements of line*: I. *An efficient deterministic algorithm*, Discrete Comput. Geom. **5** (1990), 449–483.

3. _____, *Partitioning arrangements of line*: II. *Applications*, Discrete Comput. Geom. **5** (1990), 533–573.

4. P. Agarwal and B. Aronov, *Counting facets and incidences*, Discrete Comput. Geom.(to appear).

5. P. Agarwal, B. Aronov, M. Sharir, and S. Suri, *Selecting distances in the plane*, Proc. Sixth Ann. ACM Sympos. on Computational Geometry, 1990, pp. 321–331. (Also to appear in Algorithmica.)

6. P. Agarwal, H. Edelsbrunner, O. Schwarzkopf, and E. Welzl, *Euclidean minimum spanning trees and bichromatic closest pairs*, Discrete Comput. Geom. **6** (1991), 407–422.

7. P. Agarwal, M. van Krevald, and M. Overmars, *Intersection queries for curved objects* Proc. Seventh Ann. ACM Sympos. on Computational Geometry, 1991, 41–50.

8. P. Agarwal and J. Matoušek, *Ray shooting and parametric search*, Technical Report CS-91-22, Dept. Computer Science, Duke University, Durham, NC, 1991.

9. _____, *Relative neighborhood graphs in three dimensions*, Proc. Third Ann. ACM-SIAM Sympos. on Discrete Algorithms (to appear).

10. _____, *Searching with nonlinear objects* (in preparation).

11. P. Agarwal, J. Matoušek, and S. Suri, *Farthest neighbors, maximum spanning trees and related problems in higher dimensions*, Proc. Second Workshop on Algorithms and Data Structures, 1991, 105–116. (Also to appear in Comp. Geom.: Theory & Applications)

12. P. Agarwal and M. Sharir, *Red-blue intersections with applications to collision detection*, SIAM J. Comput.**19** (1990), 297–322.

13. _____, *Counting circular arc intersections*, Proc. Seventh Ann. ACM Sympos. on Computational Geometry, 1991.

14. _____, *Applications of a new partitioning scheme*, Proc. Second Workshop on Algorithms and Data Structures, 1991.

15. P. Agarwal, M. Sharir, and P. Shor, *Sharp bounds on Davenport-Schinzel sequences*, J. Combin. Theory Ser. A **52** (1989), 228–274.

16. P. Agarwal, M. Sharir, and S. Toledo, *Applications of parametric searching in geometric optimization*, Proc. Third Ann. ACM-SIAM Sympos. on Discrete Algorithms, 1992 (to appear).

17. B. Aronov, H. Edelsbrunner, L. Guibas, and M. Sharir, *Improved bounds on the complexity of many faces in arrangements of segments*, Combinatorica (to appear).

18. B. Aronov, P. Erdős, W. Goddard, D. Kleitman, M. Klugerman, J. Pach, and J. Schulman, *Crossing families*, Proc. Seventh Ann. ACM Sympos. on Computational Geometry, 1991, pp. 351–356.

19. B. Aronov and M. Sharir, *On the zone of a surface in a hyperplane arrangement*, Proc. Second Workshop on Algorithms and Data Structures, 1991, pp. 13–19.

20. J. Beck, *On the lattice property of the plane and some problems of Dirac, Motzkin and P. Erdős in combinatorial geometry*, Combinatorica **3** (1983), 281–297.

21. J. Bentley and T. Ottman, *Algorithms for reporting and counting geometric intersections*, IEEE Trans. Comput. C–28 (1979), 643–647.

22. J. Bentley and M. Shamos, *Divide-and-conquer in multidimensional space*, Proc. Eighth Ann. ACM Sympos. on Theory of Computing, 1976, pp. 220–230.

23. M. de Berg, D. Halperin, M. Overmars, J. Snoeyink, and M. van Kreveld, *Efficient ray shooting and hidden surface removal*, Proc. Seventh Ann. ACM Sympos. on Computational Geometry, 1991, pp. 51–60.

24. B. Berger and J. Rompel, *Simulating* $(\log^c n)$*-wise independence in NC*, Proc. 30th Ann. IEEE Sympos. on Foundations of Computer Science, 1989, pp. 2–7.

25. B. Berger, J. Rompel, and P. Shor, *Efficient NC algorithms for set cover with applications to learning and geometry*, Proc. 30th Ann. IEEE Sympos. on Foundations of Computer Science, 1989 pp. 54–59.

26. A. Blummer, A. Ehrenfeucht, D. Haussler, and M. Warmuth, *Classifying learnable geometric concepts with the Vapnik-Chervonenkis dimension*, J. Assoc. Comput. Mach. **36** (1989), 926–965.

27. R. Canham, *A theorem on arrangements of lines in the plane*, Israel J. Math. **7** (1969), 393–397.

28. B. Chazelle, *Reporting and counting segment intersections*, J. Comput. System Sci. **32** (1986), 156–182.

29. _____, *Tight bounds on the stabbing number of trees in Euclidean plane*, Technical Report CS-TR-155-58, Dept. Computer Science, Princeton University, Princeton, NJ, May 1988.

30. _____, *Lower bounds on the complexity of polytope range searching*, J. Amer. Math. Soc. **2** (1989), 637–666.

31. B. Chazelle and H. Edelsbrunner, *An optimal algorithm for intersecting line segments in the plane*, Proc. 29th Ann. IEEE Sympos. on Foundations of Computer Science, 1989, pp. 590–600.

32. B. Chazelle, H. Edelsbrunner, L. Guibas, R. Pollack, R. Seidel, J. Snoeyink, and M. Sharir, *Counting and cutting cycles of lines and rods in space*, Proc. 31st Ann. IEEE Sympos. on Foundations of Computer Science, 1990, pp. 242–251.

33. B. Chazelle, *An optimal convex hull algorithm and new results on cuttings*, Proc. 32nd Ann. IEEE Sympos. on Foundations of Computer Science, 1991.

34. B. Chazelle, H. Edelsbrunner, L. Guibas, and M. Sharir, *Lines in space: Combinatorics, algorithms and applications*, Proc. 21st Ann. ACM Sympos. on Theory of Computing, 1989.

35. _____, *A singly-exponential stratification scheme for real semi-algebraic varieties and its applications*, Proc. 16th Internat. Colloq. on Automata, Languages and Programming, 1989, pp. 179–192.

36. B. Chazelle and J. Friedman, *A deterministic view of random sampling and its use in geometry*, Combinatorica **10** (1990), 229–249.

37. B. Chazelle and M. Sharir, *An algorithm for generalized point location and its application*, J. Symbolic Comput. **10** (1990), 281–309.

38. B. Chazelle, M. Sharir, and E. Welzl, *Quasi optimal upper bounds for simplex range searching and new zone theorem*, Proc. Sixth Ann. ACM Sympos. on Computational Geometry, 1990, pp. 23–33.

39. B. Chazelle and E. Welzl, *Quasi optimal range searching in spaces with finite VC-dimension*, Discrete Comput. Geom. **4** (1989), 467–489.

40. S. Cheng and R. Janardan, *Space-efficient ray shooting and intersection searching: Algorithms, dynamization and applications*, Proc. Second Ann. ACM-SIAM Sympos. on Discrete Algorithms, 1991, pp. 7–16.

41. F. Chung, *The number of different distances in the plane*, J. Combin. Theory Ser. A **36** (1984), 342–354.

42. _____, *Sphere-point incidence relations in higher dimensions with applications to unit distances and furthest neighbor pairs*, Discrete Comput. Geom. **4** (1989), 183–190.

43. F. Chung, E. Szemerédi, and W. Trotter, Jr., *The number of distinct distances determined by a finite point set in the plane*, Discrete Comput. Geom. (to appear).

44. K. Clarkson, *A randomized algorithm for closest-point queries*, SIAM J. Comput. **17** (1988), 830–847.

45. _____, *New applications of random sampling in computational geometry*, Discrete Comput. Geom. **2** (1987), 195–222.

46. _____, *Applications of random sampling in computational geometry. II*, Proc. Fourth Ann. ACM Sympos. on Computational Geometry, 1988, pp. 1–11.

47. K. Clarkson, H. Edelsbrunner, L. Guibas, M. Sharir, and E. Welzl, *Combinatorial complexity bounds for arrangements of curves and surfaces*, Discrete Comput. Geom. **5** (1990), 99–160.

48. K. Clarkson and P. Shor, *Applications of random sampling in computational geometry. II*, Discrete Comput. Geom. **4** (1989), 387–421.

49. G. Collins, *Quantifier elimination for real closed fields by cylinderical algebraic decomposition*, Lecture Notes in Computer Science, vol. 33, Springer-Verlag, Berlin, 1975, pp. 134–183.

50. D. Dobkin and H. Edelsbrunner, *Space searching for intersecting objects*, J. Algorithms **8** (1987), 348–361.

51. H. Edelsbrunner, *Algorithms in Combinatorial Geometry*, Springer-Verlag, Heidelberg, 1987.

52. H. Edelsbrunner, L. Guibas, J. Hershberger, R. Seidel, M. Sharir, J. Snoeyink, and E. Welzl, *Implicitly representing arrangements of lines or segments*, Discrete Comput. Geom. **4** (1989), 433–466.

53. H. Edelsbrunner, L. Guibas, J. Pach, R. Pollack, R. Seidel, and M. Sharir, *Arrangements of curves in the plane—Topology, combinatorics, and algorithms*, Proc. 15th Internat. Colloq. on Automata, Languages and Programming, Lecture Notes in Computer Science, vol. 318, Springer-Verlag, New York, 1988, pp. 214–229.

54. H. Edelsbrunner, L. Guibas, and M. Sharir, *The complexity of many faces in arrangement of lines and of segments*, Discrete Comput. Geom. **5** (1990), 161–196.

55. _____, *The complexity of many cells in arrangements of planes and related problems*, Discrete Comput. Geom. **5** (1990), 197–216.

56. H. Edelsbrunner, J. O'Rourke, and R. Seidel, *Constructing arrangements of lines and hyperplanes with applications*, SIAM J. Comput. **15** (1986), 341–363.

57. H. Edelsbrunner and M. Sharir, *A hyperplane incidence problem with applications to counting distances*, Proc. Internat. Sympos. on Algorithms, Lecture Notes in Computer Science, vol. 450, Springer-Verlag, Berlin and New York, 1990.

58. H. Edelsbrunner and E. Welzl, *Constructing belts in two-dimensional arrangements with applications*, SIAM J. Comput. **15** (1986), 271–284.

59. _____, *On the maximal number of edges of many faces in an arrangement*, J. Combin. Theory Ser. A **41** (1986), 159–166.

60. _____, *Halfplanar range search in linear space and $O(n^{0.695})$ query time*, Inform. Process. Lett. **23** (1986), 289–293.

61. P. Erdős, *On a set of distances of n points*, Amer. Math. Monthly **53** (1946), 248–250.

62. _____, *On sets of distances of n points in Euclidean space*, Publ. Math. Inst. Hungar. Acad. Sci. **5** (1960), 165–169.

63. _____, *On some applications of graph theory to geometry*, Canad. J. Math. **19** (1967), 968–971.

64. P. Erdős and J. Pach, *Variations on the theme of repeated distances*, Combinatorica **10** (1990), 261–269.

65. M. L. Fredman, *A lower bound on the complexity of orthogonal range queries*, J. Assoc. Comput. Mach. **28** (1981), 696–705.

66. L. Guibas, M. Overmars, and M. Sharir, *Ray shooting, implicit point location, and related queries in arrangements of segments*, Technical Report 433, Dept. Computer Science, New York University, New York, NY, March 1989.

67. _____, *Counting and reporting intersections in arrangements of line segments*, Technical Report 434, Dept. Computer Science, New York University, New York, NY, March 1989.

68. S. Hart and M. Sharir, *Nonlinearity of Davenport-Schinzel sequences and of generalized path compression schemes*, Combinatorica **6** (1986), 151–177.

69. D. Haussler and E. Welzl, *ε-nets and simplex range queries*, Discrete Comput. Geom. **2** (1987), 127–151.

70. J. Komlós, J. Pach, and G. Woeginger, *Almost tight bounds for epsilon-nets*, Discrete Comput. Geom. (to appear).

71. C. Lo, J. Matoušek, and W. Steiger, *Ham-sandwich cuts in two and three dimensions* (in preparation).

72. C. Lo and W. Steiger, *An optimal-time algorithm for ham-sandwich cuts in the plane*, Proc. Second Canadian Conf. on Computational Geometry, 1990.

73. L. Lovász, *On the ratio of integral and fractional cover*, Discrete Mathematics **13** (1975), 383–390.

74. J. Matoušek, *Construction of ε-nets*, Discrete Comput. Geom. **5** (1990), 427–448.

75. _____, *Spanning trees with low crossing numbers*, Inform. Theoret. Appl. (1991), 103–123.

76. _____, *Cutting hyperplane arrangements*, Discrete Comput. Geom. **6** (1991), 385–406.

77. _____, *More on cutting arrangements and spanning trees with low stabbing number*, Technical Report B-90-2, Fachbereich Mathematik, Freie Universitäat, February 1990.

78. _____, *Approximations and optimal geometric divide-and-conquer*, Proc. 23rd Ann. ACM Sympos. on Theory of Computing, pp. 506–511.

79. _____, *Efficient partition trees*, Proc. Seventh Ann. ACM Sympos. on Computational Geometry, 1991, pp. 1–9.

80. _____, *Halfspace range reporting*, Proc. 32nd Ann. IEEE Sympos. on Foundations of Computer Science, 1991, pp. 207–215.

81. _____, *Linear optimization queries*, manuscript, 1991.

82. J. Matoušek, R. Seidel, and E. Welzl, *How to net a little: small ε-nets for disks and halfspaces*, Proc. Sixth Ann. ACM Sympos. on Computational Geometry, 1990, pp. 16–22. (Also to appear in J. Algorithms)

83. J. Matoušek, E. Welzl, and L. Wernisch, *Discrepancy and ε-approximations for bounded VC-dimension*, Proc. 32nd Ann. IEEE Sympos. on Foundations of Computer Science, 1991, pp. 424–430. (Also to appear in Combinatorica).

84. N. Megiddo, *Applying parallel computation algorithms in design of serial algorithms*, J. Assoc. Comput. Mach. **30** (1983), 852–865.

85. _____, *Linear programming in linear time when the dimension is fixed*, J. Assoc. Comput. Mach. **31** (1984), 114–127.

86. K. Mehlhorn, *Multi-dimensional searching and computational geometry*, Springer-Verlag, Heidelberg, 1984.

87. R. Motwani, J. Naor, and M. Naor, *The probabilistic method yields deterministic parallel algorithms*, Proc. 30th Ann. IEEE Sympos. on Foundations of Computer Science, 1989, 8–13.

88. K. Mulmuley, *A fast planar partition algorithm.* I, J. Symbolic Comput. **10** (1990), 253–280.

89. K. Mulmuley and S. Sen, *Dynamic point location in arrangements of hyperplanes*, Proc. Seventh Ann. ACM Sympos. on Computational Geometry, 1991, pp. 132–141.

90. M. Overmars, H. Schipper, and M. Sharir, *Storing line segments in partition trees*, BIT **30** (1990), 385–403.

91. M. Overmars and M. Sharir, *Output sensitive hidden surface removal algorithms*, Proc. 30th Ann. IEEE Sympos. on Foundations of Computer Science, 1989, pp. 598–603.

92. J. Pach and M. Sharir, *Repeated angles in the plane and related problems*, Technical Report 492, New York University, New York. (Also to appear in J. Combin. Theory Ser. A.)

93. J. Pach and G. Woeginger, *Some new bounds for epsilon-nets*, Proc. Sixth Ann. ACM Sympos. on Computational Geometry, 1990, pp. 10–15.

94. M. Pellegrini, *Ray shooting and isotopy classes of lines in 3-D space*, Proc. Second Workshop on Algorithms and Data Structures, 1991.

95. F. Preparata and M. Shamos, *Computational geometry: An introduction*, Springer-Verlag, Heidelberg, 1985.

96. P. Raghavan, *Probabilistic construction of deterministic algorithms: approximating packing integer programs*, J. Comput. System Sci. **37** (1988), 130–143.

97. J. Reif and S. Sen, *Optimal randomized parallel algorithms for computational geometry*, Proc. 16th Internat. Conf. on Parallel Processing, 1987.

98. _____, *Polling: A new randomized sampling technique for computational geometry*, Proc. 21st Ann. ACM Sympos. on Theory of Computing, 1989.

99. N. Sarnak and R. Tarjan, *Planar point location using persistent search trees*, Comm. ACM **29** (1986), 669–679.

100. M. Shamos, *Computational Geometry*, Ph.D. thesis, Yale University, New Haven, CT, 1978.

101. D. M. H. Sommerville, *Analytical Geometry in Three Dimensions,* Cambridge, 1951.

102. J. Spencer, *Ten Lectures on the Probabilistic Method*, CBMS-NSF, SIAM, 1987.

103. J. Spencer, E. Szemerédi, and W. Trotter, Jr., *Unit distances in the Euclidean plane*, Graph Theory and Combinatorics (B. Bollobás, ed.) Academic Press, NY, 1984, pp. 293–303.

104. J. Stolfi, *Primitives for Computational Geometry,* Ph.D. dissertation, Stanford University, Standford, CA, 1989.

105. E. Szemerédi and W. Trotter, Jr., *Extremal problems in discrete geometry*, Combinatorica **3** (1983), 381–392.

106. V. N. Vapnik and A. Chervonenkis, *On the uniform convergence of relative frequencies of events to their probabilities*, Theory Probab. Appl. **16** (1971), 264–280.

107. E. Welzl, *Partition trees for triangle counting and other range searching problems*, Proc. Fourth Ann. ACM Sympos. on Computational Geometry, 1988, pp. 23–33.

108. D. Willard, *Polygon retrieval*, SIAM J. Comput. **11** (1982), 149–165.

109. A. Yao, *Space-time tradeoff for answering range queries*, Proc. 14th Ann. ACM Sympos. on Theory of Computing, 1982, pp. 128–136.

110. A. Yao and F. Yao, *A general approach to geometric queries*, Proc. 15th Ann. ACM Sympos. on Theory of Computing, 1985, pp. 163–168.

111. F. Yao, *A 3-space partition and its applications*, Proc. 15th Ann. ACM Sympos. on Theory of Computing, 1983, pp. 258–263.

112. J. Conway, H. Croft, P. Erdős, and M. Guy, *On the distribution of angles determined by coplanar points*, J. London Math. Soc. (2) **19** (1979), 137–143.

113. M. van Kreveld and M. de Berg, *Finding squares and rectangles in sets of points*, Tech. Rept. RUU-CS-89-10, Dept. Comp. Science, Univ. of Utrecht, 1989.

114. G. Purdy, *Repeated angles in E^4*, Discrete Comput. Geom. **3** (1988), 73–75.

115. J. Matoušek, *Range searching with efficient hierarchical cuttings*, manuscript, 1991.

116. P. Agarwal and J. Matoušek, *Dynamic half-space range searching and its applications*, manuscript, 1991.

117. E. Szemerédi and W. Trotter, Jr., *A combinatorial distinction between the Euclidean and projective planes*, Europ. J. Combinatorics **4** (1983), 385–394.

DEPARTMENT OF COMPUTER SCIENCE, DUKE UNIVERSITY, DURHAM, NORTH CAROLINA 27706

DIMACS Series in Discrete Mathematics
and Theoretical Computer Science
Volume 6, 1991

On the Convex Hull of the Integer Points in a Disc

ANTAL BALOG AND IMRE BÁRÁNY

ABSTRACT. Let P_r denote the convex hull of the integer points in the disc of radius r. We prove that the number of vertices of P_r is essentially $r^{2/3}$ as $r \to \infty$.

1. Introduction

Take a disc of radius r in the plane and consider P_r, the convex hull of the integer points inside the disc. How many vertices will P_r have?

Motivation for this equation comes from several sources. First, in integer programming, one wants to know the number of solutions, when c varies, to the problem $\max c \cdot x$ subject to $x \in K$ where K is a convex body in R^d. The answer is the number of vertices of $\operatorname{conv}(K \cap \mathbb{Z}^d)$. A relevant result in integer *linear* programming is the following. Let $P \subset R^d$ be a polyhedron given by the inequalities $a_i \cdot x \leqq \alpha_i$ $(i = 1, \ldots, m)$ with $a_i \in \mathbb{Z}^d$ and $\alpha_i \in \mathbb{Z}$. The *size* of P, $\operatorname{size}(P)$ is defined as the number of bits necessary to encode it as a binary string, i.e.,

$$\operatorname{size}(P) = \sum_{i=1}^{m} \left(\sum_{j=1}^{d} \lceil \log(|a_{ij}| + 1) \rceil + \lceil \log(|\alpha_i| + 1) \rceil \right).$$

Then, as it is shown in [5], the number of vertices of $\operatorname{conv}(\mathbb{Z}^d \cap P)$ is at most $2m^d[12d^2\operatorname{size}(P)]^{d-1}$. A construction in [3] shows that this result is best possible.

1991 *Mathematics Subject Classification.* Primary 52C05; Secondary 11H06, 11P21.

Both authors are on leave from the Mathematical Institute of the Hungarian Academy of Sciences, 1364 Budapest, Pf 127, Hungary.

The first author was supported by NSF grant DMS-8610730.

The second author was supported by the Program in Discrete Mathematics and its Application at Yale and NSF Grant CCR-8901484.

A second motivation comes from classical results. Write B^d for the d-dimensional Euclidean ball. Van der Corput proved in 1922 [6] that

$$(1) \qquad |\mathbb{Z}^2 \cap rB^2| = r^2 \pi + O(r^{2/3 - \varepsilon})$$

with $\varepsilon = 0.01$. Since then there have been a lot of (minor) improvements in ε, probably the last coming from Iwaniec and Mozzochi (see [8]), generalized by Huxley [8]. He proves that if D is a convex body in R^2 with $\mathscr{C}^3$ boundary and positive curvature at every point of the boundary, then

$$(2) \qquad |\mathbb{Z}^2 \cap rD| = r^2 \operatorname{Area} D + O(r^{7/11 + \varepsilon}).$$

Another classical result is due to Jarnik [9]. He showed that if Γ is a strictly convex curve in the plane whose length is s, then

$$(3) \qquad |\mathbb{Z}^2 \cap \Gamma| \leqq \frac{3}{\sqrt[3]{2\pi}} s^{2/3} + O(s^{1/3}).$$

If Γ is $\mathscr{C}^3$, then the exponent $\frac{2}{3}$ can be reduced to $\frac{3}{5}$ in (3). This is a result due to Swinnerton-Dyer [13] and Schmidt [12]. Jarnik gave an example of a strictly convex curve Γ whose length is s and whose radius of curvature is less than $7s$ at every point such that

$$|\mathbb{Z}^2 \cap \Gamma| \geq \frac{3}{\sqrt[3]{2\pi}} s^{2/3} + O(s^{1/3}).$$

(1) has been extended to higher dimensions:

$$|\mathbb{Z}^3 \cap rB^3| = r^3 \operatorname{vol}(B^3) + O(r^{4/3}),$$
$$|\mathbb{Z}^4 \cap rB^4| = r^4 \operatorname{vol}(B^4) + O(r^2 \log r),$$
$$|\mathbb{Z}^d \cap rB^d| = r^d \operatorname{vol}(B^d) + O(r^{d-2}), \quad \text{for } d > 4.$$

Here the first equality is due to Vinogradov [15] and Chen [4], the other two to Walfisz [14]. What we will need here is the weaker

$$(4) \qquad |\mathbb{Z}^d \cap rB^d| = r^d \operatorname{vol}(B^d) + o(r^{d(d-1)/(d+1)}),$$

valid for all $d \geq 2$.

Another motivation is the following. Let $x_1, \ldots, x_n$ be points chosen randomly, independently, and uniformly from B^d. Then $K_n = \operatorname{conv}\{x_1, \ldots, x_n\}$ is a random polytope. It is known (see, for instance, Schneider's survey paper [11]) that the expected number of vertices of K_n is $\operatorname{const}(d) n^{(d-1)/(d+1)}$. Now if one chooses r so that $r^d \operatorname{vol}(B^d) = n$, then in rB^d there will be essentially n integral points, and the number of vertices of $\operatorname{conv}(\mathbb{Z}^d \cap rB^d)$ must be around $n^{(d-1)/(d+1)} \approx r^{d(d-1)/(d+1)}$ if the integer points *behave* like random points in rB^d. It turns out that this is indeed the case for $d = 2$, as Theorem 1 shows.

Write $N(r, d)$ for the number of vertices of $\operatorname{conv}(\mathbb{Z}^d \cap rB^d)$ and set $N(r) = N(r, 2)$.

THEOREM 1. *For large enough* r, $c_1 r^{2/3} \leqq N(r) \leqq c_2 r^{2/3}$, *where* c_1 *and* c_2 *are absolute constants.*

From the proof we will get $c_1 \approx 0.33$ and $c_2 \approx 5.54$. It is not clear for us whether the limit $\lim_{r \to \infty} N(r) r^{-2/3}$ exists or not.

The proof of the upper bound in Theorem 1 is easier and works in any dimension:

(5)
$$N(r, d) \leqq c_d r^{d(d-1)/(d+1)}.$$

We can extend Theorem 1 to smooth enough convex bodies in R^2, using Huxley's result (2).

THEOREM 2. *If* D *is a plane convex body with* $\mathscr{C}^3$ *boundary and positive curvature, then*

$$c_1(D) r^{2/3} \leqq \# \text{ of vertices of } \operatorname{conv}(\mathbb{Z}^2 \cap rD) \leqq c_2(D) r^{2/3}$$

where the constants $c_1(D)$ *and* $c_2(D)$ *depend on the upper and lower bounds for the curvature of* D.

The proof is essentially the same, but more technical than that of Theorem 1 and will therefore be omitted.

In the proofs we will use Vinogradov's notation $\ll$ and $\ll_d$. All implied constants are effective.

2. Proof of the upper bounds

The upper bound in Theorem 1 is easier. It follows from Jarnik's result (3) but one has to make the boundary of P_r strictly convex. Actually, Jarnik's original proof applies as well giving $c_2 = 3(2\pi)^{1/3} = 5.5358\ldots$. Or one can use the following result of Andrews [1], cf. [2, 12, 10] as well. If $P \subset R^d$ is a convex polytope with integral vertices and nonempty interior, then

$$\# \text{ vertices of } P \ll_d (\operatorname{vol} P)^{(d-1)/(d+1)}.$$

This proves (5) immediately.

Now we give a simple direct proof of (5). Assume v is a vertex of $\operatorname{conv}(\mathbb{Z}^d \cap rB^d)$ and consider $M(v) = rB^d \cap (v - rB^d)$.

CLAIM 1. $\operatorname{vol} M(v) \leqq 2^d$.

Indeed, $M(v)$ is convex and centrally symmetric with respect to $v \in \mathbb{Z}^d$. By Minkowski's theorem, $\operatorname{vol} M(v) > 2^d$ would imply the existence of a point $x \in \mathbb{Z}^d \cap M(v)$, $x \neq v$. Then both x and $2v - x$ are integral and lie in rB^d so $v = \frac{1}{2}[x + (2v - x)]$ cannot be a vertex. $\square$

Assume now that v is at distance Δ from the boundary of rB^d. Clearly,

$$\operatorname{vol} M(v) > 2 \frac{\Delta}{d} (\sqrt{2r\Delta})^{d-1} \operatorname{vol} B^{d-1},$$

that gives, together with Claim 1, $\Delta \ll_d r^{(d-1)/(d+1)}$. Then, using (1) and (4)

$$N(r, d) \leq |\mathbb{Z}^d \cap rB^d| - |\mathbb{Z}^d \cap (r - \Delta)B^d| \ll_d r^{d(d-1)/(d+1)}. \quad \square$$

3. The lower bound

For the lower bound in Theorem 1 define $\Delta = 2^{-1/3} r^{-1/3}$, and set $A = A(r, \Delta) = rB^2 \setminus (r - \Delta)B^2$.

An integer point $x \in A$ is called a *vertex* if it is a vertex of P_r, and a *nonvertex* otherwise. The set of vertices will be denoted by V, the set of nonvertices by NV. For a nonvertex $x \in NV$ let $v \in V$ be the vertex nearest to x. This may not be unique, then choose any one of the nearest vertices. Draw an arrow from v to x and color this arrow green if it goes clockwise and blue if it goes counter-clockwise. We may assume that there are at least as many green arrows as blue ones. Denote the set of green arrows by G. Clearly, $|NV| \leq 2|G|$.

Observe that, if $\overrightarrow{vx} \in G$, then $\|v - x\| \leq \sqrt{2r\Delta}$. This is so because, as $x \in NV$, there must be a vertex of P_r in the cap (of rB^2) that has minimal area and contains x, and for any point y in that cap $\|x - y\| \leq \sqrt{(2r - \Delta)\Delta} < \sqrt{2r\Delta}$.

CLAIM 2. If $\overrightarrow{vx} \in G$ and $\overrightarrow{vy} \in G$, then v, x, y are collinear.

PROOF. An easy computation shows that the triangle with vertices v, x, y has area less than $\frac{1}{2}$. (This is where $\Delta = 2^{-1/3} r^{-1/3}$ is needed.) But any lattice triangle has area at least $\frac{1}{2}$ so v, x, y must be collinear.

This means that for fixed $v \in V$ there is a longest green arrow $\overrightarrow{vx}$ (with $x = x(v)$, say) containing all other green arrows starting at v. Fix now a primitive vector $p \in \mathbb{Z}^2$ (i.e., a vector $p \neq 0$ with relative prime components) and consider $S(p)$, the sum of all vectors $x(v) - v$ coming from a longest green arrow $\overrightarrow{vx}(v)$ that is parallel to p and points in the same direction.

CLAIM 3. $\|S(p)\| \ll r^{1/3}$.

We postpone the proof to the end of this section.

Clearly, $\|S(p)\|/\|p\|$ is equal to the number of green arrows that are parallel to p and point in the same direction. Now let $\{p_1, \ldots, p_m\}$ be the set of all primitive vectors with $S(p) \neq 0$. Evidently, $|V| \geq m$. On the other hand, by Claim 3

$$|G| = \sum_{i=1}^{m} \frac{\|S(p_i)\|}{\|p_i\|} \ll r^{1/3} \sum_{i=1}^{m} \frac{1}{\|p_i\|}.$$

Here $\sum_{i=1}^{m} \|p_i\|^{-1}$ will be the largest when $\{p_1, \ldots, p_m\}$ is the set of the m shortest primitive vectors in $\mathbb{Z}^2$. Then, as it is well known [7] and actually easy to see

$$\sum_{i=1}^{m} \frac{1}{\|p_i\|} \ll \sqrt{m} \leq \sqrt{|V|}.$$

Now by (1)

$$r^{2/3} \ll |A \cap \mathbb{Z}^2| = |V| + |NV| \leq |V| + 2|G| \ll |V| + r^{1/3}\sqrt{|V|},$$

which clearly implies the lower bound.

It is perhaps worth stating separately what we used in the last part of the proof: In the disc ρB^2, $o(\rho^2)$ diameters contain only $o(\rho^2)$ of the integer points in ρB^2.

PROOF OF CLAIM 3. Consider the lattice lines

$$l_i = \left\{ x \in R^2 : x = tp + i\frac{p^{\perp}}{\|p\|^2}, \ t \in R \right\}$$

where $i = 1, 2, \ldots$ and $p^{\perp}$ is the vector obtained from p by a $90°$ counter-clockwise rotation. (Here p is a primitive vector, again.) For each longest green arrow $\overrightarrow{vx}(v)$ where $x(v) = v + k(v)p$ $(k(v) = 1, 2, 3, \ldots)$ there is a line l_i such that the segment connecting v and $x(v)$ is contained in $A \cap l_i$. This intersection consists of either one or two segments but in both cases we have

$$\|x(v) - v\| = k(v)\|p\| \leq L_i := \text{half the length of } A \cap l_i.$$

More generally, let $l(h)$ denote the line parallel to p and at distance $r - h$ from the origin (so $0 < h < r$). Write $L(h)$ for the half-length of the intersection $A \cap l(h)$. Then

$$L(h) = \sqrt{(2r - h)h} - \sqrt{(2r - h - \Delta)|h - \Delta|_+}$$

where $|h - \Delta|_+ = h - \Delta$ if $h \geq \Delta$ and 0 otherwise. Clearly $l_i = l(h_i)$ with $h_i = r - i/\|p\|$. We must have

$$\|p\| \leq k(v)\|p\| \leq L_i.$$

The inequality $\|p\| \leq L(h)$ implies an upper bound for h, namely,

$$h \leq H := (1 + O(r^{-2/3}))\frac{2r\Delta^2}{\|p\|^2},$$

so that $H \ll r^{1/3}$. This shows that for $h \in [0, H]$

$$L(h) \ll \sqrt{2r}(\sqrt{h} - \sqrt{|h - \Delta|_+}).$$

Now

$$\|S(p)\| \leq \Sigma\{L_i : 0 \leq h_i \leq H\} \leq \|p\| + \|p\| \int_0^H L(h)\,dh + \max_{0 \leq h \leq H} L(h),$$

because the sum ΣL_i can be considered as an approximation to the integral

$\int_0^H L(h)\,dh$. Evidently $\max L(h) \le \sqrt{2r\Delta}$ and $\|p\| < \sqrt{2r\Delta}$. Then

$$\int_0^H L(h)\,dh \ll \sqrt{2r}\int_0^H (\sqrt{h} - \sqrt{|h-\Delta|}_+)\,dh$$
$$= \sqrt{2r}\frac{2}{3}(H^{3/2} - |H-\Delta|^{3/2})$$
$$\ll \frac{r\Delta^2}{\|p\|}.$$

So indeed,

$$\|S(p)\| \ll \sqrt{2r\Delta} + r\Delta^2 + \sqrt{2r\Delta} \ll r^{1/3}. \quad \square$$

References

1. G. E. Andrews, *A lower bound for the volumes of strictly convex bodies with many boundary points*, Trans. Amer. Math. Soc. **106** (1963), 270–279.
2. V. I. Arnold, *Statistics of integral convex polytopes*, Functional Anal. Appl. **14** (1980), 1–3. (Russian)
3. I. Bárány, R. Howe, and L. Lovász, *On integer points in polyhedra: A lower bound*, Combinatorica (to appear).
4. J. R. Chen, *The lattice points in a circle*, Sci. Sinica Ser. A **12** (1963), 633–649.
5. W. Cook, M. Hartman, R. Kannan, and C. McDiarmid, *On integer points in polyhedra*, Combinatorica (to appear).
6. J. G. van der Corput, *Verscharfung der Abschätzung beim Teilerproblem*, Math. Ann. **87** (1922), 39–65.
7. F. Fricker, *Einführung in die Gitterpunktlehre*, Birkhäuser, Basel, Boston, and Stuttgart, 1982.
8. M. N. Huxley, *Exponential sums and lattice points*, Proc. London Math. Soc. (3) **60** (1990), 471–502.
9. V. Jarnik, *Über Gitterpunkte and konvex Kurven*, Math. Z. **24** (1925), 500–518.
10. S. B. Konyagin and K. A. Sevastyanov, *Estimation of the number of vertices of a convex integral polyhedron in terms of its volume*, Functional Anal. Appl. **18** (1984), 13–15. (Russian)
11. R. Schneider, *Random approximation of convex sets*, Microscopy **151** (1988), 211–227.
12. W. M. Schmidt, *Integer points on curves and surfaces*, Monatsh. Math. **99** (1985), 45–72.
13. H. P. F. Swinnerton-Dyer, *The number of lattice points on a convex curve*, J. Number Theory **6** (1974), 128–135.
14. I. M. Vinogradov, *On the number of integer points in a sphere*, Izv. Akad. Nauk SSSR Ser. Mat. **27** (1963), 957–968. (Russian)
15. A. Walfisz, *Gitterpunkte in mehrdimensionalischen Kugeln*, Panstwowe Wydawnictwo Naukowe, Warszawa, 1957.

(Antal Balog) School of Mathematics, Institute for Advanced Study, Princeton, New Jersey 08540

(Imre Bárány) Cowles Foundation, Yale University, New Haven, Connecticut 06520

Courant Institute of Mathematical Sciences, New York University, New York, 10012

DIMACS Series in Discrete Mathematics
and Theoretical Computer Science
Volume **6**, 1991

Horizon Theorems for Lines and Polygons

MARSHALL BERN, DAVID EPPSTEIN,
PAUL PLASSMANN, AND FRANCES YAO

ABSTRACT. We give tight bounds on the complexity of the cells of a line
arrangement that are cut by another line or by a convex polygon. These
quantities are useful for the analysis of various geometric algorithms.

1. Introduction

A number of results in the analysis of algorithms depend on bounds on
the complexity of *zones* in an arrangement; that is, given an arrangement of
n lines in the plane, and some figure in the same plane, we wish to know the
sum of the numbers of sides of the cells in the arrangement that are cut by
that figure. The basic result in this area is the so-called *horizon theorem*: the
complexity of the cells supported by one side of any line in the arrangement
is at most $5n$ [**1, 2, 4**]. This result can be used to prove an $O(n^2)$ bound on
the time needed to construct the arrangement. The horizon theorem has also
been used in some recent work on hidden surface removal and constructive
solid geometry [**6**]; in these cases the figure cutting the arrangement is a convex
k-gon, and we wish to know the complexity of the cells touching the k-gon
on the inside. For fixed k the previous bound shows that this is $O(n)$.

We give a number of results on the complexity of zones in an arrangement
of lines:

(i) The maximum number of sides in all cells supported by a single line
is at most $9.5n + O(1)$, improving a previous bound of $10n$. We
give an example to show that the new bound is tight up to $O(1)$. (We
have recently learned that Edelsbrunner et al. have proved the same
bounds with different techniques [**3**].)

1991 *Mathematics Subject Classification.* Primary 68U05; Secondary 52A10, 52B05, 68Q25.
Research partially supported by the Computational Mathematics Program of the National
Science Foundation under grant DMS-8706133.

(ii) We give examples showing that this bound does not generalize to other related configurations: the complexity of cells supported by two sides of two parallel lines, either between the two lines or on the outsides of the lines, can be at least $10n + O(1)$. By previously known results, this is also an upper bound.

(iii) The maximum complexity of the cells touching the inside of a triangle is at most $10.5n + O(1)$ and at least $10n + O(1)$.

(iv) The complexity of the cells touching the inside of a convex k-gon, is at most $11n + (3/2)k^2$.

(v) The complexity of the cells touching the inside of a convex k-gon is $O(n\alpha(n, k))$, assuming k is $O(n)$. This gives a tighter bound than (iv) for k larger than about $\sqrt{n}$ and improves previous bounds of $O(n\alpha(n))$ [5] and $O(nk)$ [6]. Here $\alpha(n)$ and $\alpha(n, k)$ are one-variable and two-variable inverse Ackermann functions, respectively.

These results also hold for *pseudoline arrangements*. A pseudoline arrangement is a collection of curves, in which each pair of curves intersects at most once (at a crossing, rather than at a tangency). A convex k-gon cutting a pseudoline arrangement is assumed to cut each curve at most twice. In the final section of the paper, we give a sixth result. We adapt a construction that gives n line segments with lower envelope complexity $\Omega(n\alpha(n))$ [10, 8] to show that the complexity of an arrangement of pseudolines cut by a convex k-gon may be as large as $\Omega(n\alpha(n, k))$. This construction proves that the upper bound in (v) is tight for pseudolines up to a constant factor. It is unknown whether the bound in (v) is tight for straight lines.

2. Tight bounds for both sides of a line

Let $\mathscr{A}$ be an arrangement of lines in the plane. Following Edelsbrunner [2], we define a 1-*border* (respectively, 0-*border*) of $\mathscr{A}$ to be a side (vertex) of a polygonal cell of $\mathscr{A}$. A 1-border can be thought of as a pair, consisting of a line segment and a cell. The *zone* of a line of $\mathscr{A}$ is the set of 0- and 1-borders that bound cells supported by that line. The complexity of a zone is its cardinality.

The number of 0-borders in the zone of a line or a convex polygon is exactly the number of 1-borders minus one for each unbounded cell. Since we are interested in the maximum possible complexity of a zone, and since there are constructions for zones achieving the maximum 1-border complexity that have only four unbounded cells, we can treat the 0-border complexity as essentially equal to the 1-border complexity. Therefore, we consider only 1-borders in all our complexity bounds.

We further assume that the $n + 1$ lines of $\mathscr{A}$ are in general position. We refer to the $(n + 1)$-st line h_0 whose zone is under consideration as the *horizon* line, and assume that it is horizontal. For a line a of $\mathscr{A}$, $a \neq h_0$,

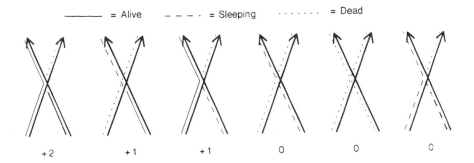

FIGURE 1. Transition rules for left sides of lines.

we distinguish the two *sides* of a, left and right. A 1-border contained in a belongs to one side or the other.

First we review an argument due to Edelsbrunner et al. [2, 4] that gives an upper bound of $5n - 1$ on the number of 1-borders that lie on one side of horizon h_0 (that, say, lie in the closed half-plane above h_0). We conceptually sweep a horizontal line h vertically away from h_0. During the sweep, each side of each line of $\mathscr{A}$, other than h_0, is in one of three states: *alive*, *sleeping*, or *dead*. Each side starts in the alive state, and transition rules determine the states resulting as h passes through line intersections. Intuitively, a side is alive if its current intersection with sweep line h is visible to horizon h_0; a side is sleeping if its intersection is currently invisible, but may become visible again as h continues; and a side is dead if it is currently invisible and will remain invisible for the rest of the sweep.

Figure 1 illustrates the six transition rules as h passes through the intersection of two lines. We show the states of only the left sides of the lines, since left sides and right sides interact independently and symmetrically, that is, the new state of a's left side depends only on its old state and the old state of b's left side.

As the sweep proceeds we count the number of 1-borders in h_0's zone that are not contained in h_0; a 1-border is added to the count when its upper endpoint is passed. (1-borders that extend to infinity are thrown in after the sweep has passed all intersections.) The numbers below each transition rule show the net gain in the count of left-side 1-borders.

Valuing an alive side at 2, a sleeping side at 1, and a dead side at 0, we see that the net gain in the count is never more than the loss in value. Since there are n left sides, all initially alive, the total number of left-side 1-borders not contained in h_0 is at most $2n$. Adding $2n$ for right sides and $n + 1$ for 1-borders contained in h_0, and then subtracting 2 since at least two sides remain alive forever (the right side with the smallest positive slope and the left side with the smallest negative slope) gives $5n - 1$. It is not hard to create an example that shows that this value is tight for the number of 1-borders above the horizon.

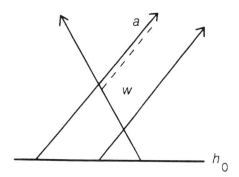

FIGURE 2. Traversing a dead wedge.

By simply doubling this value a bound of $10n - 2$ can be obtained for the total number of 1-borders, above and below, in the zone of h_0 [1, 2, 4]. We now prove a tighter bound. Our strategy is to show that many sleeping sides never return to the alive state and hence leave unused value in the accounting scheme.

We must simultaneously consider both the above and below halfplanes. From now on, the word *side* means a side of a ray with its vertex on h_0. Each line of $\mathscr{A}$ other than h_0 has upper right, upper left, lower right, and lower left sides. We write $a \times b$ for the intersection of lines a and b, and we say $a \times b$ is *above* $c \times d$ if $a \times b$ has the larger y-coordinate.

DEFINITION 1. A line a of $\mathscr{A}$, $a \neq h_0$, is *full* if all four of its sides make a transition from alive to dead.

A line that is not full must have at least one side that either makes a transition from sleeping to dead or goes off to infinity sleeping or alive. Thus a line that is not full has unused value at least 1. We say that a side *does not wake up* if it makes a transition from sleeping to dead or goes off to infinity sleeping. We say that line a *kills* a given side of line b if at $a \times b$, that side of b makes a transition from alive or sleeping to dead.

Call a region w of the plane a *dead wedge* if w is the intersection of two closed halfplanes bounded by lines of $\mathscr{A}$, and w does not intersect h_0. Assume that w lies above h_0, and name the two rays that bound w right and left in the obvious way. The following lemma is immediate and applies analogously to the other three types of sides. See Figure 2.

LEMMA 1. *Assume that at some point in the sweep, the upper right side of line a lies within dead wedge w and is sleeping. Then either a intersects the right ray of w or a's upper right side does not wake up.* □

We now look at a full line f in detail. The four sides of f each make an alive-to-dead transition. Define lines a, b, c, and d to be the lines which kill, respectively, the lower right, upper left, upper right, and lower left sides of f. Then a and b must form larger angles than f with horizon h_0 (where the angle is counterclockwise between h_0 and the other line), while

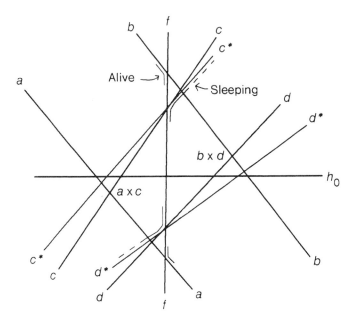

FIGURE 3. Line f is full.

c and d form smaller angles. Assume further that $b \times f$ is above $c \times f$. See Figure 3.

LEMMA 2. *Under the assumptions above:*
(a) *Intersection $a \times f$ is below $d \times f$.*
(b) *Intersection $a \times c$ lies below horizon h_0.*
(c) *Intersection $b \times d$ lies above horizon h_0.*

PROOF. Assume the opposite of (a). Then since the lower left side of f must be alive just above $f \times d$, intersection $a \times c$ must lie above h_0. But then the upper left side of f cannot be alive just below $f \times b$.

Observation (b) follows from the fact that f's upper left side must be alive just below $f \times b$. Similarly, observation (c) follows from the fact that f's lower right must be alive just above $f \times a$. $\square$

Thus Figure 3 gives the only possible arrangement of lines a, b, c, d, f, and h_0, assuming that $b \times f$ is above $c \times f$. The opposite assumption, that is, c above b gives the mirror image of Figure 3.

LEMMA 3. *There is a line c^* satisfying the following:* (1) c^* *intersects f below $f \times b$ and at or above $f \times c$,* (2) c^* *intersects b at or below $b \times c$,* (3) *the upper right side of c^* is alive just below $c^* \times b$ and is sleeping just above $c^* \times b$, and* (4) c^* *crosses h_0 to the right of $h_0 \times a$ and at or to the left of $h_0 \times c$.*

PROOF. Line c is just such a line unless its upper right side is killed somewhere between its intersections with f and b. The line that kills the upper right side of c is suitable unless this line is itself killed. Following this chain of killers leads to a suitable c^*. Notice that (4) above holds for every line along the chain, since f's upper left side is alive just below $f \times b$. □

LEMMA 4. *There is a line* d^* *satisfying the following:* (1) d^* *intersects* f *above* $f \times a$ *and at or below* $f \times d$, (2) d^* *intersects* a *at or above* $a \times d$, (3) *the lower left side of* d^* *is alive (sleeping) just above (below) its intersection with* $d^* \times a$, *and* (4) d^* *crosses* h_0 *to the left of* $h_0 \times b$ *and at or to the right of* $h_0 \times d$.

PROOF. Symmetric to the proof of Lemma 3. □

LEMMA 5. *Assume* c^* *satisfies* (1)–(4) *of Lemma* 3 *and* d^* *satisfies* (1)–(4) *of Lemma* 4. *Then either the upper right side of* c^* *or the lower left side of* d^* *does not wake up.*

PROOF. Lines c^* and d^* intersect either above or below h_0. If they intersect below, then c^* does not intersect the right ray of the dead wedge defined by b and d^* above the horizon. If they intersect above, then d^* does not intersect the left side of the dead wedge defined by a and c^* below the horizon. Lemma 1 now implies the result. □

The lemmas above show how to associate a full line f with two sleeping sides, i.e., the upper right of c^* and the lower left of d^*, such that one of them does not wake up. For the mirror image case, that is, when $c \times f$ is above $b \times f$, the associated sides are an upper left and a lower right.

Define a mapping ur from full lines to upper right sides that maps a full line f to its associated side of c^*. Notice that c^* is well defined by the procedure given in the proof of Lemma 3. Similarly define a mapping ll from full lines to lower left sides that maps a full line f to its associated side of d^*. For the mirror image case, there are mappings ul and lr.

Now assume that f and g are distinct full lines such that $ur(f) = ur(g)$ and these sides are contained in line c^*. Rename f and g if necessary so that $c^* \times f$ is above $c^* \times g$. See Figure 4.

LEMMA 6. *If* $ur(f) = ur(g)$ *and* $c^* \times f$ *is above* $c^* \times g$, *then* $ll(g) \neq ll(f)$ *and* $ll(g)$ *does not wake up.*

PROOF. Let a_f and b_f denote the lines that kill f's lower right and upper left sides, and d_f^* denote the line containing $ll(f)$. Analogously name a_g, b_g, and d_g^*. In Figure 4, the "diamond" formed by g's four associated lines is shaded.

Notice that $g \times f$ must be above $c^* \times f$ and below $b_f \times f$, because f is alive just below $f \times b_f$. Next, $b_g \times g$ must be above $c^* \times g$ and at or below $f \times g$. In Figure 4, $b_g = f$. Then the fact that $b_g \times d_g^*$ is above h_0 (as in Lemma 2(c)) implies that $d_g^* \times h_0$ is to the left of $f \times h_0$. Hence $d_g^* \neq d_f^*$. Now since c^* is alive just below $b_f \times c^*$, d_g^* must cross c^* above

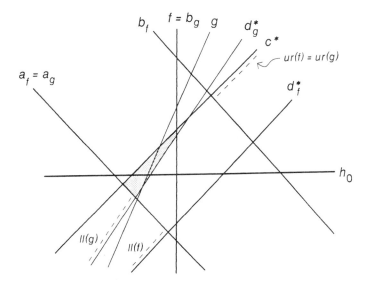

FIGURE 4. Full lines f and g are such that $ur(f) = ur(g)$.

h_0 and below $b_f \times c^*$. Then the lower left side of d_g^* does not wake up below $d_g^* \times a_g$, since to do so it would have to cross c^* below the horizon h_0. □

LEMMA 7. *Statements symmetric to Lemma 6 hold for mappings* ll, ul, *and* lr. □

Roughly speaking, Lemma 6 shows that even though ur is not one-to-one, if two full lines f and g map to the same upper right side, there are still two sides that do not wake up to "blame"— $ll(g)$ and one of $ur(f)$ and $ll(f)$.

LEMMA 8. *The number of full lines is no greater than the number of sides that do not wake up.*

PROOF. Consider the full lines whose upper left sides are killed above their upper right sides, such as f in Figure 3. Number such full lines $f_1, f_2, \ldots$ clockwise, that is by decreasing angle with h_0. Either $ur(f_1)$ and $ll(f_1)$ are unique to f_1 or there are full lines that share $ur(f_1)$ as their ur images and/or lines that share $ll(f_1)$ as their ll images. If unique, then Lemma 5 implies that one of these sides does not wake up, so assume that at least one of $ur(f_1)$ and $ll(f_1)$ is not unique to f_1. Notice that a line that shares its ur (respectively ll) image with f_1 must intersect f_1 above (below) h_0, so these two sets of lines are disjoint. By Lemma 6, a line f_i that shares its ur image with f_1 has an ll image that does not wake up.

Image $ll(f_i)$ may itself be shared with another full line f_j. We assert that j must be larger than i. To see this assertion, consider Figure 4 again. Let f in the figure be f_1 and g be f_i. Recall that the arrangement of $f = f_1$, $g = f_i$, h_0, and c^* is the only possible arrangement such that

$ur(f_1) = ur(f_i)$. Now assume f_j is such that $ll(f_j) = ll(f_i)$, and f_j forms a larger angle with h_0 than f_i, i.e., $j < i$. Since $ll(f_j) = ll(f_i)$, the lower left side of f_j must be crossed by d_g^* (the line containing $ll(f_j)$) below h_0. And $h_0 \times f_j$ must be to the right of $h_0 \times c^*$, or else c^* would kill f_j's lower left side. But now the upper left side of $f_i = g$ cannot be alive above c^* as it lies in the dead wedge formed by f_j and c^*. This contradicts the fact that f_i is full.

Starting with f_1, we can define a rooted tree of full lines, in which a parent full line shares one of its images under ur and ll with each of its children. This procedure defines a tree because, by the assertion above and a symmetric counterpart, each full line adds children of larger index than its own index. Lemmas 6 and 7 guarantee that each full line added to the tree defines a new side that does not wake up. After a tree terminates we start another tree with the first f_i that has not yet participated. A symmetric argument matches mirror image full lines with ul and lr sides that do not wake up. □

THEOREM 1. *The maximum number of 1-borders in h_0's zone is at most $\lfloor 9.5n \rfloor - 1$.*

PROOF. Let the number of full lines be F, the number of sides that do not wake up be W, and the maximum number of 1-borders in h_0's zone be M. There are $2n + 2$ 1-borders contained in h_0, and the total initial value in the two sweeps is $8n$. There are two lines—the "most horizontal" lines—each of which contains two sides that go off to infinity alive.

Recall that each line that is not full has unused value at least 1, and the two most-horizontal lines—not full—have unused value 2 each. Hence, $M \le 9n + F$. A side that does not wake up has unused value 1, so we also have the inequality $M \le 10n - 2 - W$. By Lemma 8, $F \le W$. The minimum of $9n + F$ and $10n - 2 - W$ is hence no more than $\lfloor 9.5n \rfloor - 1$. □

Figure 5 gives an example with $\lfloor 9.5n \rfloor - 3$ 1-borders. Figures 6 and 7 show that Theorem 1 cannot be generalized to the complexity of cells on opposite sides of two parallel lines—in both possible cases the complexity can be as much as $10n + O(1)$. In these figures, a few straight lines are shown curved in order to fit them on the page. These examples generalize in the obvious way to other sufficiently large n.

3. Bounds for a triangle

In the remainder of this paper, we consider the case of an arrangement of lines cut by a convex k-gon. There are two kinds of cells in the zone of this configuration: those outside the k-gon, and those inside the k-gon. Note that the $10n + O(1)$ upper bound for the zone of a line also holds, with minor modifications, for the complexity of the outside cells in the k-gon zone. We shall focus only on those cells inside the k-gon.

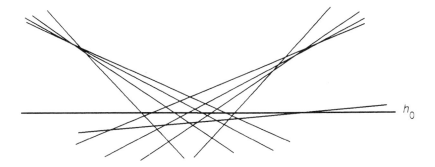

FIGURE 5. A $9.5n + O(1)$ example for two sides of one line.

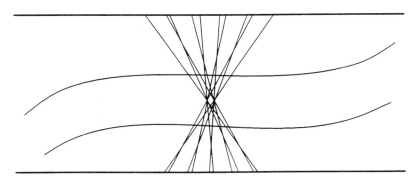

FIGURE 6. A $10n + O(1)$ example for the insides of two parallel lines.

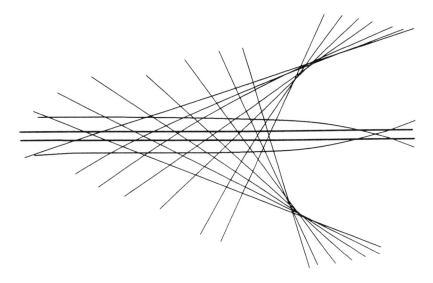

FIGURE 7. A $10n + O(1)$ example for the outsides of two parallel lines.

As with the previous bounds, it makes sense to allow the lines of the arrangement and the sides of the k-gon to be generalized to pseudolines. The requirements are: each pair of pseudolines intersects at most once, and they must cross at intersections; and each pseudoline of the arrangement crosses exactly two k-gon sides, crossing each side only once. The second requirement replaces the condition in the straight line case that the k-gon be convex.

In this section we give an upper bound of $10.5n + O(1)$ on the zone complexity of the inside of a triangle. It is easy to adapt Figure 6 to give a lower bound of $10n + O(1)$ in which no lines intersect one side of the triangle.

Let the sides of the triangle be denoted s_1, s_2, and s_3 clockwise. Let the lines cutting side s_i be N_i, with $|N_1| + |N_2| + |N_3| = 2n$. Consider any single side, say s_1. The 1-borders supported by s_1 are of two types: those contained in lines of N_1 and those contained in lines of $N_2 \cap N_3$. Let c_1 denote the lower envelope of $N_2 \cap N_3$ inside the triangle; c_1 is a convex polygonal chain with $|c_1|$ segments and endpoints on s_2 and s_3. We can similarly define chains c_2 and c_3.

Convex chain c_i is a pseudoline with respect to the lines in N_i, that is, each line of N_i intersects c_i at most once, and at that intersection crosses c_i. Consider sweeping a parallel line away from s_i as in the argument of Edelsbrunner et al. reviewed in the last section. We may assume that s_i is horizontal, so that left and right are canonically defined.

LEMMA 9. *The complexity of cells supported by s_i is at most $5|N_i| + |c_i|$.*

PROOF. First note that if $|c_i| = 0$, the lemma is true by the $5n - 1$ bound, so assume $|c_i| \geq 1$. The complexity of cells supported by s_i is the complexity of the arrangement formed by N_i and c_i, considered as a pseudoline, plus the number of corners $|c_i| - 1$. In the sweep away from s_i, pseudoline c_i adds only 2 to the complexity instead of the usual 5, because only one of its sides is visible and it does not cut s_i. Thus we have a bound of $5|N_i| + 1$. Considering c_i as a polygonal chain rather than a pseudoline adds $|c_i| - 1$. □

DEFINITION 2. A line f of $N_i \cap N_j$ is *full* if (1) it contributes a segment to c_k, $k \neq i, j$; (2) it makes 4 transitions (left and right sides on each of s_i and s_j) from alive to dead; and (3) there is a point p on the segment of f on c_k such that p is visible from s_k but not visible from s_i or s_j, that is, p lies between the two alive-to-dead transitions of the side of f facing s_k.

A 1-border can be counted either in one of the three sweeps, or as part of one of the convex chains. Requirement (3) above ensures that a full line contributes an "extra" segment to a convex chain, that is, one that is not already counted in one of the sweeps. A *nonfull* line either has unused value in one of the two sweeps, as in the last section, or it does not contribute an extra visible corner to a convex chain c_i.

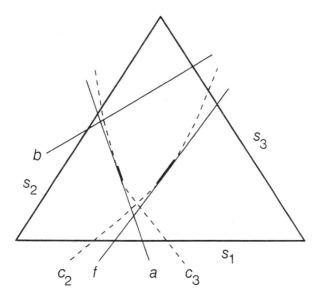

FIGURE 8. A line that kills a side of a full line cannot be full.

LEMMA 10. *Let f be a full line that contributes to c_2. Let a be the line that kills f's left side in the sweep from s_1. Then a is not full.*

PROOF. Assume a is full. Line a cannot contribute to convex chain c_2, as this would contradict the requirement that some point of f be visible from s_2 above $f \times a$. Here "above" means further along in the sweep from s_1. So assume a contributes to c_3, as shown in Figure 8. First assume a's point p_a visible from s_3 (guaranteed by (3) above) lies below $f \times a$. Now the line that kills a's left side below p_a must also kill f's left side below $f \times a$, a contradiction. So assume point p_a lies above $f \times a$. Consider the line b that kills a's left side in the sweep from s_2. For p_a to be visible from s_3, $b \times s_3$ must lie above the endpoint of c_2 on s_3. This arrangement contradicts the fact that some point of f above $f \times a$ is visible from s_2. □

LEMMA 11. *Let f be a full line that contributes to c_2, and let a be the line that kills the right side of f in the sweep from s_1. Then a is not full.*

PROOF. Assume a is full. Line a must intersect s_3, so a must contribute to c_2. The point p_a cannot lie above $a \times f$, so it must lie below $a \times f$ and above the intersection at which a's right side dies. But then the line that kills the right side of a must kill the right side of f below $f \times a$. □

LEMMA 12. *The number of full lines in N_i is at most $|N_i|/2$.*

PROOF. Consider the set of full lines F_i in N_i. By Lemmas 10 and 11 and symmetric counterparts, a line in F_i cannot kill a side of another line in F_i. Now consider only the initial segments of lines in F_i up to the lowest (in the sweep from s_i) intersection at which both sides are dead. There are no intersections among these line segments.

We now assert that it takes $|F_i|$ lines to kill all of these segments. To each initial segment f, assign the line of N_i that kills a side of f below f's upper endpoint. Assume line a is assigned twice: say a kills in sweep order the right sides of segments f and g. Then we have a contradiction to the assumption that the left side of g is alive above $g \times a$. The case that a kills two left sides is symmetric.

Finally, assume a kills in sweep order the left side of f and the right side of g. Then the line that kills the right side of f at its upper endpoint cannot escape from the triangle formed by a, g, and s_i without killing the right side of g below $g \times a$, a contradiction. The case of a right side followed by a left side is symmetric. Thus no line that kills a side of F_i is assigned twice, and we have $|F_i| \leq |N_i|/2$. □

THEOREM 2. *The complexity of the cells on the inside of a triangle, cut by* n *lines, is at most* $10.5n + O(1)$.

PROOF. Lemma 12 implies that the sum over i of the number of full lines in N_i is at most n. Since each full line cuts two sides, this means that the total number of full lines is at most $n/2$. Each line starts with value 9, that is, 2 for each live side in each of the sweeps from the sides of the triangle it cuts, and 1 more for the possibility of contributing a corner to a convex chain c_i. Each nonfull line has unused value at least 1. Thus the average used value is at most 8.5 and the total—including the $2n$ 1-borders contained in the triangle itself—is at most $10.5n + O(1)$. □

OPEN PROBLEM 1. *Reduce the gap between the lower bound of* $10n + O(1)$ *and the upper bound of* $10.5n + O(1)$ *on the complexity of the cells along the inside of a triangle.*

4. Polygons with fixed number of sides

In this section and the next we give bounds on the complexity of the zone of a k-gon in an arrangement of pseudolines.

Define $V(n, k)$ to be the maximum possible number of pairs (ℓ, s), where ℓ is a line in the arrangement and s is a side of the k-gon, where ℓ does not intersect s, but where ℓ forms the side of a cell supported by s. There may be many such cells, but still (ℓ, s) is counted only once. If ℓ forms the side of a cell supported by multiple k-gon sides, we only include the pair (ℓ, s) where s is the most clockwise of the sides supporting the cell. (However, a pair (ℓ, s'), where s' is not most-clockwise, is included if ℓ is visible to s' from another cell.) The function $V(n, k)$ is significant for the k-gon zone complexity because of the following fact.

LEMMA 13. *The complexity of the cells on the inside of a k-gon, cut by* $n \geq 1$ *lines, is at most* $10n + k + V(n, k)$.

PROOF. Let the sides of the k-gon be denoted s_1, s_2, ..., s_k. Let the lines crossing side s_i be denoted N_i. Let the lines visible to side s_i but not crossing it be denoted V_i.

If side s_i is not crossed by any line, it adds at most 1 to the total complexity of the figure; there are at most k such sides.

First assume V_i is empty. Then the cells supported by s_i are exactly those in the arrangement of only the lines in N_i, and the complexity of these cells is at most $5|N_i| - 1 \le 5|N_i| + |V_i|$.

Otherwise, the "lower envelope" of the lines in V_i (i.e., assuming s_i is horizontal and below all lines of V_i) forms a convex chain, that can be treated as a pseudoline with respect to the lines in N_i. The complexity of all cells supported by s_i is then the complexity of the arrangement formed by N_i together with this extra pseudoline, plus the complexity of the corners of the pseudoline. As in the last section, the arrangement complexity is at most $5|N_i| + 1$. The number of corners on the pseudoline is $|V_i| - 1$. Therefore in this case also the complexity of the cells supported by s_i is bounded by $5|N_i| + |V_i|$.

The complexity of the whole arrangement (counting cells supported by exactly two sides twice), is then at most

$$k + \sum_{i=1}^{k}(5|N_i| + |V_i|) = k + 5\sum_{i=1}^{k}|N_i| + \sum_{i=1}^{k}|V_i| = 10n + k + V(n, k). \quad \square$$

We now give a simple combinatorial lemma that we use to prove a bound on $V(n, k)$; this will then be used as a base case in our more complicated final bound.

LEMMA 14. *Let S_1, S_2, ..., S_m be a family of subsets of a set S with $|S| = k$, with the property that, for some constant a and all $i \ne j$, $|S_i \cap S_j| \le a$. Then $\sum_{i=1}^{m}|S_i| \le am + \binom{k}{a+1}$.*

PROOF. Let $b_i = |S_i| - a$. Then each set S_i has at least b_i $(a + 1)$-tuples, that by assumption must be distinct from the $(a + 1)$-tuples in all the other S_j. So $\sum_{i=1}^{m} b_i \le \binom{k}{a+1}$. Therefore $\sum_{i=1}^{m}|S_i| \le am + \sum_{i=1}^{m} b_i \le am + \binom{k}{a+1}$. $\square$

LEMMA 15. $V(n, k) \le 2n + k^2/2$.

PROOF. Let $m = 2n$ in Lemma 14, and let the sets S_i be the k-gon sides visible to each side of each line in the arrangement. As above, only the first side in a multi-side cell is "visible." Then, for two sets S_i and S_j that correspond to sides of distinct lines, the zone cells in the region inside both corresponding halfplanes can be ordered linearly around the k-gon. The only possible k-gon side shared by both sets is the one supporting the cell first visible to S_j in this linear order; the view from a k-gon side later in the order is blocked by the view within this cell. Thus Lemma 14 applies with $a = 1$. □

LEMMA 16. $V(n, k) \leq n + (3/2)k^2 - k$.

PROOF. Divide the lines into classes $C_{i,j}$, where line ℓ belongs to $C_{i,j}$ exactly when it crosses sides s_i and s_j of the k-gon. Then, similarly to the proof of Lemma 13, the lines in a single class visible to sides s_x with $i < x < j$ can be treated as a single pseudoline, visible only on one side. And the lines visible to sides s_x with $x < i$ or $j < x$ can be treated as another pseudoline, also visible on a side.

The number of corners on these pseudolines is at most $|C_{i,j}|$. The pseudoline can be divided up into a sequence of segments, linearly ordered by which k-gon side they are visible from; there may be gaps where the pseudoline is not visible to any side, but these can be assigned arbitrarily to either neighboring side. Then each side that sees the pseudoline sees a number of the lines composing it equal to 1 plus the number of corners in the segment assigned to that side. Therefore each corner adds at most 1 to the complexity of $V(n, k)$.

Now given two such pseudolines, as in Lemma 15, there is only one side that can be visible to both of them, and we can apply lemma 14 with $a = 1$ and $m = 2\binom{k}{2}$. Adding the pseudoline-side visibilities to the number of corners on each pseudoline gives a total bound of

$$V(n, k) \leq \sum_{i=1}^{k} \sum_{j=1}^{k} |C_{i,j}| + 2\binom{k}{2} + \binom{k}{2} \leq n + (3/2)k^2 - k. \quad \square$$

THEOREM 3. *The complexity of the cells on the inside of a k-gon, cut by $n \geq 1$ lines, is at most $11n + (3/2)k^2$.* □

This gives the tightest known bound for fixed k, of $11n + O(1)$. However the best construction known is Figure 6 (appropriately modified to fit in the k-gon), which has complexity $10n + O(1)$.

OPEN PROBLEM 2. *Reduce the gap between the lower bound of $10n + O(1)$ and the upper bound of $11n + O(1)$ on the complexity of the cells on the inside of a k-gon when k is a fixed constant.*

5. Recursive bounds for polygons

In this section we show that the maximum complexity of the zone of a k-gon, for k that is $O(n)$, is $O(n\alpha(n, k))$. In the final section of this paper we show that our bound is tight for all n and k, up to constant factors. We have recently learned that the same $O(n\alpha(n, k))$ bound can be obtained in a conceptually simpler way. First observe that the sequence of lines counted in $V(n, k)$ forms an n-letter Davenport-Schinzel sequence of order 3; that is, there can be no embedded $a \ldots b \ldots a \ldots b \ldots a$. Moreover, this sequence can be divided into k contiguous blocks, such that each block contains no repeated letters. Sharir [7] has shown an $O(n\alpha(n, k))$ bound on the length of such a Davenport-Schinzel sequence. In fact, his proof also simplifies the previous $O(n\alpha(n))$ upper bound argument for arbitrary order-3 Davenport-Schinzel sequences [5].

Our upper bound of $O(n\alpha(n, k))$ depends on the following recurrence bounding $V(n, k)$. (Sharir's proof depends on a similar, but slightly less complicated, recurrence.)

LEMMA 17. *Let $k \le b \cdot k'$. Then*

$$V(n, k) \le \max \left\{ 2n_0 + V(n_0 + b, b) + \sum_{i=1}^{b} V(n_i + 2k', k' + 1) \;\middle|\; \sum_{i=0}^{b} n_i = n \right\}.$$

PROOF. Our overall strategy is to conceptually cut the k-gon into a central b-gon along with b polygons around this b-gon, each with no more than $k' + 1$ sides. We then define a new pseudoline arrangement within each of these polygons, such that the total number of visible line-side pairs in the new arrangements is an upper bound on $V(n, k)$, the original number of visible line-side pairs in the k-gon.

Choose a configuration of n lines in a k-gon achieving the maximum value of $V(n, k)$ pairs. Bundle contiguous sides of the k-gon into b groups, with at most k' sides in each group. Define n_i, for $1 \le i \le b$, to be the number of lines in the arrangement having both ends within group i; let n_0 be the number of lines having ends in two groups. Then $\sum_{i=0}^{b} n_i = n$.

We will count the number of visibilities in the original arrangement by forming new arrangements for each group of sides. Consider the lines visible to a particular group i. These lines fall into four possible types: (1) lines starting and ending in group i, (2) lines starting in group i and ending in another group, (3) lines starting and ending in two other groups, and (4) lines starting and ending in the same other group.

The lines of type (1) will remain unchanged in the new arrangement for group i.

We subdivide the lines of type (2) according to which side of group i they cross. As in Lemma 16, we can form the lines crossing a single side into two pseudolines in the new arrangement for group i. As in that lemma, we must also count the number of corners formed on those pseudolines. The total number of pseudolines formed in group i is $2k'$; however two of those, the first and last pseudolines in the clockwise order around group i, cannot create any visibilities that are counted in $V(n, k)$. Therefore, the number of pseudolines contributing to the visibilities within group i is $2k' - 2$. The number of corners in all these pseudolines is at most equal to the number of lines of type (2); the sum of these numbers over all groups is $2n_0$.

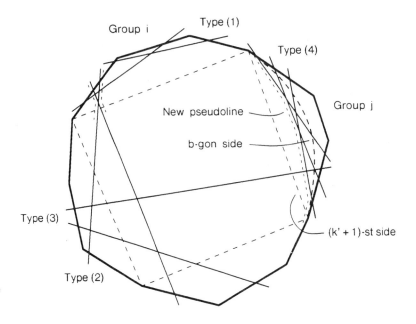

FIGURE 9. Cutting a k-gon into one b-gon and b $(k' + 1)$-gons.

We subdivide the lines of type (4) according to which other group they belong to. The lines from a single group j can be treated as a single pseudoline of type (3) as viewed from sides in group i. However, we must also count the corners on this pseudoline. These corners are counted (for all choices of i at once) by adding another side to group j to form a closed polygon with no more than $k' + 1$ sides. See Figure 9.

The lines and new pseudolines of type (3) can be treated as a single pseudoline "ceiling" when counting visibilities from group i; however, we must also count the number of corners on that ceiling. This is just the number of lines and pseudolines of type (3) composing the ceiling, which can be counted by creating a new arrangement consisting of all lines and pseudolines of type (3), cut by a b-gon corresponding to the b groups of sides. (For a geometric realization, the b-gon may have to bend out and exchange an endpoint with a new pseudoline as in Figure 9.) The number of line-side visibilities in this b-gon arrangement counts the corners on all ceilings at once. (In fact, this slightly overcounts since only one side of each new pseudoline is visible.) Thus the total number of corners for all groups is at most $V(n_0 + b, b)$.

Summarizing, we have $V(n_i + 2k', k' + 1)$ visibilities within group i and corners on the pseudoline for the type (4) lines contributed by group i to visibilities in other groups. We have $2n_0$ corners on all the pseudolines created for the type (2) visibilities. And we have $V(n_0 + b, b)$ corners on the pseudolines created for the type (3) visibilities and the pseudolines of type (3) created to count the type (4) visibilities. Adding these together gives the total bound. □

Now let us define an Ackermann function $A(i, j)$ as follows:

- $A(1, j) = 6j$.
- $A(i, j) = A(i - 1, j + 1)$ if $j \leq 6$.
- $A(i, j) = A(i - 1, A(i, j - 6))$ if $j > 6$.

Let $\alpha_i(x) = \max \{1\} \cup \{j \geq 1 \mid A(i, j + 1) \leq x\}$ and

$$\alpha(x, y) = \min \{i \geq 1 \mid A(i, \lfloor x/y \rfloor) \geq x\}.$$

Our Ackermann function is somewhat nonstandard; at the end of the paper we show that nevertheless $\alpha(x, y)$ is within a constant factor of other inverse Ackermann functions [8, 9]. Our functions have the following properties:

LEMMA 18. *For any* $k \geq 0$ *and* $\ell \geq 0$ *with* $k + l \geq 1$, $A(i, j) > A(i - k, j - \ell)$.

PROOF. We prove the lemma by induction; to prove it for (i, j) we assume that it holds for all (i', j') with $i' < i$ or with $i' = i$ and $j' < j$. Then we show that $A(i, j) > \max\{A(i - 1, j), A(i, j - 1)\}$; the full lemma for (i, j) easily follows.

- For $i = 1$, $A(i, j) > A(i, j - 1)$ immediately from the definition.
- For $j \leq 6$, $A(i, j) = A(i - 1, j + 1) > A(i - 1, j) = A(i, j - 1)$.
- For $j = 7$, $A(i, j) = A(i - 1, A(i, 1)) > A(i - 1, A(2, 1)) = A(i - 1, 12) > A(i - 1, 7) = A(i, j - 1)$.
- For $i = 2$ and $j = 7$, $A(i, j) = 72 > 42 = A(i - 1, j)$.
- For $i = 2$ and $j > 7$,

$$A(i, j) = 6(A(i, j - 6)) > 6(A(i - 1, j - 6))$$
$$= 36(j - 6) > 6j = A(i - 1, j).$$

In all the remaining inequalities to check, both left and right sides follow the recursive definition. Then $A(i, j) = A(i - 1, A(i, j - 6)) > A(i - 1, A(i, j - 7)) = A(i, j - 1)$ and $A(i, j) = A(i - 1, A(i, j - 6)) > A(i - 2, A(i - 1, j - 6)) = A(i - 1, j)$. □

COROLLARY 1. *For any* i *and* x, $\alpha_i(x) \leq \alpha_{i-1}(x)$.

LEMMA 19. *If* $i \geq 2$, *then* $\alpha_i(x) < \sqrt{x}$.

PROOF. We prove the lemma for $i = 2$; the remaining cases follow from Corollary 1. If $x \leq 72$, then $\alpha_2(x) \leq 6$, and the truth of the lemma can be seen by inspection of the possible cases. Otherwise, using induction, $\alpha_2(x) = 1 + \alpha_2(x/6) < 1 + \sqrt{x/6} < \sqrt{x}$. □

LEMMA 20. *For any* $i > 1$ *and* x, $\alpha_i(\alpha_{i-1}(x) + 1) \leq \max \{6, \alpha_i(x) - 6\}$.

PROOF. If $\alpha_i(x) \leq 6$, then $\alpha_i(\alpha_{i-1}(x) + 1) \leq 6$. Otherwise,

$$A(i, \alpha_i(\alpha_{i-1}(x) + 1) + 6 + 1) = A(i - 1, A(i, \alpha_i(\alpha_{i-1}(x) + 1) + 1))$$
$$\leq A(i - 1, \alpha_{i-1}(x) + 1) \leq x.$$

But for any y, if $A(i, y + 1) \leq x$, it follows that $y \leq \alpha_i(x)$. Hence $\alpha_i(\alpha_{i-1}(x) + 1) + 6 \leq \alpha_i(x)$. $\square$

LEMMA 21. *There is a constant $c \geq 3$ such that for any $i < \alpha(n, k)$,*
$$V(n, k) \leq 2in + 3ik\alpha_i(k) + ci(k - 3).$$

PROOF. For $i = 1$, the lemma gives the bound of Lemma 15 along with a term linear in k. So let $i > 1$. If $k \leq \sqrt{n}$, then the lemma follows from Lemma 15. Assume that $\alpha_i(k) < 12$. Then $A(i, 12) \geq k$ and $A(i, 13) \geq k^2 \geq n$. So the one-variable function $\alpha(n)$ is $O(i)$. Now $i < \alpha(n, k)$ implies that $A(i, \lfloor n/k \rfloor) < n$, which means that $n/k < 13$. By choosing c large enough, the lemma then follows from the Davenport-Schinzel bound of $n\alpha(n)$ [5].

So we may assume $i > 1$ and $\alpha_i(k) \geq 12$. Let $k' = \alpha_{i-1}(k)$, and let $b = \lceil k/k' \rceil$.

Now $A(i, k) < n$. If $k' = 1$, then $\alpha_i(k) = 1 = \alpha_{i-1}(k)$, and the bound follows easily from that for $i - 1$. Otherwise, note that $k' < b$ (by Lemma 19), $b \leq (n + 1)/2$, and

$$b\alpha_{i-1}(b) \leq bk' \leq k + k' \leq k + b.$$

By Lemma 17,

$$V(n, k) \leq 2n_0 + V(n_0 + b, b) + \sum_{j=1}^{b} V(n_j + 2k', k' + 1).$$

We inductively use the bound we are proving with $i - 1$ for $V(n_0, b)$ and with i for $V(n_j + 2k', k' + 1)$:

$$V(n, k) \leq 2n_0 + 2(i - 1)(n_0 + b) + 3(i - 1)b\alpha_{i-1}(b) + c(i - 1)(b - 3)$$
$$+ \sum_{j=1}^{b} \left(2i(n_j + 2k') + 3i(k' + 1)\alpha_i(k' + 1) + ci(k' - 2) \right).$$

We now gather terms involving n_0 and n_j,

$$V(n, k) \leq 2in + 2(i - 1)b + 3(i - 1)b\alpha_{i-1}(b) + c(i - 1)(b - 3)$$
$$+ 4ibk' + 3ib(k' + 1)\alpha_i(k' + 1) + cib(k' - 2),$$

and apply the facts that $b\alpha_{i-1}(b) \leq k + b$ and $bk' \leq k + b$,

$$V(n, k) \leq 2in + 9ib + 7ik + 3i(k + 2b)\alpha_i(k' + 1) + ci(k - 3).$$

By Lemma 20 and the fact that $\alpha_i(k) \geq 12$,

$$
\begin{aligned}
V(n,k) &\leq 2in + 9ib + 7ik + 3i(k + 2b)(\alpha_i(k) - 6) + ci(k - 3) \\
&\leq 2in - 27ib - 11ik + 6ib\alpha_i(k) + 3ik\alpha_i(k) + ci(k - 3) \\
&\leq 2in + 3ik\alpha_i(k) + ci(k - 3),
\end{aligned}
$$

since by Corollary 1, $\alpha_i(k) \leq k' \leq (k + b)/b$. □

THEOREM 4. *The complexity of the cells on the inside of a k-gon, cut by n lines, is $O(n\alpha(n,k))$.* □

6. Lower bounds for polygons

Wiernik gave a construction of n line segments with lower envelope complexity $\Omega(n\alpha(n))$ [10]. By placing such a collection of segments inside a convex n-gon and extending segments to the n-gon with curves, we can form a pseudoline arrangement with zone complexity $\Omega(n\alpha(n))$. In this section we show how to generalize this construction for any k, giving a lower bound of $\Omega(n\alpha(n,k))$.

We adapt Shor's construction for the lower-envelope complexity of line segments [8], a simplification of the construction due to Wiernik [10]. As our lower bound construction only works for pseudolines, we need not concern ourselves with some of the details of Shor's construction, such as the careful handling of line segment slopes.

Shor constructs an arrangement of positive-slope line segments $S(i,j,r)$ with nonlinear lower-envelope complexity as follows. First define an Ackermann function $F(i,j)$ by $F(1,j) = 1$, $F(i,1) = 2 \cdot F(i-1,2)$, and $F(i,j) = F(i,j-1) \cdot F(i-1,F(i,j-1))$. Arrangement $S(i,j,r)$ contains $F(i,j)$ *j-fans*. A j-fan is a set of j line segments that share the same left endpoint. Within each j-fan the segments can be numbered $1,2,\ldots j$, such that the slope of each segment is at least r times the slope of the preceding one, where r is a real number at least 3.

Arrangement $S(i,j,r)$ is defined recursively. $S(1,j,r)$ is a single j-fan. $S(i,1,r)$ is constructed from $S(i-1,2,r)$ by translating the larger-slope segment of each 2-fan by a tiny distance ϵ, so that the smaller-slope segment has endpoint somewhat to the left of the larger-slope segment. Each segment is then a 1-fan and the lower-envelope complexity doubles.

In the general inductive step, we generate many copies of $S(i,j-1,r)$ and a single copy of $S^* = S(i-1, F(i,j-1), r^*)$, where r^* is determined by the geometry of $S(i,j-1,r)$. Arrangement S^* is flattened and then tilted by an affine transformation so that all slopes of segments in S^* are very close to 1. This transformation leaves the lower envelope complexity unchanged. There will be some tiny $\epsilon > 0$ such that if each segment of S^* is translated a distance no greater than ϵ, then the only changes in the lower envelope occur at the left endpoints of fans.

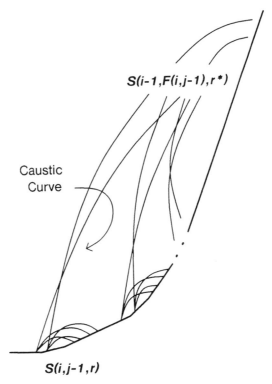

$S(i-1,F(i,j-1),r^*)$

Caustic
Curve

$S(i,j-1,r)$

FIGURE 10. Pseudoline arrangement with nonlinear zone complexity.

Next each copy of $S(i, j - 1, r)$ is first flattened so that all slopes are less than $1/r$ and then shrunk to be smaller than ϵ. We place each copy of $S(i, j - 1, r)$ next to a fan left endpoint of S^*, and then perturb the segments of S^* so that one segment of S^* joins each $(j - 1)$-fan in the nearby copy of $S(i, j - 1, r)$. The trick is that segments of S^* are perturbed so that the largest-slope segment of each fan will have rightmost left endpoint, next largest next rightmost, and so on. With suitable choice of r^*, each $F(i, j - 1)$-fan emerges from its copy of $S(i, j - 1, r)$ and then forms a "caustic curve," that is, an arrangement with $F(i, j - 1)$ subsegments in its lower envelope. This caustic curve lies to the right of all right endpoints of segments in the copy of $S(i, j - 1, r)$. See Figure 10.

The complexity of the lower envelope of $S(i, j, r)$ is the sum of the complexities of $F(i - 1, F(i, j - 1))$ copies of $S(i, j - 1, r)$, one copy of S^*, and $F(i - 1, F(i, j - 1))$ caustic curves of complexity $F(i, j - 1)$. The solution is $ijF(i, j)$ as confirmed by the following equality:

$$\begin{aligned}
ijF(i, j) = {} & F(i - 1, F(i, j - 1)) \cdot i(j - 1)F(i, j - 1) \\
& + (i - 1)F(i, j - 1) \cdot F(i - 1, F(i, j - 1)) \\
& + F(i - 1, F(i, j - 1)) \cdot F(i, j - 1).
\end{aligned}$$

We now show how to extend the segments of $S(i, j, r)$ to pseudolines and fit them into a polygonal chain with $F(i, j) + 1$ sides, that is, one side for each fan plus one final side. One or two additional sides can be added at the end to close the polygon. First, $S(1, j, r)$ fits into a chain with two sides. The left endpoint lies on a horizontal side, and right endpoints are extended with curves to intersect a second side with arbitrary positive slope. The curves are such that they do not intersect each other. Second, $S(i, 1, r)$ fits into a chain with $F(i, 1) + 1 = 2F(i - 1, 2) + 1$ sides by simply subdividing each edge, except the last, of the chain holding $S(i - 1, 2, r)$, so that each fan starts on its own side. In the general step, we need $F(i - 1, F(i, j - 1))$ copies of chains with $F(i, j - 1) + 1$ sides to hold all the copies of $S(i, j - 1, r)$. The last side of a chain for each copy doubles as the first side of the chain for the next copy, as shown in Figure 10. One final side is added to receive the extensions of the segments of S^*.

We now confirm that for each choice of the number of pseudolines n and the number of polygon sides k, this construction matches the upper bound of the last section. Define an inverse Ackermann function $\phi(n, k) = \min\{i \geq 1 \mid (n/k)F(i, \lfloor n/k \rfloor) \geq n\}$. The construction above gives zone complexity $ijF(i, j)$ for an arrangement of $jF(i, j)$ pseudolines and a polygon with $F(i, j) + O(1)$ sides. Setting $j = \lfloor n/k \rfloor$ and $i = \phi(n, k)$, the construction gives zone complexity $n\phi(n, k)$ for an arrangement of n pseudolines in a $(k + O(1))$-gon. We now show that for $n/k \geq 7$, $\phi(n, k)$ is within a constant of $\alpha(n, k)$, the function defined in the last section. These inverse Ackermann functions are also within constants of more usual inverse Ackermann functions, for example the one used by Tarjan [9].

LEMMA 22. *If* $n/k \geq 7$, $\phi(n, k) - 1 \leq \alpha(n, k) \leq 4 \cdot \phi(n, k)$.

PROOF. For each choice of i and j, $F(i + 1, j) \geq A(i, j)$, so $\alpha(n, k) \geq \phi(n, k) - 1$. For each i and each $j \geq 7$, $A(2i, 2j) \geq A(2i, j + 7) \geq jF(i, j)$, so $(1/2)\alpha(n, k) \leq \phi(n, 2k) \leq 2\phi(n, k)$. □

THEOREM 5. *The maximum complexity of the cells on the inside of a k-gon, cut by n pseudolines, is $\Theta(n\alpha(n, k))$.*

PROOF. This follows from Theorem 4 along with Lemma 22 for $n \geq 7k$ and the $n\alpha(n)$ lower bound for $n < 7k$. □

Both Wiernik's construction and our lower bound construction use pseudolines, and it seems difficult to modify them to use straight lines instead. We close by posing the following open problem.

OPEN PROBLEM 3. *Find a construction for a straight line arrangement cut by a polygon having superlinear zone complexity, or show that no such construction exists.*

Acknowledgments

We thank Mike Paterson and Peter Shor for many valuable discussions.

References

1. B. Chazelle, L. J. Guibas, and D. T. Lee, *The power of geometric duality*, BIT **25**, (1985), 76–90.

2. H. Edelsbrunner, *Algorithms in combinatorial geometry*, Springer-Verlag, Berlin and New York, 1987.

3. H. Edelsbrunner, J. W. Jaromczyk, and G. Swiatek, *The complexity of the common intersection of n double wedges*, preprint, 1989.

4. H. Edelsbrunner, J. O'Rourke, and R. Seidel, *Constructing arrangements of lines and hyperplanes with applications*, SIAM J. Comput. **15** (1986), 341–363.

5. S. Hart and M. Sharir, *Nonlinearity of Davenport-Schinzel Sequences and of Generalized Path Compression Schemes*, Combinatorica **6**, (1986), 151–177.

6. M. S. Paterson and F. F. Yao, *Binary partitions with applications to hidden-surface removal and solid modeling*, Discrete Comput. Geom. **5** (1990), 485–504.

7. M. Sharir, Unpublished proof, personal communication with P. Shor, 1989.

8. P. Shor, *Geometric realizations of superlinear Davenport Schinzel sequences*, preprint, 1988.

9. R. E. Tarjan, *Data Structures and Network Algorithms*, SIAM, Philadelphia, 1983.

10. A. Wiernik and M. Sharir, *Planar realization of nonlinear Davenport-Schinzel sequences by segments*, Discrete Comput. Geom. **3** (1988), 15–47.

(M. BERN, D. EPPSTEIN, AND F. YAO) XEROX PALO ALTO RESEARCH CENTER, PALO ALTO, CALIFORNIA 94304

(P. PLASSMAN) CENTER FOR APPLIED MATHEMATICS, CORNELL UNIVERSITY, ITHACA, NEW YORK 14853, MATHEMATICS AND COMPUTER SCIENCE DIVISION, ARGONNE NATIONAL LABORATORY, ARGONNE, ILLINOIS 60439

DIMACS Series in Discrete Mathematics
and Theoretical Computer Science
Volume **6**, 1991

On the Perimeter of a Point Set in the Plane

VASILIS CAPOYLEAS AND JÁNOS PACH

ABSTRACT. Let $\{c_0, \ldots, c_n\}, \{c_0', \ldots, c_n'\}$ be two point sets in the plane satisfying $|c_i - c_j| \le |c_i' - c_j'|$ for all i and j. By a theorem of Sudakov and Alexander, the perimeter of the convex hull of $\{c_0, \ldots, c_n\}$ does not exceed the perimeter of the convex hull of $\{c_0', \ldots, c_n'\}$. We give a simple proof of this result and establish a similar theorem in the case when the Euclidean distance is replaced by the maximum norm. We point out the close relationship between these questions and a longstanding open problem due to Thue Poulsen, Kneser, and Hadwiger.

1. The union of balls

More than thirty-five years ago E. Thue Poulsen [**14**], M. Kneser [**9**], and H. Hadwiger [**7**] proposed the following conjecture which has attracted the interest of many geometers but is still open. Let $\{C_0, \ldots, C_n\}, \{C_0', \ldots, C_n'\}$ be two collections of disks of radius r in the plane. Let c_i and c_i' denote the center of C_i and C_i', respectively, and assume that $|c_i - c_j| \le |c_i' - c_j'|$ for all i and j.

Then

$$(1) \qquad \operatorname{Area}\left(\bigcup_{i=0}^{n} C_i\right) \le \operatorname{Area}\left(\bigcup_{i=0}^{n} C_i'\right).$$

W. Habicht (see [**9**]) and B. Bollobás [**3**] settled the special case when the system $\{C_0, \ldots, C_n\}$ can be continuously transformed into $\{C_0', \ldots, C_n'\}$ so that during the transformation the mutual distances between the centers do not decrease.

1991 *Mathematics Subject Classification.* Primary 52A40; Secondary 52A25, 51K99, 05B40.

This work was carried out while the first author was at Rutgers University. Partially supported by NSF Grant CCR-8902522.

Research of the second author was supported by NSF Grant CCR-89-01484, Hungarian Science Foundation Grant OTKA-1814 and the DIMACS Center.

Assume now that (1) is true, and let the radii r of the disks tend to infinity. It is easy to see that

$$\text{Area}\left(\bigcup_{i=0}^{n} C_i\right) = r^2\pi + r \text{ Per conv}\{c_0, \dots, c_n\}$$
$$+ \text{ Area conv}\{c_0, \dots, c_n\} + O(1/r),$$

where Per and conv stand for the perimeter and the convex hull, respectively. Similarly,

$$\text{Area}\left(\bigcup_{i=0}^{n} C_i'\right) = r^2\pi + r \text{ Per conv}\{c_0', \dots, c_n'\}$$
$$+ \text{ Area conv}\{c_0', \dots, c_n'\} + O(1/r).$$

Now (1) immediately yields

(2) $$\text{Per conv}\{c_0, \dots, c_n\} \leq \text{Per conv}\{c_0', \dots, c_n'\}.$$

The aim of the present paper is to give a simple elementary proof of this weaker assertion, which was first established by Sudakov [13] and rediscovered by Alexander [1]. Our approach is similar to the one followed in Alexander's paper, but it avoids using Schläfli's formula. Both proofs are based on a simple property of simplices (Lemma 1 in the next section).

THEOREM 1. *Let $\{c_0, \dots, c_n\}, \{c_0', \dots, c_n'\}$ be two point sets in the plane satisfying $|c_i - c_j| \leq |c_i' - c_j'|$ for all i and j. Then the perimeter of the convex hull of $\{c_0, \dots, c_n\}$ does not exceed the perimeter of the convex hull of $\{c_0', \dots, c_n'\}$.*

Let us first recall some basic notions and results from the theory of convex bodies. Let B^n denote the n-dimensional unit ball, and let S^{n-1} be the boundary of B^n. For any convex set $K \subseteq \mathbf{R}^n$, let $K + rB^n$ denote the *parallel body* of K with radius r, i.e., the set of all points of the space whose distance from at least one element of K is at most r. Let Vol_n stand for the n-dimensional volume. It is well known (see, e.g., Bonnesen–Fenchel [2], Busemann [4], Leichtweiss [10]) that $\text{Vol}_n(K + rB^n)$ can be expressed as a polynomial of degree n in r,

(3)
$$\text{Vol}_n(K + rB^n) = W_0(K) + \binom{n}{1} W_1(K)r$$
$$+ \binom{n}{2} W_2(K)r^2 + \cdots + W_n(K)r^n.$$

The coefficient $W_m(K)$ is called the mth mean projection measure (Quermassintegral) of K. $W_0(K) = \text{Vol}_n(K)$, $\binom{n}{1}W_1(K)$ is the surface area of K, $W_n(K) = \text{Vol}_n B^n = \kappa_n$. In general, apart from a factor depending only on n, $W_m(K)$ is the average of $\mathrm{Vol}_{n-m} K(F)$ over all $(n-m)$-dimensional subspaces $F \subseteq \mathbf{R}^n$, where $K(F)$ denotes the orthogonal projection of K on F. In particular,

$$(4) \qquad W_{n-1}(K) = \frac{1}{2n} \int_{S^{n-1}} \mathrm{width}_n(K, x) \, dx,$$

where $\mathrm{width}_n(K, x) = \mathrm{Vol}_1 K(x)$ is the distance between the two supporting hyperplanes of K perpendicular to the unit vector $x \in S^{n-1}$.

Let us turn now to the proof of Theorem 1. Imagine that the plane is embedded in $\mathbf{R}^n$. Put $K = \mathrm{conv}\{c_0, \dots, c_n\}$, $K' = \mathrm{conv}\{c_0', \dots, c_n'\}$. Furthermore, let B_i and B_i' denote the n-dimensional balls of radius r centered at c_i and c_i' respectively. Obviously, $\bigcup_{i=0}^n B_i \subseteq K + rB^n$.

CLAIM 1. Let $\mathrm{diam}\, K = \max_{0 \le i, j \le n} |c_i - c_j|$. If $r \ge \mathrm{diam}\, K$, then

$$K + \left(r - \frac{(\mathrm{diam}\, K)^2}{r} \right) B^n \subseteq \bigcup_{i=0}^n B_i.$$

PROOF. Let $x \in K + (r - (\mathrm{diam}\, K)^2/r)B^n$, i.e, there exists $k \in K$ such that $|x - k| \le r - (\mathrm{diam}\, K)^2/r$. If x is not a vertex of K, then one can choose c_i so that $\angle xkc_i \le \pi/2$. But then

$$|x - c_i| \le \sqrt{|x - k|^2 + |k - c_i|^2} \le \sqrt{\left(r - \frac{(\mathrm{diam}\, K)^2}{r} \right)^2 + (\mathrm{diam}\, K)^2} \le r$$

provided that $r \ge \mathrm{diam}\, K$. Hence, $x \in B_i$. $\square$

CLAIM 2. Let B_i and B_i' denote the n-dimensional balls of radius r centred at c_i and c_i' respectively. Then, $\mathrm{Vol}_n(\bigcup_{i=0}^n B_i) \le \mathrm{Vol}_n(\bigcup_{i=0}^n B_i')$.

Now we show how Theorem 1 can be deduced from these observations, and postpone the proof of Claim 2 until the next section.

Combining Claims 1 and 2, we obtain that

$$\mathrm{Vol}_n \left(K + \left(r - \frac{(\mathrm{diam}\, K)^2}{r} \right) B^n \right) \le \mathrm{Vol}_n(K' + rB^n).$$

Substituting (3), this implies

$$W_n(K) \left(r - \frac{(\mathrm{diam}\, K)^2}{r} \right)^n + W_{n-1}(K) \left(r - \frac{(\mathrm{diam}\, K)^2}{r} \right)^{n-1} + O(r^{n-2})$$

$$\le W_n(K')r^n + W_{n-1}(K')r^{n-1} + O(r^{n-2}).$$

Using the fact that $W_n(K) = W_n(K') = \kappa_n$, and taking the limits as r tends to infinity, we get

$$(5) \qquad W_{n-1}(K) \le W_{n-1}(K').$$

However, K is a *planar* convex set, so (4) can be rewritten as

$$W_{n-1}(K) = \frac{1}{2n} \int_{S^1} \int_0^{+\pi/2} \int_{(\sin\phi)S^{n-3}} \text{width}_2(K,y) \cos^2\phi \, dz \, d\phi \, dy$$

$$= \frac{(n-2)\kappa_{n-2}}{2n} \left(\int_{S^1} \text{width}_2(K,y) \, dy \right) \left(\int_0^{+\pi/2} \sin^{n-3}\phi \cos^2\phi \, d\phi \right).$$

Similarly,

$$W_{n-1}(K') = \frac{(n-2)\kappa_{n-2}}{2n} \left(\int_{S^1} \text{width}_2(K',y) \, dy \right) \left(\int_0^{+\pi/2} \sin^{n-3}\phi \cos^2\phi \, d\phi \right).$$

Hence, it follows immediately from (5) that

$$\text{Per } K = \frac{1}{2} \int_{S^1} \text{width}_2(K,y) \, dy \le \frac{1}{2} \int_{S^1} \text{width}_2(K',y) \, dy = \text{Per } K',$$

completing the proof of Theorem 1.

2. Proof of Claim 2

The proof consists of two steps. The first one is a slight generalization of a simple fact which was also used by M. Gromov [6] to verify a special case of the following "dual" counterpart of the Hadwiger-Kneser-Thue Poulsen conjecture: Let $\{B_0, B_1, \ldots, B_m\}$, $\{B_0', B_1', \ldots, B_m'\}$ be two sets of balls in n-dimensional space, and let c_i and c_i' denote the center of B_i and B_i', respectively. If $|c_i - c_j| \le |c_i' - c_j'|$ for all i and j, then

$$(6) \qquad \text{Vol}_n \left(\bigcap_{i=0}^m B_i \right) \ge \text{Vol}_n \left(\bigcap_{i=0}^m B_i' \right).$$

Gromov proved this result when the number of balls does not exceed $n + 1$, i.e., $m \le n$.

LEMMA 1. *Let* $K = \text{conv}\{c_0, \ldots, c_n\}$ *and* $K' = \text{conv}\{c_0', \ldots, c_n'\}$ *be two nondegenerate simplices in* n-*dimensional space, and assume that* $|c_i - c_j| \le |c_i' - c_j'|$ *for all* i *and* j.

Then K *can be continuously transformed into a congruent copy of* K' *in a finite number of steps so that*

  (i) *in each step we move only one vertex,*
  (ii) *the motion of this vertex is smooth,*
  (iii) *the edgelengths of the simplex never decrease.*

PROOF. Let $\mathbf{A}$ and $\mathbf{A}'$ denote the n by n matrices whose column vectors are $c_1 - c_0, c_2 - c_0, \ldots, c_n - c_0$ and $c_1' - c_0', c_2' - c_0', \ldots, c_n' - c_0'$, respectively. Put $\mathbf{D} = \mathbf{A}^T\mathbf{A}$ and $\mathbf{D}' = (\mathbf{A}')^T\mathbf{A}'$. Let $d_{ij} = |c_i - c_j|$ and

$d'_{ij} = |c'_i - c'_j|$. Then the entries of $\mathbf{D}$ can be expressed by the law of cosine, as follows.

$$\mathbf{D}_{ij} = \, < c_i - c_0, c_j - c_0 > \, = \frac{(d_{0i})^2 + (d_{0j})^2 - (d_{ij})^2}{2}.$$

Similarly,

$$\mathbf{D}'_{ij} = \, < c'_i - c'_0, c'_j - c'_0 > \, = \frac{(d'_{0i})^2 + (d'_{0j})^2 - (d'_{ij})^2}{2}.$$

In particular, both $\mathbf{D}$ and $\mathbf{D}'$ are symmetric.

For any $0 \le t \le 1$, let

$$d^t_{ij} = \sqrt{(1 - t)(d_{ij})^2 + t(d'_{ij})^2},$$

$$\mathbf{D}^t_{ij} = \frac{(d^t_{0i})^2 + (d^t_{0j})^2 - (d^t_{ij})^2}{2} = (1 - t)\mathbf{D}_{ij} + t\mathbf{D}'_{ij}.$$

A symmetric matrix $\mathbf{D}$ can be written in the form $\mathbf{D} = \mathbf{A}^T \mathbf{A}$ for some nonsingular matrix $\mathbf{A}$, if and only if $\mathbf{D}$ is positive definite. Hence $\mathbf{D} = \mathbf{D}^0$, $\mathbf{D}' = \mathbf{D}^1$ are positive definite, and so is $\mathbf{D}^t = (1 - t)\mathbf{D} + t\mathbf{D}'$. Let us express $\mathbf{D}^t$ as $(\mathbf{A}^t)^T(\mathbf{A}^t)$ for some nonsingular matrix $\mathbf{A}^t$, and let us consider the simplex K^t induced by $c^t_0 = 0$ and the column vectors $c^t_1, \dots, c^t_n$ of $\mathbf{A}^t$. Then

$$|c^t_i - c^t_j| = \sqrt{\mathbf{D}^t_{ii} + \mathbf{D}^t_{jj} - 2\mathbf{D}^t_{ij}} = d^t_{ij}$$

for all $0 < i, j \le n$, and

$$|c^t_i - c^t_0| = |c^t_i| = \sqrt{\mathbf{D}^t_{ii}} = d^t_{0i}$$

for every i.

Clearly, d^t_{ij} is a smooth nondecreasing function in $[0, 1]$.

Let us assign to any nondegenerate n-dimensional simplex the point in $\mathbf{R}^{\binom{n+1}{2}}$, whose coordinates are the squares of its edgelengths. The set $\mathscr{D}$ of those points of $\mathbf{R}^{\binom{n+1}{2}}$ which are assigned to such a simplex is obviously *open*, because these points correspond to symmetric positive definite matrices. Let p and p' denote the points corresponding to the simplices K and K', respectively. Then the points assigned to some K^t $(0 \le t \le 1)$ form a segment connecting p and p' in $\mathscr{D}$. Choose a small $\delta > 0$ such that the parallel body of the segment pp' with radius δ is entirely contained in $\mathscr{D}$. We can easily connect p and p' within $pp' + \delta B^n$ by a finite (oriented) polygon each side of which points in the positive direction of some coordinate axis. Each point of this polygon corresponds to a simplex $\overline{K}^t$ $(0 \le t \le 1)$ whose first vertex without loss of generality can be chosen to be the origin. We can assume that the second vertex of $\overline{K}^t$ is on the positive x_1-axis, the third vertex is in the $x_1 x_2$ plane with positive x_2, etc. Then $\overline{K}^t$ $(0 \le t \le 1)$

describes a continuous, piecewise smooth transformation of a congruent copy of K into a congruent copy of K' so that we move one vertex at a time and the edgelengths never decrease. □

LEMMA 2. *Let* $\{B_0, B_1, \ldots, B_m\}$ *be a collection of n-dimensional balls of arbitrary radii in* $\mathbf{R}^n$, *and let* c_i *denote the center of* B_i. *Let* $\vec{v}$ *be a vector with the property that translating* c_0 *along* $\vec{v}$, *the distance between* c_0 *and any other* c_i *does not decrease. Then* $\mathrm{Vol}_n(\bigcup_{i=0}^m B_i)$ *does not decrease during this translation.*

PROOF. Let S^+ (and S^-) denote the set of those points p on the surface of B_0, which do not belong to any other B_i, and such that the ray $R(p)$ starting at p and pointing to the direction of $\vec{v}$ does not intersect (resp. intersects) the interior of B_0.

Let $\hat{S}^+$ and $\hat{S}^-$ denote the orthogonal projections of S^+ and S^-, respectively, on a hyperplane perpendicular to $\vec{v}$. We claim that

(7) $\mathrm{Vol}_{n-1}\hat{S}^+ \geq \mathrm{Vol}_{n-1}\hat{S}^-.$

To see this, it suffices to show that for every $p \in S^-$, the ray $R(p)$ passes through a point of S^+. Suppose not. Then the (other) intersection point of $R(p)$ and the surface of B_0 belongs to some B_i $(i \neq 0)$. This would imply that the distance between c_0 and c_i decreases as we move B_0 along $\vec{v}$, contradicting our assumption.

Lemma 2 follows from (7) by integration, because translating B_0 along $\vec{v}$ by an infinitesimal amount dx, $\mathrm{Vol}_n(\bigcup_{i=0}^m B_i)$ will change by

$$(\mathrm{Vol}_{n-1}\hat{S}^+ - \mathrm{Vol}_{n-1}\hat{S}^-)\,dx. \quad \square$$

Now we are in the position to prove Claim 2. Let us perturb slightly the point sets $\{c_0, \ldots, c_n\}, \{c_0', \ldots, c_n'\} \subseteq \mathbf{R}^n$ to make them full-dimensional, without violating the conditions $|c_i - c_j| \leq |c_i' - c_j'|$ for all i and j.

Let us assume that this perturbation does not change $\mathrm{Vol}_n(\bigcup_{i=0}^n B_i)$ and $\mathrm{Vol}_n(\bigcup_{i=0}^n B_i')$ by more than ϵ.

Lemmas 1 and 2 now imply $\mathrm{Vol}_n(\bigcup_{i=0}^n B_i) - \epsilon \leq \mathrm{Vol}_n(\bigcup_{i=0}^n B_i') + \epsilon$ for any $\epsilon > 0$, and Claim 2 follows.

3. Perimeter in the maximum norm

Throughout we have been measuring distances in the usual Euclidean norm. However, it is possible that (2) is valid for a large class of norms.

Let l_∞ denote the distance induced by the maximum norm, i.e., given two points $a = (x, y), a' = (x', y') \in \mathbf{R}^2$, $l_\infty(a, a') = \max(|x - x'|, |y - y'|)$. Let Per_∞ denote the perimeter of a convex polygon measured in the l_∞-distance. We have the following theorem.

THEOREM 2. *Let* $\{c_0, \ldots, c_n\}, \{c_0', \ldots, c_n'\}$ *be two point sets in the plane satisfying* $l_\infty(c_i, c_j) \le l_\infty(c_i', c_j')$ *for all* i *and* j. *Then*

$$\text{Per}_\infty \text{ conv}\{c_0, \ldots, c_n\} \le \text{Per}_\infty \text{ conv}\{c_0', \ldots, c_n'\}.$$

PROOF. First we make an observation about the perimeter of a convex polygon in the maximum norm.

LEMMA 3. *Given a finite point set* P *in the plane, let* R *denote the smallest rectangle enclosing* P *such that the angles between the sides of* R *and the coordinate axes are equal to* $\pi/4$. *Let* a, b, c, d *be* (*not necessarily distinct*) *elements of* P *such that each side of* R *contains at least one of them. Then*

$$\text{Per}_\infty \text{ conv}(P) = \text{Per}_\infty \text{ conv}\{a, b, c, d\} = \frac{\text{Per}(R)}{\sqrt{2}}.$$

PROOF. We say that a segment is *nearly horizontal* (*vertical*) if the angle between its supporting line and the x-axis is at most (at least) $\pi/4$.

Suppose that a, b, c, d are in clockwise order, and a has the largest sum of coordinates. Then all sides of conv P between a and b, and between c and d are nearly vertical. All other sides are nearly horizontal. Thus, in the maximum norm the total length of the sides between a and b is equal to the difference between the y-coordinates of these two points, i.e., to $l_\infty(a, b)$. We can argue similarly for the sides between b and c, c and d, d and a, to obtain the first equality in the lemma. To prove the second one, notice that the part of the perimeter of R between, say, a and b, has Euclidean length $\sqrt{2}l_\infty(a, b)$. □

COROLLARY 1. *Let* P *be a finite point set in the plane,* $P^* \subseteq P$. *Then*

$$\text{Per}_\infty \text{ conv } P^* \le \text{Per}_\infty \text{ conv } P. \quad □$$

Next we show that it is sufficient to prove Theorem 2 for $n = 3$ (for 4 points). For $n < 3$ there is nothing to prove. Assume that the statement is true for $n = 3$, and let $n > 3$. Let a, b, c, d denote the points of $P = \{c_0, \ldots, c_n\}$ sitting on the boundary of the rectangle R enclosing P, as in Lemma 3. Let a', b', c', d' denote the corresponding points in $P' = \{c_0', \ldots, c_n'\}$. Suppose, in order to obtain a contradiction, that

$$\text{Per}_\infty \text{ conv } P > \text{Per}_\infty \text{ conv } P'.$$

By Lemma 3 and Corollary 1,

$$\text{Per}_\infty \text{ conv }\{a, b, c, d\} = \text{Per}_\infty \text{ conv } P > \text{Per}_\infty \text{ conv } P'$$
$$\ge \text{Per}_\infty \text{ conv }\{a', b', c', d'\},$$

contradicting our assumption that Theorem 2 is true for $n = 4$.

Suppose now that in the maximum norm distances between the elements of $P = \{a, b, c, d\}$ are at most as large as the corresponding distances within $P' = \{a', b', c', d'\}$, but Per_∞ conv $P > \mathrm{Per}_\infty$ conv P'.

We can assume without loss of generality that a', b', c', d' are in convex position. Suppose not. Then one of them, say d', is in the interior of the triangle induced by the others. The l_∞-distance of d' from a', b', and c' is either their vertical distance or their horizontal distance. Suppose without loss of generality that at most one of these three distances is the horizontal distance. Then we can move d' horizontally until it hits the boundary of the triangle $a'b'c'$, so that none of the l_∞-distances decreases.

We can also assume that a, b, c, d are in convex position. Otherwise, if, say, d is in the triangle abc, then by Corollary 1

$$\mathrm{Per}_\infty \text{ conv } \{a, b, c\} = \mathrm{Per}_\infty \text{ conv } P > \mathrm{Per}_\infty \text{ conv } P'$$
$$\geq \mathrm{Per}_\infty \text{ conv } \{a', b', c'\},$$

a contradiction.

Let $abcd$ be the cyclic order of the vertices of conv P. By symmetry, there are only two essentially different cases.

- *Case A*: The cyclic order of the vertices of conv P' is $a'b'c'd'$. Then

$$\mathrm{Per}_\infty \text{ conv } P = l_\infty(a, b) + l_\infty(b, c) + l_\infty(c, d) + l_\infty(d, a)$$
$$\leq l_\infty(a', b') + l_\infty(b', c') + l_\infty(c', d') + l_\infty(d', a')$$
$$= \mathrm{Per}_\infty \text{ conv } P'.$$

- *Case B*: The cyclic order of the vertices of conv P' is $a'c'b'd'$. By the triangle inequality for the maximum norm, $l_\infty(a, b) + l_\infty(c, d) \leq l_\infty(a, c) + l_\infty(b, d)$. Hence,

$$\mathrm{Per}_\infty \text{ conv } P = l_\infty(a, b) + l_\infty(b, c) + l_\infty(c, d) + l_\infty(d, a)$$
$$\leq l_\infty(b', c') + l_\infty(d', a') + l_\infty(a', c') + l_\infty(b', d')$$
$$= \mathrm{Per}_\infty \text{ conv } P'.$$

This completes the proof of Theorem 2. □

4. Related problems and generalizations

Theorem 1 can readily be generalized to bounded infinite sets $S \subset R^2$. A mapping $f : S \to R^2$ is said to be a *contraction* if $|f(p) - f(q)| \leq |p - q|$ for all $p, q \in S$.

THEOREM 3. *Let f be a contraction of a bounded set $S \subseteq R^2$. Then*

$$\mathrm{Per} \text{ conv } f(S) \leq \mathrm{Per} \text{ conv } S. \quad \square$$

Evidently, similar results cannot be true for the *surface area* of the convex hull of higher-dimensional sets. However, the above arguments immediately yield

THEOREM 4. *Let* f *be a contraction of a bounded set* $S \subseteq R^n$. *Then*

$$W_{n-1}(\operatorname{conv} f(S)) \leq W_{n-1}(\operatorname{conv} S),$$

where W_{n-1} *is the* $(n-1)$*st mean projection measure* (*the mean width*) *in* n *dimensions.* □

It might be interesting to notice that by slightly modifying the proof presented in §2, we can generalize Lemma 2 in two different directions.

LEMMA $2'$. *Let* $\{B_1, \ldots, B_k, B_{k+1}, \ldots, B_m\}$, $k < m$, *be a collection of* n-*dimensional balls of arbitrary radii in* $\mathbf{R}^n$, *and let* c_i *denote the center of* B_i. *Let* $\vec{v}$ *be a vector with the property that translating* $\{c_1, \ldots, c_k\}$ *along* $\vec{v}$, *the distance between any* c_i *and* c_j, $i \leq k < j$, *does not decrease. Then*

(i) $\operatorname{Vol}_n(\bigcup_{i=1}^m B_i)$ *does not decrease,*

(ii) $\operatorname{Vol}_n(\bigcap_{i=1}^m B_i)$ *does not increase,*

during this translation. □

Combining Lemma 1 and (the special case $k = 1$ of) Lemma $2'$(ii), we obtain a proof of (6) for $m \leq n$, somewhat different from the one given in [6]. It also implies the following theorem of Kirszbraun [8]: No contraction increases the circumradius of any n-dimensional compact set. A related result of Gale [5] states that the same is true for the width instead of the circumradius.

It is worth mentioning that conjecture (1) of Thue Poulsen, Kneser, and Hadwiger cannot be generalized to other Minkowski planes. More precisely, we have the following theorem.

THEOREM 5. *Let the plane be equipped with a norm* $\|.\|$ *such that the unit disk* $C = \{x \in \mathbf{R}^2 : \|x\| \leq 1\}$ *is not an ellipse.*

Then one can find points $a, b, a', b' \in \mathbf{R}^2$ *with the property that* $\|a - b\| < \|a' - b'\|$ *but*

$$\operatorname{Area}((C + a) \cup (C + b)) > \operatorname{Area}((C + a') \cup (C + b')). \quad \square$$

On the other hand, the weaker statement (2) might remain true for a large class of other norms substantially different from the Euclidean one.

In the last section we have shown that (2) holds in the plane equipped with the maximum norm (when the unit disk is a square). A related question is the following.

Given two collections of axis parallel unit squares in the plane, $\{S_1, \ldots, S_n\}$ and $\{S'_1, \ldots, S'_n\}$, such that $\text{Area}(S_i \cup S_j) \geq \text{Area}(S'_i \cup S'_j)$ for all i and j. Is it true that $\text{Area}(\cup S_j) \geq \text{Area}(\cup S'_j)$?

Some related questions with applications to particle physics are discussed by Lieb and Simon in [11] and [12].

REFERENCES

1. R. Alexander, *Lipschitzian mappings and total mean curvature of polyhedral surfaces*. I, Trans. Amer. Math. Soc. **288** (1985), 661–678.

2. T. Bonnesen and W. Fenchel, *Theorie der Konvexen Körper*, Springer-Verlag, Berlin, 1934 (reprinted by Chelsea Publ. Co., New York, 1948).

3. B. Bollobás, *Area of the union of disks*, Elem. Math. **23** (1968), 60–61.

4. H. Busemann, *Convex surfaces*, Interscience, London and New York, 1958.

5. D. Gale, *On Lipschitzian mappings of convex bodies*, Convexity Proc. Sympos. Pure Math., vol. 7, Amer. Math. Soc., Providence, RI, 1963, pp. 221–223.

6. M. Gromov, *Monotonicity of the volume of intersection of balls*, Geometrical Aspects of Functional Analysis, Lecture Notes in Math., vol. 1267, Springer-Verlag, Berlin, 1987, pp. 1–4.

7. H. Hadwiger, *Ungelöste Probleme No. 11*, Elem. Math. **11** (1956), 60–61.

8. M. Kirszbraun, *Über die zusammenziehenden und Lipschitzschen Transformationen*, Fund. Math. **22** (1934), 77–108.

9. M. Kneser, *Einige Bemerkungen über das Minkowskische Flächenmass*, Arch. Math. **6** (1955), 382–390.

10. K. Leichtweiss, *Konvexe Mengen*, Springer-Verlag, Berlin, Heidelberg, and New York, 1980.

11. E. Lieb, *Monotonicity of the molecular electronic energy in the nuclear coordinates*, J. Phys. B **15** (1982), L63–L66.

12. E. Lieb and B. Simon, *Monotonicity of the electronic contribution to the Born-Oppenheimer energy*, J. Phys. B **11** (1978), L537–L542.

13. V. Sudakov, *Gaussian random processes and measures of solid angles in Hilbert space*, Soviet Math. Dokl. **12** (2) (1971).

14. E. Thue Poulsen, *Problem 10*, Math. Scand. **2** (1954), 346.

COURANT INSTITUTE, NEW YORK UNIVERSITY, NEW YORK, NEW YORK 10012

COURANT INSTITUTE, NEW YORK UNIVERSITY, NEW YORK, NEW YORK 10012 AND MATHEMATICAL INSTITUTE OF THE HUNGARIAN ACADEMY OF SCIENCES, 1364 BUDAPEST, PF. 127

DIMACS Series in Discrete Mathematics
and Theoretical Computer Science
Volume **6**, 1991

Lines in Space—A Collection of Results

HERBERT EDELSBRUNNER

ABSTRACT. Many computational geometry problems are exceedingly more difficult if the setting is the (three-dimensional real) space $\mathbf{R}^3$ rather than the plane $\mathbf{R}^2$. Most often the reason for this striking increase in complexity is the appearance of new geometric phenomena caused by one-dimensional objects in space. The intention of recent studies on problems for lines in space is to shed light on these new phenomena and their complexities. This paper reviews some of the most important results and shows how they are related to problems in dimensions 2 and 5.

1. Introduction

In computational geometry it is a truism that problems in three-dimensional space, $\mathbf{R}^3$, are significantly more difficult and complex than the corresponding problems in the two-dimensional plane, $\mathbf{R}^2$. Of course, reality is not that simplistic, but it seems this way often enough to be a generally accepted view. As an example consider the problem of finding the *width* of a convex polygon/polytope with n vertices, that is, the smallest distance between two parallel lines/planes that bracket the polygon/polytope. In $\mathbf{R}^2$ this distance is realized by a vertex and an edge, and it is fairly easy to see how to enumerate and test $O(n)$ pairs in time $O(n)$ in order to find the width. In $\mathbf{R}^3$ the problem is more complicated because the width can either be realized by a vertex-facet pair or an edge-edge pair, and it is not clear how to limit the number of edge pairs that have to be considered. Indeed, no algorithm is known that computes the width of a convex polytope in time anywhere close to linear in n.

1991 *Mathematics Subject Classification.* Primary 05C35, 52A37, 68C05.

Key words and phrases. Computational and discrete geometry, three dimensions, lines, line segments, weavings, quadrics, cycles, hidden line elimination, Plücker coordinates, segment trees, polyhedral terrains.

Research of the author was supported by the National Science Foundation under grant CCR-8921421.

In a nut-shell, the kind of interaction between the edges as it appears in the width problem is a geometric phenomenon which cannot be observed in $\mathbf{R}^2$ and which significantly complicates matters in $\mathbf{R}^3$. More generally, the intricate interplay between one-dimensional objects in space is what makes computational geometry more difficult and interesting in $\mathbf{R}^3$ than in $\mathbf{R}^2$. Fairly recent research in the area specifically concentrated on this aspect of spatial geometry. This survey is based on this research, more specifically on the papers [4, 5, 6, 10], and summarizes some of the more important results in a fairly leisurely manner.

The first part of this survey, §§2–7, is combinatorial in nature and focuses on the notion of cycles defined by lines or line segments in space. Section 2 formally introduces the notion of a cycle and presents some motivation for why cycles seem a worthwhile object of study. The second part of this survey, §§8–15, has a more algorithmic flavor than the first part. The objects under investigation are sets of line segments and polyhedral terrains.

2. Lines and cycles

Given a finite collection of lines in space and a viewpoint, we can define a relation "$\prec$" that reflects when one line obstructs the view of another line, see Figure 2.1 for an example.

The main topic of the first part of this survey paper are combinatorial considerations related to "$\prec$". The objects under consideration are (infinite) lines and line segments (also called *rods*) in space. Typical questions asked are:

   (i) How many cycles can n lines define?
  (ii) How many cuts are necessary to eliminate all cycles?
 (iii) How fast can a cycle be detected, if there is one?

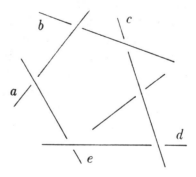

FIGURE 2.1. $a \prec b$, $b \prec c$, $c \prec d$, $d \prec e$, and $e \prec a$, therefore $a \prec b \prec c \prec d \prec e \prec a$ is a cycle.

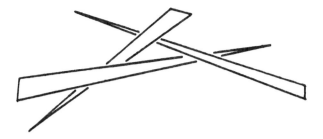

FIGURE 2.2. A cycle in "$\prec$" makes Step 1 of the Painter's algorithm impossible. A way out is to cut the objects so that all cycles are removed.

While these questions are of independent interest, there *are* problems in computational geometry that motivate the study of cycles. The *hidden surface removal problem* for a set of opaque polygonal objects, e.g. triangles in space, is the problem of drawing the view of the objects as seen from a given viewpoint. One of the common algorithm for this problem is the *Painter's algorithm*:

1. Sort the triangles consistently with "$\prec$."
2. Paint the objects according to this ordering from back to front.

Overmars and Sharir [9] have algorithms that eliminate hidden lines and surfaces by constructing the view from the front to the back. The complexity of their algorithms is sensitive to the output. A common drawback of the Painter's algorithm and the algorithms in [9] is that they assume acyclicity of "$\prec$". As illustrated in Figure 2.2, "$\prec$" can be cyclic for as few as three objects. So-called binary space partition trees are a common method for removing all cycles by effectively cutting the objects into smaller pieces; see [8, 11]. This method is related to the results presented in §§6 and 7.

The *point location problem* in space seeks to store a cell complex so that for a given query point the cell that contains it can be found as fast as possible. While in the plane there are solutions that take storage $O(n)$ and query time $O(\log n)$, where n is the total number of faces (regions, edges, and vertices) of the complex, the problem in space seems much harder. A data structure that takes storage $O(n)$ and query time $O(\log^2 n)$ can be found in [3]; it assumes, however, that the cells of the complex are acyclic with respect to the viewpoint at $(0, 0, \infty)$. Only recently, Preparata and Tamassia [15] gave an algorithm that is reasonable efficient in the general case. It takes storage $O(n \log^2 n)$ and query time $O(\log^2 n)$.

3. Weavings and perfect weavings

A *weaving* is a simple arrangement of lines (or line segments) in the plane, together with a binary "over-under" relation $\prec$ defined for them. A weaving is *perfect* if along each line the lines intersecting it alternate between "over" and "under." A weaving is *realizable* if there are lines (or line segments) in

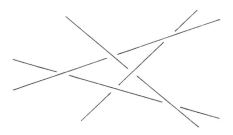

FIGURE 3.1. A perfect 4-weaving.

space that project (vertically) onto the weaving. A *weaving pattern* is a class of weavings that are combinatorially the same and it is *realizable* if at least one representative in the class is realizable. The following result is due to Pach, Pollack, and Welzl [10].

THEOREM 3.1. *Any perfect weaving of* $n \geq 4$ *lines is unrealizable.*

The case $n = 4$ can proved simply by rotating two of the four lines in space (see Figure 3.1) so that their projections remain the same, three of the four lines pairwise intersect, and the fourth line still has the strict alternation between "over" and "under" with respect to the three lines. This, of course, is impossible because the first three lines lie in a common plane and the fourth line intersects this plane in a single point. The proof for $n \geq 5$ is more difficult and can be found in [10].

4. Bipartite weavings

The roots of a quadratic form in x, y, z define a quadratic surface in space (also called *quadric*). Some members of the family of quadrics, in

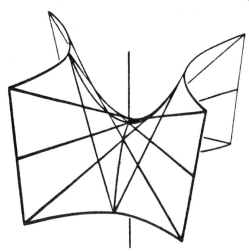

FIGURE 4.1. A hyperbolic paraboloid.

FIGURE 4.2. A perfect 4-by-4 weaving is not realizable.

particular the hyperboloid of one sheet and the hyperbolic paraboloid (see Figure 4.1), turn out to be useful in the study of weavings. Both surfaces have a number of interesting properties.

1. A line not contained in the surface meets it in at most two points.
2. There are two infinite families of lines, A and B, so that the surface is equal to $\bigcup A = \bigcup B$, the lines of each family are pairwise skew, and any two lines of different families either intersect or are parallel.
3. The surface divides space into two connected components.
4. Any three pairwise skew lines define a unique such surface that contains all three lines.

These properties imply fairly straightforwardly that for four lines in general position in space there are either two or zero other lines that meet all four.

A weaving of $m + n$ line segments is *bipartite* if the set of line segments can be split into sets $H = \{h_1, h_2, \ldots, h_m\}$ and $V = \{v_1, v_2, \ldots, v_n\}$ so that no two line segments of the same set intersect, and all line segments in H (V) intersect all line segments in V (H) in the same order. It is *perfect* if along each line segment the other line segments strictly alternate between "over" and "under," see Figure 4.2. Note that for fixed m and n there is only one perfect bipartite weaving *pattern*. The following result is taken from [10].

THEOREM 4.1. *A perfect bipartite m-by-n weaving is realizable iff* $\min\{m, n\} \leq 3$.

We will not repeat the proof of this result which is more complicated than one would like. It heavily uses properties of the two ruled surfaces mentioned above. Using either surface it is easy to construct a perfect bipartite 3-by-n weaving that is realizable. Just take three lines of one ruling family and n lines of the other family. By slightly rotating a line of the second family we can generate either an over-under-over or an under-over-under sequence, as desired. Since arbitrarily large perfect bipartite weavings are nonrealizable it seems natural to ask how contaminated with cycles $\prec$ can get. This leads to the question of counting or bounding the number of cycles.

5. Counting cycles in bipartite weavings

The question considered in this section is the maximum number of cycles, where the maximum is taken over all *realizable* m-by-n weavings. Assume for simplicity that $m = n$ and that either $h_i \prec v_j$ or $v_j \prec h_i$ for every $h_i \in H$ and $v_j \in V$.

Note that every cycle has even length and contains a subcycle of length 4. Furthermore, the maximum number of 4-cycles is $\Theta(n^4)$ (see Figure 5.1). The following lemma turns out to be rather useful in our combinatorial analysis. It can be found in Bollobás [2].

LEMMA 5.1. *If a bipartite graph on m and n nodes in each class contains no $K_{s,t}$, s of the m and t of the n nodes, then it has only $O(t^{1/s}mn^{1-1/s} + sn)$ arcs.*

5.1. An upper bound for elementary cycles. A *elementary cycle* is a 4-cycle defined by two adjacent line segments in H and two adjacent line segments in

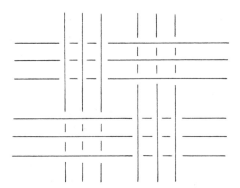

FIGURE 5.1. There are $4k$ line segments and k^4 4-cycles.

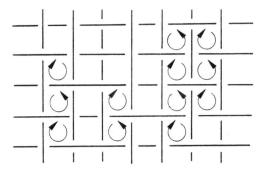

FIGURE 5.2. Cw and ccw elementary cycles in a 5-by-7 weaving.

V; it is either *cw* or *ccw* (see Figure 5.2). Every two adjacent line segments from H and from V in a perfect bipartite weaving define an elementary cycle. The following result, taken from [6], shows that in the realizable case the number of elementary cycles is far less than $(n-1)^2$, the maximum number in the nonrealizable case.

Upper Bound. *A realizable n-by-n weaving has at most* $O(n^{3/2})$ *elementary cycles.*

PROOF. For reasons of symmetry we count only ccw elementary cycles. Define a graph $(\bar{H} \cup \bar{V}, A)$ with $\bar{H} = \{(h_i, h_{i+1}) \mid 1 \le i \le n-1\}$, $\bar{V} = \{(v_j, v_{j+1}) \mid 1 \le j \le n-1\}$, and $\{(h_i, h_{i+1}), (v_j, v_{j+1})\} \in A$ if the four line segments define a ccw cycle. A $K_{2,2}$ in this graph corresponds to a perfect 4-by-4 weaving. Since such a weaving is nonrealizable, the graph cannot have a $K_{2,2}$ and has therefore at most $cn^{3/2}$ arcs. □

5.2. A lower bound for elementary cycles. A lower bound of $\Omega(n^{4/3})$ on the maximum number of cycles in a realizable n-by-n bipartite weaving can be found in [6]. The example is based on a construction of n points and n lines in the plane that define $\Theta(n^{4/3})$ point-line incidences (see [16, 7]). The points in this construction are arranged in a grid-like fashion, and the line set contains the n lines that contain the most points.

Here is a rough outline (illustrated in Figure 5.3) how this two-dimensional construction with point set P and line set L can be lifted to three-dimensions.

1. Embed the point-line construction in the yz-plane in space.
2. From each point $p \in P$ erect a line v_p normal to the yz-plane that intersects it in p.

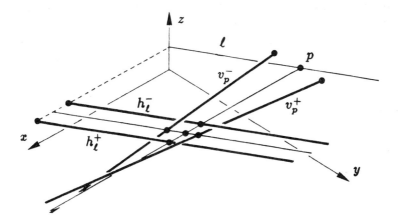

FIGURE 5.3. A point-line incidence in the yz-plane is lifted to an elementary cycle.

3. For each line $\ell \in L$ construct a line h_ℓ parallel to ℓ so that the orthogonal projection of h_ℓ onto the yz-plane is ℓ. The distance from the yz-plane is equal to the slope of ℓ. Notice that v_p and h_ℓ intersect iff $p \in \ell$.

4. Replace h_ℓ by two lines parallel to h_ℓ, one slightly in front and the other slightly behind h_ℓ, as viewed from the positive x-direction.

5. Replace v_p by two lines, one slightly to the left of v_p and with slightly negative slope, and the other slightly to the right of v_p and with slightly positive slope.

The construction can be done so that each point-line incidence becomes an elementary cycle, in the orthogonal projection of the lines onto the xy-plane.

6. Cutting cycles

As argued in §§5.1 and 5.2, a realizable n-by-n weaving can have $\Omega(n^{4/3})$ elementary cycles. This implies that sometimes $\Omega(n^{4/3})$ cuts of lines or line segments are necessary to eliminate all cycles. To get a subquadratic upper bound we first consider a topological property of cycles.

For a cycle we can look at its *polygon* and its *edges*, which are the polygon edges (see Figure 6.1).

LEMMA 6.1. *Any cycle contains a subcycle of length* 4 *whose "horizontal" edges lie on edges of the original cycle.*

To prove this lemma one can use two operations to simplify the polygon of the cycle. If the polygon has a self-intersection then a shortcut can be taken at this point leaving a shorter cycle. If the polygon is simple but not yet a 4-cycle there is a reflex vertex and the "vertical" edge at this vertex can be extended until it hits another "horizontal" edge. There is a shorter cycle either to the left or the right of this extended edge.

Another property of cycles in a bipartite weaving is that if the polygons of two ccw 4-cycles defined by common "horizontal" line segments overlap,

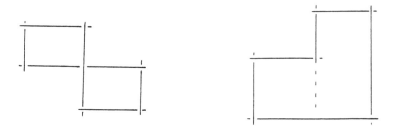

FIGURE 6.1. The left illustrates the polygon of a cycle. The right shows how an edge can be extended so as to get a shorter cycle.

then they overlap in the polygon of another ccw 4-cycle. Both results suggest that we cut 4-cycles at their "horizontal" edges and thus eliminate all cycles.

THEOREM 6.2. $O(n^{9/5})$ *cuts are sufficient to eliminate all cycles in a realizable n-by-n weaving.*

PROOF. First, set $m = n^{1/5}$ and cut along every $(m+1)$st rod of H and of V. This requires $O(n^{9/5})$ cuts and localizes each cycle to within an m-by-m bipartite weaving. Second, to eliminate the remaining cycles cut each 4-cycle at its two "horizontal" rods, unless its quadrilateral contains the quadrilateral of another 4-cycle and shares at least one of the "horizontal" rods with it.

To count we define a bipartite graph with sets of nodes A and B and prepare a $K_{2,2}$ argument. A is the set of pairs (v_i, v_j) with $1 \le j - i \le m$ and B is the set of pairs (h_i, h_j) with $1 \le j - i \le m$. So A and B are of size $O(mn)$ each. Connect a node $a \in A$ with a node $b \in B$ if the four lines define a ccw 4-cycle and the quadrilateral does not contain the quadrilateral of another ccw 4-cycle with which it shares at least one "horizontal" rod. Partition B into about $\binom{m+2}{2}$ subsets so that for each subset $j - i$ is invariant and any two nodes define nonoverlapping index intervals. The subgraph induced by A and such a subset of B has no $K_{2,2}$. If n_i denotes the size of the ith subset of B then the number of arcs/cuts in the entire graph is bounded by

$$\sum_{i=1}^{\binom{m+2}{2}} (n_i \sqrt{mn} + mn) = O(m^{3/2} n^{3/2} + m^3 n).$$

The assertion follows because $m = n^{1/5}$. □

7. A point-line incidence problem

For a weaving that is not bipartite it is still possible to associate a cycle with its polygon. In this more general case, a cycle is called *elementary* if its polygon is a face of the line arrangement. Clearly, there are at most $O(n^2)$ elementary cycles because there are at most that many faces. Currently no better upper bound on the number of elementary cycles is known. To shed some light on this problem we modify it somewhat and study lines in space and their intersection patterns.

Given a set of n lines in space, a point is called a *joint* if it lies on (at least) three noncoplanar lines of the set. It is fairly easy to see that $\Omega(n^{3/2})$ is a lower bound for the maximum number of joints defined by n lines. Take k planes in general position in space. They define $n = \binom{k}{2}$ lines and $\binom{k}{3} = \Omega(n^{3/2})$ vertices, each a joint.

The connection between this point-line incidence extremum problem and counting cycles is that a slight perturbation of the lines can make a joint an elementary 3-cycle. The following upper bound on the number of joints is taken from [6].

THEOREM 7.1. *A set of* n *lines in space defines at most* $O(n^{7/4})$ *joints.*

PROOF. Let L be the set of n lines, define $A = \{\{a, b\} \in \binom{L}{2} \mid a \cap b \neq \emptyset\}$, let $G = (L, A)$ be the intersection graph of L, and set $k = n^{1/4}$. Note that each joint is "witnessed" by at least one arc in A. The proof modifies G in three steps and finally presents the counting argument.

STEP 1. If a joint is incident to at least k lines then remove the $\binom{k}{2}$ arcs witnessing the joint. This creates at most $\frac{n(n-1)}{k(k-1)} = O(n^{3/2})$ *orphans*, that is, joints not witnessed by any arc.

STEP 2. If a plane contains at least k lines then remove the corresponding nodes, together with incident arcs. A plane contains fewer than n joints, which now possibly become orphans. This creates at most $\frac{n^2}{k} = O(n^{7/4})$ orphans.

STEP 3. If a quadric contains at least k lines then remove the corresponding nodes with incident arcs. A quadric contains fewer than $2n$ joints which now possibly become orphans. This creates at most $\frac{2n^2}{k} = O(n^{7/4})$ orphans.

The remaining graph contains no $K_{3,2k}$. Because if there are lines a_1, a_2, a_3 and $b_1, b_2, \ldots, b_{2k}$ that define a $K_{3,2k}$ then some k of the lines b_i either go through a common vertex or lie in a common plane or quadric, a contradiction. Using the lemma of §5 we get

$$|A| = O(k^{1/3} n^{5/3}) = O(n^{7/4}). \quad \square$$

8. Bichromatic problems

As mentioned in §1, for many algorithmic problems in space it is important to efficiently test the relative position of lines or rods. One example is the *ray-shooting problem*: given a collection of polytopes and a ray, determine which polytope intersects the ray closest to its starting point. We refer to Agarwal and Sharir [1] and Pellegrini [12] who use Plücker coordinates, explained in §12, for solutions to this problem. Other examples of such problems deal with so-called (*polyhedral*) *terrains*, that are continuous and piecewise linear maps from $\mathbf{R}^2$ to $\mathbf{R}$.

Given two terrains, one might want to determine whether or not they intersect, or compute their intersection, or compute their pointwise maximum; see Figure 8.1. These problems will be addressed in §§13–15. The underlying data structures, algorithms, and geometric techniques will be discussed in §§9–12. A particularly important data structure in this context is the segment tree and some of its variants as described in §§9 and 11. This structure facilitates the reduction of problems for polyhedral terrains and rods to problems

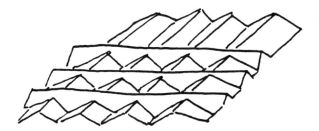

FIGURE 8.1. The pointwise maximum of two terrains.

for lines. The latter will be treated with algorithmic techniques based on the use of Plücker coordinates to represent lines.

9. The segment tree

The segment tree is a (one-dimensional) data structure storing a set S of n intervals (see e.g. [5, 14]). The n intervals have $k \leq 2n$ endpoints, and the k endpoints decompose $\mathbf{R}^1$ into $k + 1$ *atomic segments*. The *segment tree* is a minimum height, ordered binary tree with $k + 1$ leaves corresponding to the atomic segments, from left to right. The *segment* $\sigma(\kappa)$ of a node κ is the corresponding atomic segment if κ is a leaf, and $\sigma(\mu) \cup \sigma(\nu)$ if μ and ν are the children of κ. Finally, for every node μ with parent κ (if it exists) we define $L_\mu = \{I \in S \mid \sigma(\mu) \subseteq I \text{ and } \sigma(\kappa) \not\subseteq I\}$, the *list* of μ, see Figure 9.1. Here is a short list of basic properties of the segment tree of S.

1. An interval I belongs to the list of at most two nodes per level, thus it belongs to at most $2 \log_2 n + O(1)$ lists or nodes.
2. The segments $\sigma(\mu)$ of the nodes μ with $I \in L_\mu$ define a partition of I.

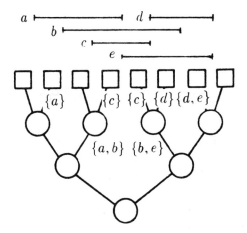

FIGURE 9.1. The segment tree defined by five intervals with a total of seven endpoints.

3. For a point p let $\rho = \mu_0, \mu_1, \ldots, \mu_h = \lambda$ be the path starting at the root ρ and ending at the leaf λ with $p \in \sigma(\lambda)$. Then $\{I \in S \mid p \in I\} = \bigcup_{i=0}^{h} L_{\mu_i}$.

Note that the lists of the nodes on the path in property 3 are disjoint, by definition. Define $L_\mu^* = \{I \in S \mid \sigma(\mu) \cap I \neq \emptyset \text{ and } \sigma(\mu) \not\subseteq I\}$.

4. $I \in L_\mu^*$ iff μ is ancestor of a node ν with $I \in L_\nu$.
5. An interval I belongs to lists L^* of at most two nodes per level of the segment tree.

Properties 1 and 5 imply that the segment tree, with lists L_μ and L_μ^* stored at nodes μ, takes storage $O(n \log n)$ to represent a set of n intervals.

10. A bichromatic intersection lemma

Let B be a set of m blue and R be a set of n red line segments in $\mathbf{R}^2$, with the property that no two line segments of the same color intersect. We use a single segment tree to store B and R. The segment tree is defined for the vertical projections of the line segments onto the x-axis, assuming no line segment is vertical. However, instead of intervals, the corresponding *line segments* are stored in the lists L and L^*. Accordingly, we extend the definition of $\sigma(\mu)$ from a one-dimensional interval to the vertical slab that intersects the x-axis in this interval. The blue and red line segments are stored in separate lists, so for each node ν we have lists B_ν, B_ν^*, R_ν, R_ν^*, defined just as L_ν and L_ν^*. The following fairly straightforward but very useful result is taken from [5].

LEMMA 10.1. *For every pair $b \cap r \neq \emptyset$, $b \in B$ and $r \in R$, there is a unique node ν with*

(i) $b \cap r \in \sigma(\nu)$, *and*
(ii) $(b \in B_\nu$ *and* $r \in R_\nu)$ *or* $(b \in B_\nu$ *and* $r \in R_\nu^*)$ *or* $(b \in B_\nu^*$ *and* $r \in R_\nu)$.

PROOF. For $p = b \cap r$ consider the path $\rho = \nu_0, \nu_1, \ldots, \nu_h = \lambda$ with $p \in \sigma(\lambda)$. Define i so that $b \in B_{\nu_i}$ and j so that $r \in R_{\nu_j}$. If $i = j$ then $\nu = \nu_i = \nu_j$, if $i < j$ then $\nu = \nu_i$, and if $i > j$ then $\nu = \nu_j$. $\square$

The three cases in (ii) are disjoint which will be crucial in the upcoming applications of the lemma. For example, consider the problems of reporting and counting all pairs $(b, r) \in B \times R$ so that $b \cap r \neq \emptyset$. The solution sketched below is based on the segment tree for $B \cup R$ (with lists B_ν, B_ν^*, R_ν, R_ν^* per node ν). The above lemma admits a reduction to reporting/counting, for each node ν, the pairs $(b, r) \in (B_\nu \times R_\nu) \cup (B_\nu \times R_\nu^*) \cup (B_\nu^* \times R_\nu)$ that intersect. The intersecting pairs in $B_\nu \times R_\nu^*$ can be computed as follows; the other two cases are similar.

STEP 1. Sort the line segments in B_ν from top to bottom. STEP 2. Use binary search to locate the endpoints of the (pieces of the) line segments in R_ν^* amids the line segments in B_ν.

The amount of storage needed for the segment tree is $O(n \log n)$, and the time is $O(n \log^2 n)$ because there are endpoints of $O(n \log n)$ (pieces of) red line segments to be located. Also, sorting the lists B_ν and R_ν costs time $O(n \log^2 n)$. Using refined algorithmic techniques, the storage can be improved to $O(n)$ (by traversing the segment tree rather than storing it), and the time can be brought down to $O(n \log n)$ (using merging and fractional cascading). Details can be found in [5].

11. Another layer

The above bichromatic line segment intersection problem can also be solved by adding another layer to the segment tree. In effect, this defines coverings $B = \bigcup_{i=1}^{k} B_i$ and $R = \bigcup_{i=1}^{k} R_i$ with the following properties.

1. $b \cap r \neq \emptyset$ iff there is a unique index $1 \leq i \leq k$ with $b \in B_i$ and $r \in R_i$.
2. All $b \in B_i$ meet all $r \in R_i$ in the same sequence, and vice versa.
3. $k = O(N \log N)$ with $N = m + n$.
4. $\sum_{i=1}^{k} |B_i| + |R_i| = O(N \log^2 N)$.

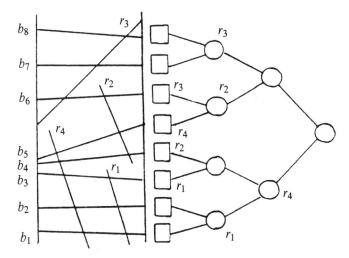

FIGURE 11.1. A slab $\sigma(\nu)$ with sets B_ν and R_ν^* is shown. The line segments in B_ν can be ordered from bottom to top, each corresponding to a leaf of a secondary segment tree. A blue line segment b is stored at each ancestor of its leaf. A line segment in R_ν^* intersects a contiguous subsequence of the line segments in B_ν and is stored in the nodes that cover this subsequence, as for ordinary segment trees.

It should be clear that after constructing the two coverings with the above properties it is easy to report or count intersecting pairs. Figure 11.1 illustrates the definition of the coverings. The significance of the coverings of B and R is that they can be used to reduce a bichromatic line segment problem into "few" and "not too big" subproblems which have the additional structure expressed in property 2 above.

12. Plücker coordinates

A (directed) line in space can be given by the homogeneous coordinates of two of its points:

$$\ell = \begin{pmatrix} \alpha_0 & \alpha_1 & \alpha_2 & \alpha_3 \\ \beta_0 & \beta_1 & \beta_2 & \beta_3 \end{pmatrix}, \quad \text{with } \alpha_0, \beta_0 > 0,$$

$$\ell' = \begin{pmatrix} \gamma_0 & \gamma_1 & \gamma_2 & \gamma_3 \\ \delta_0 & \delta_1 & \delta_2 & \delta_3 \end{pmatrix}, \quad \text{with } \gamma_0, \delta_0 > 0.$$

The eight coordinates of a line can be arranged in a 2-by-4 matrix, as above, and by taking the determinants of the six 2-by-2 submatrices we obtain the *Plücker coordinates* of the line:

$$p(\ell) = (\pi_{01}, \pi_{02}, \pi_{03}, \pi_{12}, \pi_{13}, \pi_{23}), \quad \text{with } \pi_{ij} = \det \begin{pmatrix} \alpha_i & \alpha_j \\ \beta_i & \beta_j \end{pmatrix}.$$

We call $p(\ell)$ the *Plücker point* of ℓ, given by homogeneous coordinates in projective five-dimensional space, P^5. Similarly, the π_{ij} can be interpreted as coefficients of a hyperplane:

$$v(\ell) = (\pi_{23}, -\pi_{13}, \pi_{12}, \pi_{03}, -\pi_{02}, \pi_{01}),$$

$$h(\ell) = \{p \mid \langle v(\ell), p \rangle = 0\} \quad \text{and} \quad h^+(\ell) = \{p \mid \langle v(\ell), p \rangle \geq 0\}.$$

Observe that ℓ and ℓ' intersect (or are parallel which we interpret as an intersection at infinity) iff the four defining points are coplanar, which is the case iff the determinant of the 4-by-4 matrix defined by their sixteen coordinates is zero. This determinant is obtained by plugging $p(\ell)$ into the equation of $h(\ell')$, or $p(\ell')$ into the equation of $h(\ell)$. This implies the following basic properties; see [4].

LEMMA 12.1. (1) $\ell \cap \ell' \neq \emptyset$ iff $p(\ell) \in h(\ell')$ (or, equivalently, $p(\ell') \in h(\ell)$).

(2) *If ℓ and ℓ' are oriented from left to right and ℓ is cw to ℓ' (in the projection onto the xy-plane), then ℓ lies above ℓ' iff $p(\ell) \in h^+(\ell')$.*

13. The relative position of lines

The segment tree reduction outlined in §§10 and 11 in combination with Plücker coordinates as introduced in §12 turn out to be powerful tools in attacking problems for rods in space. So let B and R be two sets of lines in $\mathbf{R}^3$, each line oriented consistently with the x-axis, so that in the orthogonal projection onto the xy-plane each line in R is cw to each line in B. Define $m = |B|$, $n = |R|$, and $N = m + n$. The following facts are taken from [4] and are useful when the above tools are put to work.

1. A line $r \in R$ lies above all lines in B iff $p(r) \in \mathscr{P}(B) = \bigcap_{b \in B} h^+(b)$.
2. $\mathscr{P}(B)$ is a convex cone in $\mathbf{R}^6$ (a polytope in P^5) with $O(m^2)$ faces.
3. There is a data structure (which we call the *envelope structure*) that stores B in storage $O(m^{2+\epsilon})$ so that for a given r we can decide in $O(\log m)$ time whether r lies above all $b \in B$.

Sets R and B are said to have the *towering property* if all $r \in R$ lie above all $b \in B$.

4. There is a randomized algorithm that tests in expected time $O(N^{4/3+\epsilon})$ whether or not R and B have the towering property.

14. Detecting the intersection of terrains

Here is a problem that can be solved by application of the segment tree in combination with Plücker coordinates. A *terrain* is a continuous function from $\mathbf{R}^2$ to $\mathbf{R}$ and it is *polyhedral* if it is piecewise linear. For two given polyhedral terrains, Σ_1 and Σ_2, with m and n edges in space, the problem is to determine whether or not they intersect.

Σ_1 is above Σ_2 iff first, every vertex of Σ_1 lies above Σ_2, second, every vertex of Σ_2 lies below Σ_1, and third, if $b \in \Sigma_1$ and $r \in \Sigma_2$ are two edges whose projections onto the xy-plane intersect then b lies above r. This characterization suggests the following high-level algorithm. Primes are used to denote orthogonal projections onto the xy-plane.

STEP 1. Locate every vertex of Σ_1' in Σ_2', and vice versa, and test points versus facets.

STEP 2. Use the two-layered segment tree for the line segment sets of Σ_1' and Σ_2' (see §11) and test the towering property for each pair $(B_i, R_i)_{i=1,2,...,k}$ using the algorithm suggested in §12.

Step 1 takes time $O(N \log N)$ using any one of the optimal point location algorithms available in the literature. Step 2 takes expected time $O(N^{4/3+\epsilon})$.

15. The pointwise maximum of terrains

In the worst case, the *pointwise maximum* (or the *upper envelope*) Σ of Σ_1 and Σ_2 has $\Theta(mn)$ edges (see Figure 8.1), and a trivial algorithm can construct it in this time. For a given instance, let k be the number of edges of Σ.

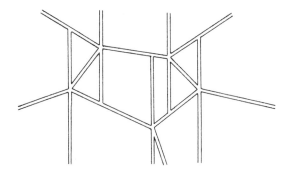

FIGURE 15.1. Projecting, decomposing into trapezoids, and shrinking.

THEOREM 15.1. *There is a randomized algorithm that constructs Σ in expected time $O(N^{3/2+\epsilon} + k \log^2 N)$ and storage $O(N)$.*

Σ can be constructed by reduction to edge-facet intersections. For one thing, Σ consists of pieces of Σ_1 and pieces of Σ_2 pasted together along curves of $\Sigma_1 \cap \Sigma_2$. Each vertex of such a curve is the intersection between an edge of Σ_1 and a facet of Σ_2, or vice versa. After one such vertex is found, the entire curve can be completed by tracing the intersection, edge-by-edge. The following representation of a polyhedral terrain Σ_1 is used to obtain the above result. Project Σ_1 onto the xy-plane, decompose the regions into trapezoids, and shrink the regions symbolically a tiny amount, as shown in Figure 15.1. Now store the nonvertical sides of the trapezoids as a collection of blue pairwise nonintersecting line segments in a two-layered segment tree with lists B_μ only. Observe that the upper and lower sides of a trapezoid are stored in exactly the same lists B_μ, and that B_μ can be viewed as an ordered list of line segments spanning $\sigma(\mu)$ or, alternatively, as an ordered list of trapezoids. Finally, store a list B_μ as an envelope structure if $|B_\mu| \leq N^{(1-\epsilon)/2}$. A list takes $O(N)$ randomized time and storage, and altogether this adds up to $O(N^{3/2+\epsilon})$. We find intersections between red edges and blue facets as follows.

1. Distribute each line segment $r \in R$ (the projection of Σ_2) to the appropriate nodes μ.
2. Query B_μ, that is,
   2.1. if μ is a leaf then test r against the only trapezoid; report if there is an intersection;
   2.2. if B_μ is not stored as an envelope structure then recurse for both children;
   2.3. if B_μ is stored as an envelope structure then
      Case 1. r is above all $b \in B_\mu$; return;
      Case 2. r is not above all $b \in B_\mu$; recurse for both children.

The overhead per $r \in R$ is the time spent at lists not stored as envelope structures, which is $O(N^{(1+\epsilon)/2})$. For each intersection we follow a path with queries, so we spend another $O(\log^2 n)$ time per intersection, as claimed in the theorem. Details can be found in [5].

REFERENCES

1. P. K. Agarwal and M. Sharir, *Applications of a new space partitioning technique*, Proc. Workshop on Algorithms and Data Structures, 1991 (to appear).

2. B. Bollobás, *Extremal graph theory*, Academic Press, London, 1978.

3. B. Chazelle, *How to search in history*, Inform. and Control **64** (1985), 77–99.

4. B. Chazelle, H. Edelsbrunner, L. J. Guibas, M. Sharir, and J. Stolfi, *Lines in space: combinatorics and algorithms*, Rept. UIUCDCS-R-90-1569, Dept. Comput. Sci., Univ. Illinois, Urbana, IL, 1990.

5. B. Chazelle, H. Edelsbrunner, L. J. Guibas, and M. Sharir, *Algorithms for bichromatic line segment problems and polyhedral terrains*, Rept. UIUCDCS-R-90-1578, Dept. Comput. Sci., Univ. Illinois, Urbana, IL, 1990.

6. B. Chazelle, H. Edelsbrunner, L. J. Guibas, R. Pollack, R. Seidel, M. Sharir, and J. Snoeyink, *Counting and cutting cycles of lines and rods in space*, Rept. UIUCDCS-R-90-1608, Dept. Comput. Sci., Univ. Illinois, Urbana, IL, 1990.

7. H. Edelsbrunner, *Algorithms in combinatorial geometry*, Springer-Verlag, Heidelberg, 1987.

8. H. Fuchs, Z. Kedem, and B. Naylor, *On visible surface generation by a priori tree structures*, Computer Graphics (SIGGRAPH'80 Conference Proceedings), pp. 124–133.

9. M. H. Overmars and M. Sharir, *Output-sensitive hidden surface removal*, Proc. 30th Ann. IEEE Sympos. Found. Comput. Sci. 1989, pp. 598–603.

10. J. Pach, R. Pollack, and E. Welzl, *Weaving patterns of lines and line segments in space*, Lecture Notes in Comput. Sci., vol. 450, Springer-Verlag, 1990, pp. 439–446.

11. M. S. Paterson and F. F. Yao, *Binary partitions with applications to hidden-surface removal and solid modelling*, Proc. 5th Ann. Sympos. Comput. Geom. 1989, pp. 23–32.

12. M. Pellegrini, *Ray shooting and isotopy classes of lines in 3-dimensional space*, Proc. Workshop on Algorithms and Data Structures, 1991 (to appear).

13. J. Plücker, *On a new geometry of space*, Philos. Trans. Royal Soc. London **155** (1865), 725–791.

14. F. P. Preparata and M. I. Shamos, *Computational geometry—an introduction*, Springer-Verlag, New York, 1985.

15. F. P. Preparata and R. Tamassia, *Efficient point location in a convex spatial cell complex*, Rept. CS-89-47, Dept. Comput. Sci., Brown Univ., Providence, RI, 1989.

16. E. Szemerédi and W. T. Trotter, *Extremal problems in discrete geometry*, Combinatorica **3** (1983), 381–392.

DEPARTMENT OF COMPUTER SCIENCE, UNIVERSITY OF ILLINOIS AT URBANA-CHAMPAIGN, URBANA, ILLINOIS 61801

DIMACS Series in Discrete Mathematics
and Theoretical Computer Science
Volume **6**, 1991

Singularities of Minimal Surfaces and Networks and Related Extremal Problems in Minkowski Space

Z. FÜREDI, J. C. LAGARIAS, AND F. MORGAN

ABSTRACT. This paper describes results on two questions about points in a Minkowski space arising from the study of minimal surfaces and networks with singularities. Let Φ denote a norm on $\mathbb{R}^n$ having unit ball $\mathscr{B}$. The first question concerns the maximal number of vectors in a Φ-equilateral set both for general norms and for strictly convex norms. A new proof is given of the known result that a Φ-equilateral set has cardinality at most 2^n for a general norm. There exists a strictly convex norm having a Φ-equilateral set of cardinality at least $(1.02)^n$, for large n. The second question concerns the maximal number of Φ-unit vectors such that $\Phi(\mathbf{x}_i + \mathbf{x}_j) \leq 1$ whenever $i \neq j$, both with and without the side condition $\Sigma_{i=1}^m \mathbf{x}_i = \mathbf{0}$. Here, exponentially large sets exist without the side condition; with it there are at most $2n$ vectors in the set.

1. Introduction

Soap films, grain boundaries in materials, and crystals all tend to minimize energy, and often have interesting singularities. The study of such singularities leads to various auxiliary problems, including questions in combinatorial geometry concerning arrangements of unit vectors in Minkowski space; [16–19; 23, Problem 11; 24–25].

A Minkowski space $(\mathbb{R}^n, \Phi)$ is just $\mathbb{R}^n$ with distances measured using a norm Φ. A norm Φ is completely determined by its unit ball

$$\mathscr{B} = \{\mathbf{x} : \Phi(\mathbf{x}) \leq 1\},$$

which is a bounded convex body with nonempty interior, centrally symmetric around $\mathbf{0}$. The *dual norm* Φ^* has unit ball

$$\mathscr{B}^* = \{\mathbf{y} : \langle \mathbf{x}, \mathbf{y} \rangle \leq 1 \text{ for all } \mathbf{x} \in \mathscr{B}\},$$

and $\Phi^{**} = \Phi$. A norm Φ is said to be *strictly convex* (also called *rotund*) if the boundary of $\mathscr{B}$ contains no line segment; it is said to be *smooth*

1980 *Mathematics Subject Classification* (1985 *Revision*). Primary 52A40, 49F10.

The third author was partially supported by the National Science Foundation and the Institute for Advanced Study.

(also called *differentiable*) if each point on the boundary of $\mathscr{B}$ has a unique supporting hyperplane. These are dual notions: a norm Φ is strictly convex if and only if Φ^* is smooth [5].

Lawlor and Morgan [13] recently obtained results on the structure of singularities of cones and of networks minimizing energies given by certain functionals involving general norms Φ on $\mathbb{R}^n$. They show that sets of Φ-equidistant points yield singular hypersurfaces C that minimize an energy $\int_C \Phi^*(\mathbf{n})$, where $\mathbf{n}$ is the unit normal to C, subject to certain boundary conditions.

THEOREM I. *Let* Φ *be a norm on* $\mathbb{R}^n$ *with dual norm* Φ^*. *Suppose there are points* $\mathbf{p}_1, \ldots, \mathbf{p}_m \in \mathbb{R}^n$ *such that*

$$\Phi(\mathbf{p}_j - \mathbf{p}_i) = 1 \quad \text{for } i \neq j.$$

Let $C \subset \mathscr{B}(\mathbf{0}, 1)$ *be a hypersurface that divides the Euclidean unit ball* $\mathscr{B}(\mathbf{0}, 1)$ *into regions* $R_1, \ldots, R_m$ *separated by pieces of hyperplanes* H_{ij} *with unit normals* $\mathbf{n}_{ij}$ *dual to* $\mathbf{p}_j - \mathbf{p}_i$ *(i.e.,* $\mathbf{n}_{ij} \cdot (\mathbf{p}_j - \mathbf{p}_i) = \Phi(\mathbf{n}_{ij})$). *Then* C *minimizes* $\sum_{i,j} \Phi^*(H_{ij})$ *among hypersurfaces (closed sets that are* C^1 *manifolds almost everywhere) that separate the fixed boundary regions* $R_i \cap S(\mathbf{0}, 1)$ *on the sphere* $S(\mathbf{0}, 1)$ *from each other in* $\mathscr{B}(\mathbf{0}, 1)$.

Lawlor and Morgan also prove a result concerning the singularity at $\mathbf{0}$ of a network C in $\mathbb{R}^n$ that minimizes $\int_C \Phi^*(\mathbf{t})$, where $\mathbf{t}$ is the unit tangent to C.

THEOREM II. *Let* Φ *be a strictly convex norm on* $\mathbb{R}^n$, *and let* Φ^* *denote the dual norm. Let* $\mathbf{a}_1, \ldots, \mathbf{a}_k \in \mathbb{R}^n$, *normalized so that* $\Phi^*(\mathbf{a}_j) = 1$, *and let* $\mathbf{b}_1, \ldots, \mathbf{b}_k$ *denote the unique dual vectors such that* $\Phi(\mathbf{b}_j) = 1$ *and* $\mathbf{a}_j \cdot \mathbf{b}_j = 1$. *Then the network* C *consisting of rays from the origin to* $\mathbf{a}_1, \ldots, \mathbf{a}_k$ *is* Φ^*-*minimizing if and only if every subcollection* J *of the* $\mathbf{b}_j$ *has a sum in the* Φ-*norm of length at most one, i.e.,*

$$(1.1) \qquad\qquad \Phi\left(\sum_{j \in J} \mathbf{b}_j\right) \leq 1,$$

and in addition

$$(1.2) \qquad\qquad \mathbf{b}_1 + \cdots + \mathbf{b}_k = \mathbf{0}.$$

If Φ *is not strictly convex, the dual vectors* $\mathbf{b}_j$ *are not necessarily uniquely determined, but if* (1.1) *and* (1.2) *hold for some choice of such* $\mathbf{b}_j$, *then* C *is* Φ^*-*minimizing.*

We call the condition (1.2) a *balancing condition*. For the Euclidean norm on $\mathbb{R}^2$, these conditions include the well-known conditions for a Steiner point in a Euclidean length-minimizing network: three edges must meet at $120°$ angles. Hwang [11] surveys results on the Steiner problem, and Alfaro et al.

[1] and Levy [14] obtain results on two-dimensional singularities for general norms.

Determining the maximal complexity of possible singularities leads to the following extremal problems for unit vectors in Minkowski space.

(1) EQUILATERAL SET PROBLEM. What is the maximum cardinality of a Φ-equilateral set $\{\mathbf{x}_i\} \subset \mathbb{R}^n$, i.e., a set such that $\Phi(\mathbf{x}_i - \mathbf{x}_j) = 1$ whenever $i \neq j$?

(2) SUMS OF UNIT VECTORS PROBLEM. What is the maximal number m in a set $S = \{\mathbf{x}_i\}$ of Φ-unit vectors $\Phi(\mathbf{x}_i) = 1$ such that $\Phi(\mathbf{x}_i + \mathbf{x}_j) \leq 1$ whenever $i \neq j$ and $\sum_{i=1}^m \mathbf{x}_i = \mathbf{0}$?

Here, (2) has weakened condition (1.1) of Theorem II to requiring a bound only on sums of *pairs* of vectors in the set S; we will see, presently, that this does not affect the answer. In the context of Theorem II, we are especially interested in the case that Φ is strictly convex.

These are natural problems, and versions of them have been raised repeatedly as pure questions in combinatorial geometry. For example, Kusner [12] lists a number of such problems concerning equilateral sets. Variants of these problems include putting extra restrictions on the norm (e.g., smoothness) and, in the case of (2), by adding or removing the balancing condition $\sum_{i=1}^m \mathbf{x}_i = \mathbf{0}$.

The purpose of this paper is to give new results and new proofs of old results on these problems. We also describe known results and state some open problems. It turns out that there are upper bounds exponential in n for both problems (Theorems 2.1 and 3.5), and exponential lower bounds for most variants of these problems (Theorems 2.4 and 3.6); however, the balancing condition in (2) is shown to imply sharp upper and lower bounds linear in n (Theorems 3.1 and 3.4).

These extremal problems turn out to have interesting connections to several other problems in combinatorial geometry. These include the problem of characterizing convex bodies that can be perfectly packed with identical smaller copies of themselves, and Hadwiger's problem of finding the maximal number of translates of a body $\mathscr{B}$ that all intersect $\mathscr{B}$ and have disjoint interiors. These also include problems of finding large sets of points $\{\mathbf{x}_i\}$ and $\{\mathbf{y}_j\}$ in $\mathbb{R}^n$ whose inner products $\langle \mathbf{x}_i, \mathbf{y}_j \rangle$ are constrained in various ways, such as the Inner Product Problem discussed in §3, in which all angles $\mathbf{x}_i \mathbf{0} \mathbf{y}_j$ are required to be at least $\frac{\pi}{2}$.

2. Equilateral sets

The Equilateral Set Problem for a general norm was settled by Petty [20].

THEOREM 2.1 [20]. *Any set $S = \{\mathbf{x}_i\}$ of points in $(\mathbb{R}^n, \Phi)$ such that*

$$\Phi(\mathbf{x}_i - \mathbf{x}_j) = 1 \quad for \ i \neq j$$

has cardinality $|S| \leq 2^n$. Equality is attained only when the unit ball of Φ is affinely equivalent to the n-cube.

Petty [20] deduced this theorem from results of Danzer and Grünbaum [3] on sets of pairwise antipodal points. We give an alternate proof.

PROOF OF THEOREM 2.1. The upper bound 2^n is a simple consequence of the

ISODIAMETRIC INEQUALITY. *Let* Φ *be a norm on* $\mathbb{R}^n$ *with unit ball* $\mathscr{B}$ *and let* $\mathscr{K}$ *be a closed body of* Φ-*diameter* ≤ 2. *Then*

$$(2.1) \qquad\qquad \mathrm{Vol}(\mathscr{K}) \leq \mathrm{Vol}(\mathscr{B}),$$

with equality if and only if $\mathscr{K} = \mathscr{B}$.

This inequality is a generalization of the so-called Bieberbach inequality. In the case of the Euclidean norm, Bieberbach proved it for $\mathbb{R}^2$ and Urysohn [26] for $\mathbb{R}^n$. Mel'nikov [15] proved it for general finite-dimensional normed spaces. Burago and Zalgaller [2, 11.2.1, p. 93] give a short proof of the isodiametric inequality using the Brunn-Minkowski theorem.

To deduce Theorem 2.1, observe that if $\{\mathbf{x}_1, \ldots, \mathbf{x}_k\}$ is a set of Φ-equidistant points, then the interiors of the Φ-balls $\frac{1}{2}\mathscr{B} + \mathbf{x}_i$ are disjoint, and $\mathscr{W} = \bigcup_{i=1}^{k}(\frac{1}{2}\mathscr{B} + \mathbf{x}_i)$ has diameter ≤ 2. Hence,

$$(2.2) \qquad\qquad \mathrm{vol}(\mathscr{B}) \geq \mathrm{vol}(\mathscr{W}) = k2^{-n}\,\mathrm{vol}(\mathscr{B}).$$

Thus, $k \leq 2^n$.

For the case of equality $k = 2^n$, one must have equality in (2.2), and the isodiametric inequality requires that

$$\mathscr{B} = \bigcup_{i=1}^{k}(\tfrac{1}{2}\mathscr{B} + \mathbf{x}_i),$$

i.e., $\mathscr{B}$ is perfectly packed with translates of $\frac{1}{2}\mathscr{B}$. We use the following result of Groemer [9, Hilfssatz 2], which says that for arbitrary full-dimensional convex bodies $\mathscr{K}$, this last property is already sufficient to force $\mathscr{K}$ to be affinely equivalent to an n-cube. Here, $\mathscr{K}$ is not assumed to be centrally symmetric.

THEOREM 2.2 [9]. *Let* $\mathscr{K}$ *be a bounded convex body in* $\mathbb{R}^n$ *that is the closure of its interior, such that for some finite* $t > 1$, *the body* $t\mathscr{K}$ *can be perfectly packed by translates of* $\mathscr{K}$. *Then* $\mathscr{K}$ *is affinely equivalent to an n-cube, and* t *is an integer. The packing is unique and extends to a lattice packing of* $\mathbb{R}^n$.

The remaining part of Theorem 2.1 follows immediately. Since the packing is unique, any equilateral set S of cardinality 2^n must be affinely equivalent to the set of vertices of the n-cube, under the same affine equivalence taking $\mathscr{B}$ to an n-cube. □

We remark that Gritzmann [7, Theorem 3.4] has proved a stronger variant of Theorem 2.2; see also [8].

THEOREM 2.3 [8]. *Let $\mathscr{L}$ be a bounded convex body in $\mathbb{R}^n$ that is the closure of its interior and that can be perfectly packed by translates of a convex body $\mathscr{K}$. Then there are complementary subspaces of dimension k and $n-k$ and convex bodies $\mathscr{K}_1$ and $\mathscr{K}_2$ in these subspaces such that $\mathscr{K}_1$ is affinely equivalent to a k-cube, with $\mathscr{K} = \mathscr{K}_1 + \mathscr{K}_2$ and $\mathscr{L} = t\mathscr{K}_1 + \mathscr{K}_2$, where t is a positive integer and $+$ denotes Minkowski sum.*

Now we consider the Equilateral Set Problem for strictly convex norms. For $n = 2$, a sharp bound is easily seen to be 3. For $n = 3$, Petty [20] observes that results of Grünbaum [10] imply that a strictly convex norm in $\mathbb{R}^3$ has at most five vectors at unit distances from each other. Lawlor and Morgan [13, §3.5] give a smooth and strictly convex norm where five vectors occur. The exact answer for strictly convex norms is not known for any $n \geq 4$.

For n dimensions, there is an exponential lower bound for strictly convex norms.

THEOREM 2.4. *There is a strictly convex norm Φ in $\mathbb{R}^n$ and a set $S = \{\mathbf{x}_i\}$ of Φ-unit vectors such that*

$$\Phi(\mathbf{x}_i - \mathbf{x}_j) = 1 \quad \text{if } i \neq j,$$

which has cardinality $|S| \geq (1.02)^n$, for all $n \geq n_0$.

To prove this theorem, we will construct a suitable norm. The fundamental fact used to do this is that for the Euclidean norm there exist exponentially large sets of "nearly orthogonal" unit vectors.

LEMMA 2.5. *For any fixed $\delta > 0$, there is a constant $f(\delta) > 0$ such that for all sufficiently large n there exists a set $Y = \{\mathbf{y}_i\}$ in $\mathbb{R}^n$ such that*

(2.3a) $\langle \mathbf{y}_i, \mathbf{y}_i \rangle = 1 \quad$ for all i,

(2.3b) $|\langle \mathbf{y}_i, \mathbf{y}_j \rangle| \leq \delta \quad$ whenever $i \neq j$,

which has cardinality $|Y| \geq (1 + f(\delta))^n$. In particular for $\delta = \frac{1}{6}$, one may take $f(\delta) > .02$.

PROOF. Without loss of generality, $\delta \leq \frac{1}{2}$. We use a simple random construction. Draw m independent uniform samples $\{\mathbf{z}_i\}$ from the set of all ± 1 vectors in $\mathbb{R}^n$. Any such sample having the property that

$$|\langle \mathbf{z}_i, \mathbf{z}_j \rangle| \leq \delta n \quad \text{whenever } i \neq j,$$

gives a set $Y = \{\frac{1}{\sqrt{n}}\mathbf{z}_i\}$ satisfying (2.3). Now for any pair of samples $(\mathbf{z}_i, \mathbf{z}_j)$, one has

$$\text{Prob}\{|\langle \mathbf{z}_i, \mathbf{z}_j \rangle| > \delta n\} = 2^{-n+1} \sum_{0 \leq j \leq (1/2-\delta)n} \binom{n}{j} \leq 2^{1+(H(1/2-\delta)-1)n},$$

where $H(t) = -t \log_2 t - (1-t) \log_2(1-t)$ is the binary entropy function; [**27**, Theorem 1.4.5]. Since there are $\binom{m}{2}$ such pairs, one has

$$\text{Prob}\{\text{all } |\langle \mathbf{z}_i, \mathbf{z}_j \rangle| \leq \delta n\} \geq 1 - \binom{m}{2} 2^{1+(H(1/2-\delta)-1)n}.$$

This probability is nonzero if we choose $m = |Y| = 2^{1/2(1-H(1/2-\delta))n}$. Since $H(\frac{1}{2} - \delta) < 1$, the desired bound follows.

For $\delta = \frac{1}{6}$, $H(\frac{1}{2} - \delta) \leq .918$, and one may take $f(\frac{1}{6}) > .02$. $\quad\square$

PROOF OF THEOREM 2.4. By Lemma 2.5, for all large enough n there exists a set $Y = \{\mathbf{w}_0, \ldots, \mathbf{w}_m\}$ of $m+1$ Euclidean unit vectors in $\mathbb{R}^n$ having

(2.4) $|\langle \mathbf{w}_i, \mathbf{w}_j \rangle| \leq \frac{1}{6}$ whenever $i \neq j$,

with $m \geq (1.02)^n$.

Consider the centrally symmetric polytope $\mathscr{B}_1$ which is the convex hull of all $\mathbf{w}_i - \mathbf{w}_j$ with $i \neq j$. We show that all points $\mathbf{w}_i - \mathbf{w}_j$ lie on the boundary of $\mathscr{B}_1$; and, furthermore, that each point can serve as its own dual vector in the norm determined by $\mathscr{B}_1$.

CLAIM 1. For $i \neq j$ and small enough $\eta > 0$, the hyperplane

$$H_{ij}(\eta) = \{\mathbf{x} : \langle \mathbf{x}, \mathbf{w}_i - \mathbf{w}_j \rangle = \|\mathbf{w}_i - \mathbf{w}_j\|^2 - \eta\}$$

separates $\mathbf{w}_i - \mathbf{w}_j$ from $\mathbf{0}$, and from all $\mathbf{w}_k - \mathbf{w}_l$ with $(k, l) \neq (i, j)$.

The proof of Claim 1 starts with the estimate

$$\langle \mathbf{w}_i - \mathbf{w}_j, \mathbf{w}_i - \mathbf{w}_j \rangle = 2 - 2\langle \mathbf{w}_i, \mathbf{w}_j \rangle \geq \tfrac{10}{6}.$$

Separation of $\mathbf{w}_i - \mathbf{w}_j$ from $\mathbf{0}$ and $\mathbf{w}_j - \mathbf{w}_i$ is clear. For i, j, k, l, all distinct

$$\langle \mathbf{w}_k - \mathbf{w}_l, \mathbf{w}_i - \mathbf{w}_j \rangle \leq \tfrac{4}{6},$$

while if exactly one of k, l equals i or j, then

$$\langle \mathbf{w}_k - \mathbf{w}_l, \mathbf{w}_i - \mathbf{w}_j \rangle \geq \tfrac{9}{6},$$

proving Claim 1.

Now let $\mathscr{B}_2$ be the intersection of the half-spaces containing $\mathbf{0}$ cut out by all the hyperplanes $H_{ij}(0)$. Then $\mathscr{B}_1 \subseteq \mathscr{B}_2$, and Claim 1 implies, on letting $\eta \to 0$, that the boundaries of $\mathscr{B}_1$ and $\mathscr{B}_2$ intersect exactly in the points $\mathbf{w}_i - \mathbf{w}_j$ for $i \neq j$ and that each of these points lies in the relative interior of a facet of $\mathscr{B}_2$.

It suffices to show that there exists a strictly convex, centrally symmetric body $\mathscr{B}$ with $\mathscr{B}_1 \subset \mathscr{B} \subset \mathscr{B}_2$. If so, then the strictly convex norm Φ determined by $\mathscr{B}$ has

$$\Phi(\mathbf{w}_i - \mathbf{w}_j) = 1 \quad \text{whenever } i \neq j,$$

since they are on the boundaries of both $\mathscr{B}_1$ and $\mathscr{B}_2$. If we then set $\mathbf{x}_i = \mathbf{w}_i - \mathbf{w}_0$ for $1 \leq i \leq m$, then

(2.5) $\Phi(\mathbf{x}_i) = \Phi(\mathbf{x}_i - \mathbf{x}_j) = 1$ whenever $i \neq j$,

which proves the theorem.

Thus, it remains to show

CLAIM 2. Let $\mathscr{B}_1$ and $\mathscr{B}_2$ be convex polytopes with $\mathscr{B}_1 \subseteq \mathscr{B}_2$, whose boundaries intersect in a finite number of points, all in the relative interior of facets of $\mathscr{B}_2$. Then there exists a strictly convex body $\mathscr{B}$ with $\mathscr{B}_1 \subset \mathscr{B} \subset \mathscr{B}_2$. If $\mathscr{B}_1$ and $\mathscr{B}_2$ are centrally symmetric about $\mathbf{0}$, then such a $\mathscr{B}$ exists that is centrally symmetric about $\mathbf{0}$.

To prove Claim 2, let $\mathscr{C}$ be a closed halfspace with the bounding hyperplane H, and suppose first that $\mathscr{B}_1$ is a convex polytope which is contained in $\mathscr{C}$ and which touches H only at a single point $\mathbf{q}$. Then for all sufficiently large radii r, the spherical ball of radius r contained in $\mathscr{C}$ that is tangent to H at $\mathbf{q}$ contains $\mathscr{B}_1 - \{\mathbf{q}\}$ in its interior. If $\mathscr{B}_1$ is strictly inside $\mathscr{C}$, then for an arbitrary point $\mathbf{q}$ on H, one can find such a ball in $\mathscr{C}$ containing $\mathscr{B}_1$. Next, given $\mathscr{B}_1 \subseteq \mathscr{B}_2$ satisfying the given hypotheses, one constructs such balls associated to pairs $(\mathscr{C}, \mathbf{q})$ for each hyperplane $\mathscr{C}$ containing a facet of $\mathscr{B}_2$. Finally, take $\mathscr{B}$ to be the intersection of the balls associated to the various pairs $(\mathscr{C}, \mathbf{q})$; it is strictly convex. In the centrally symmetric case, these balls can be chosen in symmetric pairs for opposite facets $(\mathscr{C}, \mathbf{q})$ and $(-\mathscr{C}, -\mathbf{q})$ so that $\mathscr{B}$ is centrally symmetric. $\square$

The proof of Claim 2 actually constructs a body $\mathscr{B}$ that is uniformly convex, a property that is stronger than strict convexity. A norm Φ is *uniformly convex* if there exists $\beta > 0$ such that $\Phi(\mathbf{x}) - \beta \|\mathbf{x}\|$ is also a norm. Claim 2 probably remains true with its conclusion further strengthened to require that $\mathscr{B}$ be both of class C^∞ and uniformly convex. Any strengthenings of the conditions of the norm constructed in Claim 2 automatically carry over to corresponding strengthenings of Theorems 2.4 and 3.6.

The following problem is open.

CONJECTURE 2.6. There is a constant $\gamma > 0$ such that, for any strictly convex norm Φ in $\mathbb{R}^n$, any equilateral set S for Φ has

$$|S| \leq (2 - \gamma)^n.$$

This is analogous to a conjecture of Erdös and Füredi [6], which states: If S is a finite set of points in $\mathbb{R}^n$ such that any angle determined by three of its points is less than $\frac{\pi}{2}$, then its cardinality $|S| \leq (2 - \gamma)^n$.

3. Sums of unit vectors

Given $S = \{\mathbf{x}_i : 1 \leq i \leq m\}$ with $\Phi(\mathbf{x}_i) = 1$, we begin by relating the bounded sums condition

(3.1) $\Phi(\mathbf{x}_i + \mathbf{x}_j) \leq 1$

to a condition involving only the Euclidean norm. Consider the dual norm Φ^*, specified by the unit ball

$$(3.2) \qquad \mathscr{B}^* = \{\mathbf{x}^* : \langle \mathbf{x}^*, \mathbf{y} \rangle \le 1 \text{ for all } \mathbf{y} \in \mathscr{B}\},$$

where $\langle \mathbf{x}, \mathbf{y} \rangle = \sum_{i=1}^{n} x_i y_i$ is the Euclidean inner product on $\mathbb{R}^n$. For each vector $\mathbf{x}$ with $\Phi(\mathbf{x}) = 1$, there exists at least one dual vector $\mathbf{x}^*$ with $\Phi^*(\mathbf{x}^*) = 1$ and

$$(3.3) \qquad \langle \mathbf{x}, \mathbf{x}^* \rangle = 1,$$

e.g., $\mathbf{x}^*$ points in a direction normal to a tangent hyperplane to the unit ball $\mathscr{B}$ of Φ at its boundary point $\mathbf{x}$. Choose for each $\mathbf{x}_i \in S$ a corresponding $\mathbf{x}_i^* \in \mathscr{B}^*$ satisfying (3.3), and set $S^* = \{\mathbf{x}_i^* : 1 \le i \le m\}$. We claim that if S satisfies (3.1), then

$$(3.4) \qquad \langle \mathbf{x}_i, \mathbf{x}_j^* \rangle \le 0 \quad \text{whenever } i \ne j.$$

Indeed by (3.1) and $\mathbf{x}_j^* \in \mathscr{B}^*$,

$$\langle \mathbf{x}_i, \mathbf{x}_j^* \rangle + 1 = \langle \mathbf{x}_i + \mathbf{x}_j, \mathbf{x}_j^* \rangle \le \Phi(\mathbf{x}_i + \mathbf{x}_j) \le 1,$$

which proves (3.4).

This leads us to consider the auxiliary problem

INNER PRODUCT PROBLEM. Let X and Y be sets in $\mathbb{R}^n$ with $|X| = |Y| = m$, such that

$$(3.5a) \qquad \langle \mathbf{x}_i, \mathbf{y}_i \rangle > 0 \quad \text{for } 1 \le i \le m.$$
$$(3.5b) \qquad \langle \mathbf{x}_i, \mathbf{y}_j \rangle \le 0 \quad \text{whenever } i \ne j.$$

Bound m, under given side conditions on X and Y.

The condition (3.5b) is a condition on angles: all angles $\mathbf{x}_i 0 \mathbf{y}_j$ are at least $\frac{\pi}{2}$.

Upper bounds for the Inner Product Problem (with side conditions) will imply the same upper bounds for the Sums of Unit Vectors Problem (with side conditions); the converse is not necessarily true. We consider, in particular, the following side conditions:

(i) EUCLIDEAN NORM CASE: $X = Y$.
(ii) WEAK BALANCING CONDITION: $\mathbf{0}$ is in the relative interior of the convex hull of $X = \{\mathbf{x}_i : 1 \le i \le m\}$.

Note that the balancing condition

$$\sum_{i=1}^{m} \mathbf{x}_i = \mathbf{0}$$

implies that the weak balancing condition (ii) holds.

We first consider the Sums of Unit Vectors Problem, assuming the weak balancing condition.

THEOREM 3.1. *Let* Φ *be a norm on* $\mathbb{R}^n$. *If* $S = \{\mathbf{x}_i\}$ *is a set of* Φ-*unit vectors in* $\mathbb{R}^n$ *such that*

$$(3.6) \qquad \Phi(\mathbf{x}_i + \mathbf{x}_j) \leq 1 \quad \text{for } i \neq j,$$

which satisfies the weak balancing condition, then it has cardinality $|S| \leq 2n$. *Equality can be attained only when the set* S *is linearly equivalent to the set* $\{\pm\mathbf{e}_i : 1 \leq i \leq n\}$, *where* $\mathbf{e}_i$ *are the unit vectors in each coordinate direction.*

In fact the equality $|S| = 2n$ is still attained under the stronger hypotheses

$$\Phi\left(\sum_{\mathbf{x}_i \in J} \mathbf{x}_i\right) \leq 1 \quad \text{for all } J \subseteq S, \qquad \sum_{i=1}^{n} \mathbf{x}_i = \mathbf{0},$$

if Φ has the n-cube as unit ball and $S = \{\pm\mathbf{e}_i : 1 \leq i \leq n\}$. Under these stronger hypotheses, it can be proved that the case of equality $|S| = 2n$ holds only when the unit ball of Φ is affinely equivalent to the n-cube.

Theorem 3.1 is an immediate corollary of a similar upper bound for the Inner Product Problem.

THEOREM 3.2. (a) *Given sets* X *and* Y *in* $\mathbb{R}^n$ *with cardinalities* $|X| = |Y| = m$ *such that*

$$(3.7a) \qquad \langle \mathbf{x}_i, \mathbf{y}_i \rangle > 0 \quad \text{for } 1 \leq i \leq m,$$
$$(3.7b) \qquad \langle \mathbf{x}_i, \mathbf{y}_j \rangle \leq 0 \quad \text{whenever } i \neq j,$$

where X *satisfies the condition that* $\mathbf{0}$ *is in the relative interior of the convex hull of* X. *Then* $m \leq 2n$, *and equality can hold only if the* $\mathbf{x}_i$ *can be renumbered so that* $\mathbf{x}_{n+i} = -\lambda_i \mathbf{x}_i$ *with* $\lambda_i > 0$ *for* $1 \leq i \leq n$.

(b) *If in addition*

$$\langle \mathbf{x}_i, \mathbf{y}_j \rangle < 0 \quad \text{whenever } i \neq j,$$

then $m \leq n+1$.

Our proof of Theorem 3.2 is based on an observation of I. Bárány, improving on the original proof of the authors'. It uses

STEINITZ'S THEOREM. (1) *Let* S *be a finite set in* $\mathbb{R}^n$ *whose convex hull is a body* $\mathscr{K}$ *of dimension* r, *such that* $\mathbf{0}$ *lies in the relative interior of* $\mathscr{K}$. *Then there is a subset* T *of* S *of cardinality* $t \leq 2r$, *say* $T = \{\mathbf{x}_i : 1 \leq i \leq t\}$, *whose convex hull is of dimension* r, *such that* $\mathbf{0}$ *is a strict convex combination of* T, *i.e.,*

$$\mathbf{0} = \sum_{i=1}^{t} \lambda_i \mathbf{x}_i,$$

with all $\lambda_i > 0$, $\sum_{i=1}^{t} \lambda_i = 1$.

(2) *If there exists no subset* T *with* $t \leq 2r - 1$ *having the above property, then necessarily* T *consists of exactly* $2r$ *vectors, which are collinear in pairs.*

The first part of Steinitz's theorem is due to Steinitz and is discussed at length in Danzer, Grünbaum, and Klee [**4**, §3], and the second part is due to Robinson [**22**, Lemma 2a].

PROOF OF THEOREM 3.2. The weak balancing condition implies by Steinitz's theorem that there is a subset T of $t \leq 2n$ vectors of X, say $T = \{\mathbf{x}_i : 1 \leq i \leq t\}$, with

$$(3.8) \qquad \mathbf{0} = \sum_{i=1}^{t} \lambda_i \mathbf{x}_i \quad \text{all } \lambda_i > 0,$$

and $\dim(T) = \dim(X) = r$. If $|X| \geq t + 1$ then

$$\mathbf{x}_{t+1} = \sum_{i=1}^{t} \gamma_i \mathbf{x}_i,$$

since $\mathbf{x}_{t+1}$ is in the subspace spanned by T. By adding a large enough multiple of (3.8) to this equation, we obtain

$$\mathbf{x}_{t+1} = \sum_{i=1}^{t} \beta_i \mathbf{x}_i$$

with all $\beta_i > 0$. But $\langle \mathbf{x}_{t+1}, \mathbf{y}_{t+1} \rangle > 0$ by hypothesis, while

$$\langle \mathbf{x}_{t+1}, \mathbf{y}_{t+1} \rangle = \sum_{i=1}^{t} \beta_i \langle \mathbf{x}_i, \mathbf{y}_{t+1} \rangle \leq 0$$

by (3.7b), a contradiction. Hence, $T = X$ and $|X| \leq t \leq 2n$.

If $|X| = 2n$, then the case (2) of equality in Steinitz's theorem requires that the vectors in X be collinear in pairs. This proves the second assertion in (a).

For part (b) we use Caratheodory's theorem in place of Steinitz's theorem: Every vector in the convex hull of X is a convex combination of at most $n + 1$ vectors in X. Thus, the weak balancing condition implies that

$$\mathbf{0} = \sum_{i=1}^{n+1} \alpha_i \mathbf{x}_i,$$

with all $\alpha_i \geq 0$, $\sum_{i=1}^{n+1} \alpha_i = 1$, after renumbering X if necessary. If $|X| \geq n + 2$, then

$$0 = \left\langle \sum_{i=1}^{n+1} \alpha_i \mathbf{x}_i, \mathbf{y}_{n+2} \right\rangle = \sum_{i=1}^{n+1} \alpha_i \langle \mathbf{x}_i, \mathbf{y}_{n+2} \rangle < 0,$$

a contradiction proving part (b). □

A similar easy upper bound $2n$ holds for the Inner Product Problem assuming the Euclidean norm side condition $X = Y$, without any balancing condition.

THEOREM 3.3. *Given a set* X *of cardinality* $|X| = m$ *in* $\mathbb{R}^n$ *such that*

(3.9a) $\langle \mathbf{x}_i, \mathbf{x}_i \rangle > 0 \quad \text{for } 1 \leq i \leq m,$

(3.9b) $\langle \mathbf{x}_i, \mathbf{x}_j \rangle \leq 0 \quad \text{whenever } i \neq j,$

then $m \leq 2n$.

PROOF. This is proved by induction on the dimension n, the case $n = 1$ being obvious. Given X in $\mathbb{R}^n$, let $\mathbf{w}_i$ denote the projection of $\mathbf{x}_i$ onto the $(n-1)$-dimensional subspace perpendicular to $\mathbf{x}_1$. Then $\mathbf{w}_1 = \mathbf{0}$, and at most one other $\mathbf{w}_i = \mathbf{0}$, which occurs only if there is some $\mathbf{x}_i = -\lambda \mathbf{x}_1$ with $\lambda > 0$. Now for $i \geq 2$, (3.9b) implies that $\mathbf{x}_i = \mathbf{w}_i - \lambda_i \mathbf{x}_1$ with $\lambda_i \geq 0$, hence for $i \neq j$

$$\langle \mathbf{x}_i, \mathbf{x}_j \rangle = \langle \mathbf{w}_i, \mathbf{w}_j \rangle + \lambda_i \lambda_j \|\mathbf{x}_1\|^2 \leq 0.$$

Thus, the set of nonzero $\mathbf{w}_i$ satisfies the hypotheses (3.9) in $\mathbb{R}^{n-1}$, so there are at most $2n - 2$ of them, and the induction step follows.   □

Note that in (3.9a) we can rescale $\mathbf{x}_i$ to require $\langle \mathbf{x}_i, \mathbf{x}_i \rangle = 1$ without changing (3.9b). Then (3.9) asserts that all open spherical caps of angular measure $\frac{\pi}{4}$ about each $\mathbf{x}_i$ are disjoint. In this reformulation the inequality $|X| \leq 2n$ is derived in Rankin [21], exactly as above.

For a strictly convex norm the weak balancing condition gives the stronger bound $n + 1$ in place of $2n$.

THEOREM 3.4. *Let* Φ *be a strictly convex norm in* $\mathbb{R}^n$. *Let* $S = \{\mathbf{x}_i\}$ *be a set of* Φ-*unit vectors in* $\mathbb{R}^n$ *such that*

$$\Phi(\mathbf{x}_i + \mathbf{x}_j) \leq 1 \quad \text{for } i \neq j,$$

which satisfies the weak balancing condition. Then S *has cardinality* $|S| \leq n + 1$.

PROOF. The condition of strict convexity of Φ sharpens the condition (3.4) to

(3.10) $\langle \mathbf{x}_i, \mathbf{x}_j^* \rangle < 0 \quad \text{whenever } i \neq j.$

Now the result follows from Theorem 3.2(b).   □

Next, we consider the Sums of Unit Vectors Problem without any balancing condition and give an exponential upper bound.

THEOREM 3.5. *Consider any norm* Φ *in* $\mathbb{R}^n$ *and any set* $S = \{\mathbf{x}_i\}$ *of* Φ-*unit vectors satisfying*

(3.11) $\Phi(\mathbf{x}_i + \mathbf{x}_j) \leq 1 \quad \text{whenever } i \neq j.$

Then S *has cardinality* $|S| \leq 3^n - 1$.

PROOF. By the triangle inequality

$$\Phi(\mathbf{x}_i - \mathbf{x}_j) \geq \Phi(2\mathbf{x}_i) - \Phi(\mathbf{x}_i + \mathbf{x}_j) \geq 1,$$

whenever $i \neq j$, and all $\Phi(\mathbf{x}_i) \geq 1$ also. Hence, the interiors of all the sets $\mathbf{x}_i + \frac{1}{2}\mathscr{B}$ and of $\frac{1}{2}\mathscr{B}$ are pairwise disjoint. Since these sets all lie in $\frac{3}{2}\mathscr{B}$, volume considerations yield $|S| + 1 \leq 3^n$. □

The proof of Theorem 3.5 shows that the sets $\{2\mathbf{x}_i + \mathscr{B}\}$ satisfy the conditions of *Hadwiger's problem*, which is that of bounding the maximum number of translates of a closed convex body $\mathscr{B}$ that intersect $\mathscr{B}$, but which have pairwise disjoint interiors. The bound 3^n is sharp for Hadwiger's problem on taking $\mathscr{B}$ to be affinely equivalent to the n-cube, and Theorem 2.2 shows that this is the only case of equality. Since (3.11) does not hold for the n-cube, improvement is possible in the upper bound of Theorem 3.5. See Danzer, Grünbaum, and Klee [4, p. 149] for history and results on Hadwiger's problem.

In fact, exponential size sets can occur in the Sums of Unit Vectors Problem when no balancing condition is present, even with strict convexity imposed.

THEOREM 3.6. *There exists a strictly convex norm* Φ *in* $\mathbb{R}^n$ *and a set* $S = \{\mathbf{x}_i\}$ *of* Φ-*unit vectors satisfying*

$$(3.12) \qquad \Phi(\mathbf{x}_i + \mathbf{x}_j) < 1 \quad \text{whenever } i \neq j,$$

which has cardinality $|S| \geq (1.02)^n$.

PROOF. The proof uses similar ideas to Theorem 2.4. By Lemma 2.5, for all large enough n, there exists a set of $Y = \{\mathbf{w}_i\}$ in $\mathbb{R}^{n-1}$ having $|Y| \geq (1.02)^n$ and

$$(3.13) \qquad |\langle \mathbf{w}_i, \mathbf{w}_j \rangle| \leq \frac{1}{6} \quad \text{whenever } i \neq j.$$

View $\mathbb{R}^{n-1}$ embedded in $\mathbb{R}^n$ as the first $n-1$ coordinates, take the unit vector $\mathbf{e} = (0, 0, \ldots, 0, 1)$ orthogonal to all $\mathbf{w}_i$, and set

$$\mathbf{x}_i = \mathbf{w}_i + \lambda \mathbf{e}, \qquad 1 \leq i \leq m,$$

where $\lambda > 0$ is arbitrary. Now let $\mathscr{B}_1$ be the convex hull of all the vectors $\pm \mathbf{x}_i$ and $\pm 1.01(\mathbf{x}_i + \mathbf{x}_j)$ where $i \neq j$. Clearly each $\mathbf{x}_i + \mathbf{x}_j \in \text{Int}(\mathscr{B}_1)$ whenever $i \neq j$, and we assert that

$$(*) \qquad\qquad \text{All } \mathbf{x}_i \text{ are on the boundary of } \mathscr{B}_1.$$

To show this we use the dual vectors

$$\mathbf{y}_i = \mathbf{w}_i - \frac{1}{5\lambda}\mathbf{e}, \qquad 1 \leq i \leq m.$$

One has

$$\langle \mathbf{x}_i, \mathbf{y}_i \rangle = \langle \mathbf{w}_i + \lambda \mathbf{e}, \mathbf{w}_i - \frac{1}{5\lambda}\mathbf{e} \rangle = 1 - \frac{1}{5}\langle \mathbf{e}, \mathbf{e} \rangle = \frac{4}{5}.$$

Then $(*)$ is an immediate consequence of the following:

CLAIM. *The hyperplane* $H_i = \{\mathbf{x} : \langle \mathbf{x}, \mathbf{y}_i \rangle = \frac{4}{5} - \frac{1}{40}\}$ *separates* $\mathbf{x}_i$ *from* $\mathbf{0}$, *from* $-\mathbf{x}_i$, *from all other* $\pm \mathbf{x}_j$, *and from all* $\pm 1.01(\mathbf{x}_k + \mathbf{x}_l)$ *having* $k \neq l$.

The claim is obvious for $\mathbf{0}$ and $-\mathbf{x}_i$, while for $j \neq i$,

$$\langle \mathbf{x}_j, \mathbf{y}_i \rangle = \langle \mathbf{w}_j + \lambda \mathbf{e}, \mathbf{w}_i - \tfrac{1}{5\lambda}\mathbf{e} \rangle = \langle \mathbf{w}_j, \mathbf{w}_i \rangle - \tfrac{1}{5} \leq -\tfrac{1}{30}$$

and

$$\langle -\mathbf{x}_j, \mathbf{y}_i \rangle = \langle -\mathbf{w}_j - \lambda \mathbf{e}, \mathbf{w}_i - \tfrac{1}{5\lambda}\mathbf{e} \rangle = \langle -\mathbf{w}_j, \mathbf{w}_i \rangle + \tfrac{1}{5} \leq \tfrac{1}{6} + \tfrac{1}{5}.$$

Also for $j \neq i$,

$$\langle \mathbf{x}_i + \mathbf{x}_j, \mathbf{y}_i \rangle = \tfrac{4}{5} + \langle \mathbf{x}_j, \mathbf{y}_i \rangle \leq \tfrac{4}{5} - \tfrac{1}{30},$$
$$\langle -(\mathbf{x}_i + \mathbf{x}_j), \mathbf{y}_i \rangle = -\tfrac{4}{5} + \langle -\mathbf{x}_j, \mathbf{y}_i \rangle \leq -\tfrac{2}{5},$$

and, finally, for i, k, l all distinct,

$$\langle \mathbf{x}_k + \mathbf{x}_l, \mathbf{y}_i \rangle = \langle \mathbf{x}_k, \mathbf{y}_i \rangle + \langle \mathbf{x}_l, \mathbf{y}_i \rangle \leq 0$$
$$\langle -(\mathbf{x}_k + \mathbf{x}_l), \mathbf{y}_i \rangle = \langle -\mathbf{x}_k, \mathbf{y}_i \rangle + \langle -\mathbf{x}_l, \mathbf{y}_i \rangle \leq 2(\tfrac{1}{6} + \tfrac{1}{5}) \leq \tfrac{4}{5} - \tfrac{2}{30}.$$

Multiplying all these inequalities by 1.01 keeps all inner products $\leq \tfrac{4}{5} - \tfrac{1}{40}$; hence, the claim is proved.

Thus, the norm Φ_1 determined by $\mathscr{B}_1$ has the desired properties of the theorem, except that it is not strictly convex. To finish the proof, take $\mathscr{B}_2$ to be the intersection of the closed half-spaces $Q_i = \{\mathbf{x} : \langle \mathbf{x}, \mathbf{y}_i \rangle \leq \tfrac{4}{5}\}$. Now $\mathscr{B}_1 \subseteq \mathscr{B}_2$, and their common boundary points are exactly $\{\pm \mathbf{x}_i : 1 \leq i \leq m\}$, so Claim 2 of Theorem 2.4 applies to give a body $\mathscr{B}$ that is strictly convex and symmetric about $\mathbf{0}$, with $\mathscr{B}_1 \subseteq \mathscr{B} \subseteq \mathscr{B}_2$. The norm Φ determined by $\mathscr{B}$ has the required property. $\square$

For the Inner Product Problem with no side conditions, there is *no upper bound at all on* m when $n \geq 3$. To see this, write $\mathbb{R}^3 = \{(x, y, z)\}$ and take for any fixed m the set $X = \{\mathbf{x}_i : 1 \leq i \leq m\}$ where all $\mathbf{x}_i$ lie in the plane $z = 1$ and form an equilateral m-gon centered at $(0, 0, 1)$, say $\mathbf{x}_i = (\cos \tfrac{2\pi j}{m}, \sin \tfrac{2\pi j}{m}, 1)$ for $1 \leq j \leq m$. Now we can find m lines $\{l_i : 1 \leq i \leq m\}$ lying in the plane $z = 1$, which separate each $\mathbf{x}_i$ from all the other $\mathbf{x}_i$, e.g.,

$$l_i : x(\cos \tfrac{2\pi j}{m}) + y(\sin \tfrac{2\pi j}{m}) = 1 - \varepsilon,$$

for small enough positive ε; see Figure 3.1.

Now let $Y = \{\mathbf{y}_i\}$, where $\mathbf{y}_i$ is a unit vector perpendicular to the plane H_i determined by the line l_i and the point $(0, 0, 0)$, and lying on the same side of H_i as $\mathbf{x}_i$ does. Then (3.7a)–(3.7b) clearly hold since H_i separates $\mathbf{x}_i$ from all $\{\mathbf{x}_j : j \neq i\}$. (In this example, the hyperplane $z = \tfrac{1}{2}$ separates all points in X from $\mathbf{0}$.)

Finally, we pose the problem of whether the stronger condition (1.1) on sums of unit vectors implies a polynomial upper bound on their number, when no balancing condition is present.

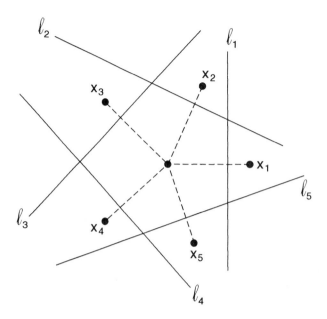

FIGURE 3.1. Lines separating each vertex of a regular m-gon from the other vertices are used in showing the necessity of same side condition in the Inner Product Problem.

PROBLEM 3.7. For a general norm $(\mathbb{R}^n, \Phi)$ and a set S of Φ-unit vectors, does the condition

$$\Phi\left(\sum_{j \in J} \mathbf{x}_j\right) \leq 1$$

for all $J \subseteq S$ imply a polynomial bound $p(n)$ on $|S|$?

Acknowledgments

We are indebted to G. D. Chakerian, V. Klee, E. Lutwak, L. A. Shepp, and P. Winkler for helpful comments and references. We are also grateful to I. Bárány for helpful discussions and for permitting us to include his proof of Theorem 3.2.

REFERENCES

1. M. Alfaro, M. Conger, K. Hodges, A. Levy, R. Kochar, L. Kuklinski, Z. Mahmood, and K. von Haam, *The structure of singularities in Φ-minimizing networks in $\mathbb{R}^n$*, Pacific J. Math. **149** (1991), 201–210.
2. Y. D. Burago and V. A. Zalgaller, *Geometric inequalities*, Springer-Verlag, New York, 1988.
3. L. Danzer and B. Grünbaum, *Über zwei Probleme bezüglich konvexen Körper von P. Erdös und V. L. Klee*, Math. Z. **79**, (1962), 95–99.
4. L. Danzer, B. Grünbaum, and V. Klee, *Helly's theorem and its relatives*, Convexity, Proc. Sympos. Pure Math, vol. VII, Amer. Math. Soc., Providence, RI, 1963, pp. 101–181.
5. M. M. Day, *Normed linear spaces*, Springer-Verlag, Berlin, 1958.

6. P. Erdös and Z. Füredi, *The greatest angle among n points in the d-dimensional Euclidean space*, Ann. Discrete Math. **17** (1983), 275–283.

7. P. Gritzmann, *Finite packungen und Überdeckungen*, Habilitationsschrift, Universität Siegen, 1984.

8. P. Gritzmann and J. Wills, *Finite packing and covering*, Studia Sci. Math. Hungar. **21** (1986), 149–162.

9. H. Groemer, *Abschätzungen für die Anzahl der Konvexen Körper, die einen konvexen Körper berühren*, Monatsh. Math. **65** (1961), 74–81.

10. B. Grünbaum, *Strictly antipodal sets*, Israel J. Math. **1** (1963), 5–10.

11. F. K. Hwang, *A primer of the Euclidean Steiner Problem*, Proc. NATO Advanced Research Workshop on Topological Network Design, 1991, (to appear).

12. Robert B. Kusner, R. K. Guy, *An olla-podrida of open problems, often oddly posed*, Amer. Math. Monthly **90** (1983), 196–199.

13. G. Lawlor and F. Morgan, *Paired calibrations applied to soap films, immiscible fluids, and surfaces and networks minimizing other norms*, preprint, 1991.

14. A. Levy, *Energy-minimizing networks meet only in threes*, J. Undergrad. Math. **22** (1990), 53–59.

15. M. S. Mel'nikov, *Dependence of volume and diameter of sets in n-dimensional Banach space*, Uspekhi Math. Nauk. **18** (1963), 165–170. (Russian)

16. F. Morgan, *The cone over the Clifford torus in $\mathbb{R}^4$ is Φ-minimizing*, Math. Ann. **289** (1991), 341–354.

17. ____, *Minimal surfaces, crystals and norms on $\mathbb{R}^n$*, Proc. 7th Ann. Sympos. on Computational Geometry, ACM Press, 1991, pp. 204–213.

18. ____, *Minimal surfaces, crystals, shortest networks and undergraduate research*, Math. Intelligencer, 1992 (to appear).

19. ____, *Riemannian geometry: A beginner's guide*, Jones & Bartlett (to appear).

20. C. M. Petty, *Equilateral sets in Minkowski spaces*, Proc. Amer. Math. Soc. **29** (1971), 369–374.

21. R. A. Rankin, *The closest packing of spherical caps in n dimensions*, Proc. Glasgow Math. Assoc. **2** (1955), 139–144.

22. C. V. Robinson, *Spherical theorems of Helly type and congruence indices of spherical caps*, Amer. J. Math. **64** (1942), 260–272.

23. B. Simon, *Fifteen problems in mathematical physics*, Perspectives in Mathematics, Birkhauser-Verlag, Basel, 1984, pp. 423–454.

24. J. Taylor, *The structure of singularities in soap-bubble-like and soap-film-like minimal surfaces*, Ann. of Math. (2) **103** (1976), 489–539.

25. ____, *Crystalline variational problems*, Bull. Amer. Math. Soc. **84** (1978), 568–588.

26. P. Urysohn, *Mittlere Breite und Volumen der konvexen Körper in m-dimensionalen Raume*, Mat. Sb. SSSR **31** (1924), 477–486.

27. J. H. van Lint, *Introduction to coding theory*, Springer-Verlag, New York, 1982.

DEPARTMENT OF MATHEMATICS, MASSACHUSETTS INSTITUTE OF TECHNOLOGY, CAMBRIDGE, MASSACHUSETTS 02139

AT&T BELL LABORATORIES, MURRAY HILL, NEW JERSEY 07974
E-mail address: jcl@research.att.com

DEPARTMENT OF MATHEMATICS, WILLIAMS COLLEGE, WILLIAMSTOWN, MASSACHUSETTS 01267
E-mail address: fmorgan@williams.edu

DIMACS Series in Discrete Mathematics
and Theoretical Computer Science
Volume 6, 1991

Wu-Ritt Characteristic sets
and Their Complexity

G. GALLO AND B. MISHRA

ABSTRACT. The concept of a characteristic set of an ideal was originally introduced by J. F. Ritt, in the late forties, and later, independently rediscovered by Wu Wen-Tsün, in the late seventies. Since then Wu-Ritt characteristic sets have found wide applications in Symbolic Computational Algebra, Automated Theorem Proving in Elementary Geometries, and Computer Vision. In this paper, we survey the original Wu-Ritt process, and its applications: particularly, in geometry. We also present optimal algorithms for computing a characteristic set with simple-exponential sequential and polynomial parallel time complexities. These algorithms were first devised by the present authors by using an effective version of Hilbert's Nullstellensatz, due to D. Brownawell and J. Kollár.

1. Introduction

The history of the origin and subsequent developments of *Wu-Ritt characteristic sets* is somewhat peculiar. The concept of a characteristic set was discovered in the late forties by J.F. Ritt (see his now classic book *Differential Algebra* [**34**]) in an effort to extend some of the constructive algebraic methods to differential algebra. However, these concepts languished in near oblivion until the last decade when the Chinese mathematician Wu Wen-Tsün [**35, 36, 37**] realized its power in the case where Ritt's techniques are specialized to commutative algebra. In particular, he exhibited its effectiveness (largely through empirical evidence) as a powerful tool for Mechanical Theorem proving. This proved to be a turning point; a renewed interest in the subject has contributed to a better understanding of the power and complexity of the Ritt's techniques.

For a system of algebraic or differential equations, its characteristic set is a certain effectively-constructible triangular set of equations that preserves

1991 *Mathematics Subject Classification*. Primary 68C20, 68C25; Secondary 13A15, 14A10.
 Supported by NSF grants CDA-90-18673 and CCR-9002819, ONR grant N00014-89-J3042, and an Italian CNR fellowship.

111

many of the interesting geometric properties of the original system [10, 17, 18, 25, 34]. However, until very recently, the constructivity of Ritt's characteristic set had not been explicitly demonstrated [18]; the original *Wu-Ritt process*, first devised by Ritt [34], subsequently modified by Wu [35, 36, 37] and widely implemented [10], computes only an *extended* characteristic set. Furthermore, the *Wu-Ritt process*, as it is, has a nonelementary[1] worst-case time complexity and thus, in principle, infeasible.

It has been speculated that the reason the Wu-Ritt process has failed to receive its deserving place in constructive algebra was largely because it has been eclipsed by the spectacular success of Buchberger's "Gröbner Bases Algorithm." A lively debate about the respective power of these two algorithms has sprung up. Considerable amounts of experimental data are available, comparing the time-complexity of various implementations of the Wu-Ritt process with Gröbner Bases algorithms [7, 9, 10, 27, 30, 31]. However, while the worst-case complexity of Gröbner Bases algorithms are fairly well-understood at this point [2, 3, 4, 15, 16, 19, 23, 28, 29, 32, 38], the computational complexity of Wu-Ritt characteristic sets had remained open until very recently [18].

In this paper, we survey the original Wu-Ritt process, and its applications: particularly, in geometry. We also present optimal algorithms for computing a characteristic set with simple-exponential sequential and polynomial parallel time complexities. These algorithms were first devised by the present authors by using an effective version of Hilbert's Nullstellensatz, due to D. Brownawell and J. Kollár [6, 26].

2. Characteristic sets

Let $k[x_1, \ldots, x_n]$ denote the ring of polynomials in n indeterminates, with coefficients in a field k. Consider a fixed ordering on the set of indeterminates; without loss of generality, we may assume that the given ordering is the following:

$$x_1 \prec x_2 \prec \cdots \prec x_n.$$

DEFINITION 2.1. CLASS AND CLASS DEGREE. Let $f \in k[x_1, \ldots, x_n]$ be a multivariate polynomial with coefficients in k. An indeterminate x_j is said to be *effectively present in* f, if some monomial in f with nonzero coefficient contains a (strictly) positive power of x_j. For $1 \leq j \leq n$, *degree of* f *with respect to* x_j, $\deg_{x_j}(f)$, is defined to be the maximum degree of the indeterminate x_j in f. The *degree of* f is defined to be $\deg(f) = \sum_{j=1}^{n} \deg_{x_j}(f)$.

---

[1]For a discussion of the nonelementary computational problems, see pp. 419–423 of the algorithms text by Aho, Hopcroft, and Ullman [1]. Roughly, a problem is said to have a nonelementary complexity, if its complexity cannot be bounded by a function that involves only a fixed number of iterations of the exponential function.

The *class* and the *class degree* (cdeg) of a polynomial $f \in k[x_1, \ldots, x_n]$ with respect to a given ordering is defined as follows:

    (i) If *no indeterminate x_j is effectively present in f*, (i.e., $f \in k$) then, by convention, class$(f) = 0$ and cdeg$(f) = 0$; otherwise,

    (ii) If x_j *is effectively present in f, and no $x_i \succ x_j$ is effectively present in f* (i.e., $f \in k[x_1, \ldots, x_j] \setminus k[x_1, \ldots, x_{j-1}]$) then class$(f) = j$ and cdeg$(f) = \deg_{x_j}(f)$.

Thus, with each polynomial $f \in k[x_1, \ldots, x_n]$, we can associate a pair of integers, its *type*:

$$\text{type} : k[x_1, \ldots, x_n] \to \mathbb{N} \times \mathbb{N}$$
$$: f \mapsto \langle \text{class}(f), \text{cdeg}(f) \rangle. \quad \square$$

DEFINITION 2.2. ORDERING ON THE POLYNOMIALS. Given two polynomials f_1 and $f_2 \in k[x_1, \ldots, x_n]$, we say f_1 is of *lower rank than* f_2, $f_1 \prec f_2$, if either (i) class$(f_1) <$ class(f_2), or (ii) class$(f_1) =$ class(f_2) and cdeg$(f_1) <$ cdeg(f_2). This is equivalent to saying that the polynomials are ordered according to the lexicographic order on their types:

$$f_1 \prec f_2 \quad \textit{if and only if} \quad \text{type}(f_1) <_{\text{lex}} \text{type}(f_2).$$

Note that there are distinct polynomials f_1 and f_2 that are not comparable under the preceding order. In this case, type$(f_1) =$ type(f_2), and f_1 and f_2 are said to be of the *same rank*, $f_1 \sim f_2$. $\quad \square$

Thus, a polynomial f of class j and class degree d can be written as

$$(1) \qquad f = I_d(x_1, \ldots, x_{j-1})x_j^d + I_{d-1}(x_1, \ldots, x_{j-1})x_j^{d-1}$$
$$+ \cdots + I_0(x_1, \ldots, x_{j-1}),$$

where $I_l(x_1, \ldots, x_{j-1}) \in k[x_1, \ldots, x_{j-1}]$, for $l = 0, 1, \ldots, d$.

DEFINITION 2.3. INITIAL POLYNOMIAL. Given a polynomial f of class j and class degree d, its *initial polynomial*, in(f), is defined to be the polynomial $I_d(x_1, \ldots, x_{j-1})$ as in equation (1). $\quad \square$

LEMMA 2.1. PSEUDO-DIVISION LEMMA. *Consider two polynomials f and $g \in k[x_1, \ldots x_n]$, with* class$(f) = j$. *Then there exist two polynomials q and r, and an integer α such that the following equation holds:*

$$(2) \qquad \qquad \text{in}(f)^\alpha g = qf + r,$$

where $\deg_{x_j}(r) < \deg_{x_j}(f)$ *and* $\alpha \leq \deg_{x_j}(g) - \deg_{x_j}(f) + 1$.
If α is assumed to be the smallest possible power satisfying equation (1) then q and r are uniquely determined. In this case, the polynomial r is said to be the pseudo-remainder *of g with respect to f, and denoted by,* prem(g, f). *We say a polynomial g is* reduced *with respect to f if $g =$ prem(g, f). $\quad \square$

DEFINITION 2.4. ASCENDING SET. A sequence of polynomials $\mathscr{F} = \langle f_1, f_2, \ldots, f_r \rangle \subseteq k[x_1, \ldots, x_n]$ is said to be an *ascending set* (or *chain*), if one of the following two conditions holds:

(i) $r = 1$ and f_1 is not identically zero;
(ii) $r > 1$, and $0 < \mathrm{class}(f_1) < \mathrm{class}(f_2) < \cdots < \mathrm{class}(f_r) \leq n$,
and each f_i is reduced with respect to the preceding polynomials, f_j's $(1 \leq j < i)$.

Every ascending set is finite and has at most n elements. The *dimension* of an ascending set $\mathscr{F} = \langle f_1, f_2, \ldots, f_r \rangle$, dim $\mathscr{F}$, is defined to be $(n - r)$. □

Thus, with each ascending set $\mathscr{F}$ we can associate an $(n + 1)$-vector, its *type*,

$$\text{type: Family of Ascending Sets} \to (\mathbb{N} \cup \{\infty\})^{n+1},$$

where ∞ is assumed to be greater than any integer. For all $0 \leq i \leq n$, the ith component of the vector is

$$\text{type}(\mathscr{F})[i] = \begin{cases} \mathrm{cdeg}(g), & \text{if}\,(\exists!\, g \in \mathscr{F})\ \mathrm{class}(g) = i, \\ \infty, & \text{otherwise.} \end{cases}$$

DEFINITION 2.5. ORDERING ON THE ASCENDING SETS. Given two ascending sets

$$\mathscr{F} = \langle f_1, \ldots, f_r \rangle \quad \text{and} \quad \mathscr{G} = \langle g_1, \ldots, g_s \rangle,$$

we say $\mathscr{F}$ is of *lower rank* than $\mathscr{G}$, $\mathscr{F} \prec \mathscr{G}$, if one of the following two conditions is satisfied:

(i) There exists an index $i \leq \min\{r, s\}$ such that

$$(\forall 1 \leq j < i)\quad f_j \sim g_j \quad \text{and} \quad f_i \prec g_i;$$

(ii) $r > s$ and $(\forall 1 \leq j \leq s)\ f_j \sim g_j$.

Note that there are distinct ascending sets $\mathscr{F}$ and $\mathscr{G}$ that are not comparable under the preceding order. In this case $r = s$, and $(\forall 1 \leq j \leq s)\ f_j \sim g_j$, and $\mathscr{F}$ and $\mathscr{G}$ are said to be of the *same rank*, $\mathscr{F} \sim \mathscr{G}$. □

Hence,

$$\mathscr{F} \prec \mathscr{G} \quad \textit{if and only if} \quad \text{type}(\mathscr{F}) <_{\text{lex}} \text{type}(\mathscr{G}).$$

Thus the map, *type*, is a partially ordered homomorphism from the family of ascending sets to $(\mathbb{N} \cup \{\infty\})^{n+1}$, where $(\mathbb{N} \cup \{\infty\})^{n+1}$ is ordered by the lexicographic order. Hence, the family of ascending sets endowed with the ordering "$\prec$" is a well-ordered set.

DEFINITION 2.6. CHARACTERISTIC SET. Let I be an ideal in $k[x_1, \ldots, x_n]$. Consider the family of all ascending sets, each of whose components is in I,

$$S_I = \{ \mathscr{F} = \langle f_1, \ldots, f_r \rangle : \mathscr{F} \text{ is an ascending set and } f_i \in I, 1 \le i \le r \}.$$

A minimal element in S_I (with respect to the $\prec$ order on ascending sets) is said to be a *characteristic set* of the ideal I. □

We remark that if $\mathscr{G}$ is a characteristic set of I then

$$n \ge |\mathscr{G}| \ge n - \dim I.$$

Since $\mathscr{G}$ is an ascending set, by definition, $n \ge |\mathscr{G}|$. The other inequality can be shown as follows: Consider some arbitrary ordering of variables and assume that $|\mathscr{G}| = k$ and the class variables are $v_k, \ldots, v_1$. Let the remaining variables be called $u_1, \ldots, u_l$. We claim that u's must all be independent. Then $n - k = l \le \dim I$ and $|\mathscr{G}| = k \ge n - \dim I$.

The proof of the claim: Suppose that the claim is false. Then I must contain a polynomial $f(u_1, \ldots, u_l)$ which is of class, say u_j. Also, f is reduced with respect to those polynomials of $\mathscr{G}$ with lower ranks. Then one can add f to $\mathscr{G}$ to get an ascending set of lower rank, which is impossible, by the definition of characteristic set.

Also, observe that, for a given ordering of the indeterminates, characteristic set of an ideal is not necessarily unique. However, any two characteristic sets of an ideal must be of the same rank.

The following technical lemma, concerning a system of univariate polynomials, will have further applications in obtaining degree and complexity bounds; for a direct (but lengthy) proof[2] see [18].

LEMMA 2.2. *Let* $H = \{h_1, h_2, \ldots, h_n\} \subseteq k[x_1, x_2, \ldots, x_n]$ *be a set of univariate monic polynomials, one in each variable:*

$$h_j(x_j) \in k[x_j] \setminus k \quad and \quad \text{in}(h_j) = 1.$$

Then $\mathscr{H} = \langle h_1, h_2, \ldots, h_n \rangle$ *is a characteristic set of* (H) *with respect to the following ordering of the indeterminates:*

$$x_1 \prec x_2 \prec \cdots \prec x_n. \quad \square$$

LEMMA 2.3. GENERALIZED PSEUDO-DIVISION LEMMA. *Consider an ascending set* $\mathscr{F} = \langle f_1, f_2, \ldots, f_r \rangle \subseteq k[x_1, \ldots, x_n]$ *and a polynomial* $g \in k[x_1, \ldots, x_n]$. *Then there exists a sequence of polynomials (called*

---

[2] A somewhat indirect proof of the above lemma may be obtained by first observing that H is a Gröbner basis of the ideal (H) (with respect to any admissible ordering), which immediately implies that

$$(\forall g \in (H)) (\forall 1 \le j \le n) [\deg_{x_j}(g) \ge \text{cdeg}(h_j)].$$

Since, class$(h_j) = j$, and since it has the minimal possible class degree, indeed, $\mathscr{H}$ is a characteristic set of (H).

a pseudo-remainder sequence), g_0, g_1, $\ldots$, $g_r = g$, *such that for each* $1 \leq i \leq r$, *the equation*

$$(\exists! \, q_i') \, (\exists! \, \alpha_i) \quad [\mathrm{in}(f_i)^{\alpha_i} g_i = q_i' f_i + g_{i-1}]$$

holds, where g_{i-1} *is reduced with respect to* f_i *and* α_i *assumes the smallest possible power achievable. Thus, the pseudo-remainder sequence is uniquely determined. Moreover, each* g_{i-1} *is reduced with respect to* f_i, f_{i+1}, $\ldots$, f_r.

$$(3) \qquad \mathrm{in}(f_r)^{\alpha_r} \, \mathrm{in}(f_{r-1})^{\alpha_{r-1}} \cdots \mathrm{in}(f_1)^{\alpha_1} g = \sum_{i=1}^{r} q_i f_i + g_0.$$

The polynomial $g_0 \in k[x_1, \ldots, x_n]$ *is said to be the* (generalized) *pseudo-remainder of* g *with respect to the ascending set* $\mathscr{F}$, $g_0 = \mathrm{prem}(g, \mathscr{F})$. *By the earlier observations,* g_0 *is uniquely determined, and reduced with respect to* f_1, f_2, $\ldots$, f_r. *We say a polynomial* g *is* reduced *with respect to an ascending set* $\mathscr{F}$ *if* $g = \mathrm{prem}(g, \mathscr{F})$. $\square$

For an ascending set $\mathscr{F}$, we use the following notation to describe the set of all polynomials that are *pseudo-divisible* by $\mathscr{F}$:

$$\mathscr{M}(\mathscr{F}) = \{g \in k[x_1, \ldots, x_n]: \mathrm{prem}(g, \mathscr{F}) = 0\}.$$

PROPOSITION 2.4. *Let* I *be an ideal in* $k[x_1, \ldots, x_n]$. *Then the ascending set* $\mathscr{G} = \langle g_1, \ldots, g_r \rangle$ *is a characteristic set of* I *if and only if*

$$(\forall f \in I) \quad [\mathrm{prem}(f, \mathscr{G}) = 0]. \quad \square$$

3. Properties of characteristic sets

The main properties of characteristic sets are summarized in the next theorem. First, we need some additional notations.

For any set of polynomials $F = \{f_1, \ldots, f_r\} \subseteq k[x_1, \ldots, x_n]$, we write $\mathscr{Z}(F)$, to denote its *zero set*:

$$\mathscr{Z}(F) = \{\langle p_1, \ldots, p_n \rangle \in k^n: (\forall f \in F) \, f(p_1, \ldots, p_n) = 0\}.$$

By F^{∞}, we shall denote the set of all *finite products* of

$$\mathrm{in}(f_1), \ldots, \mathrm{in}(f_r),$$

and, for any ideal I in $k[x_1, \ldots, x_n]$, we use the notation $I: F^{\infty}$ for

$$\{h : (\exists f \in F^{\infty}) \, [hf \in I]\}.$$

It is easily seen that $I: F^{\infty}$ is itself an ideal. Note that, for an ascending set $\mathscr{F} = \langle f_1, \ldots, f_r \rangle$,

$$\mathrm{prem}(h, \mathscr{F}) = 0 \quad \Rightarrow \quad h \in (\mathscr{F}): \mathscr{F}^{\infty}.$$

THEOREM 3.1. *Let I be an ideal in $k[x_1, \ldots, x_n]$ generated by $F = \{f_1, \ldots, f_s\}$. Let $\mathscr{G} = \langle g_1, \ldots, g_r \rangle$ be a characteristic set of I, and let $J = (g_1, \ldots, g_r) \subseteq I$ be the ideal generated by the elements of $\mathscr{G}$. Then*

(i) *$J \subseteq I \subseteq J : \mathscr{G}^\infty$;*
(ii) *$\mathscr{Z}(\mathscr{G}) \setminus (\bigcup_{i=1}^{r} \mathscr{Z}(\text{in}(g_i))) \subseteq \mathscr{Z}(I) \subseteq \mathscr{Z}(\mathscr{G})$;*
(iii) *$I = prime\ ideal \quad \Rightarrow \quad I \cap \mathscr{G}^\infty = \emptyset$ and $I = J : \mathscr{G}^\infty$.*

PROOF. First, make the following two observations:

(i) $J = (g_1, \ldots, g_r) \subseteq I$ [because $g_i \in I$ by definition].
(ii) $\text{in}(g_i) \notin I$ [because $\text{prem}(\text{in}(g_i), \mathscr{G}) = \text{in}(g_i) \neq 0$].

The first assertion follows from the first observation and the fact that

$$(\forall f \in I) \quad [\text{prem}(f, \mathscr{G}) = 0];$$

thus, for some element $\text{in}(g_r)^{\alpha_r} \text{in}(g_{r-1})^{\alpha_{r-1}} \cdots \text{in}(g_1)^{\alpha_1} = g \in \mathscr{G}^\infty$,

$$gf = \sum_{i=1}^{r} q_i g_i \in J.$$

The second assertion follows from the previous assertion and the elementary properties of the zero sets of polynomials. Thus,

$$\mathscr{Z}(I) \subseteq \mathscr{Z}(J) = \mathscr{Z}(\mathscr{G}).$$

Hence, it suffices to prove that

$$\mathscr{Z}(\mathscr{G}) \setminus \left(\bigcup_{i=1}^{r} \mathscr{Z}(\text{in}(g_i)) \right) \subseteq \mathscr{Z}(I).$$

Let $p \in k^n$ be a point in the zero set $\mathscr{Z}(\mathscr{G}) \setminus (\bigcup_{i=1}^{r} \mathscr{Z}(\text{in}(g_i)))$. Let $f \in I$. Then there exists a $g = \prod_{i=1}^{r} \text{in}(g_i)^{\alpha_i} \in \mathscr{G}^\infty$ such that $gf \in J$. Thus, $g(p)f(p) = 0$. But by assumption, $g(p) \neq 0$. Then $f(p) = 0$, as is to be shown.

To see the last assertion, observe that if I is a prime ideal then

• If some $g \in \mathscr{G}^\infty$ belongs to I then so does one of the factors of g, say $\text{in}(g_i)$. But this contradicts the second observation made in the beginning of this proof.
• Consider an $f \in J : \mathscr{G}^\infty$. By definition, there exists a $g \in \mathscr{G}^\infty$ such that

$$gf \in J \subseteq I \quad \text{and} \quad g \notin I.$$

From the primality of I, conclude that $f \in I$. That is, $J : \mathscr{G}^\infty = I$.

$\square$

As an immediate consequence of the above statement, we have

$$\text{prem}(f, \mathscr{G}) = 0 \quad \Leftrightarrow \quad f \in I,$$

provided that I is prime.

REMARK 3.1.   (i) The inclusions in assertion (ii) of Theorem 3.1 can
        be strict. Consider the ideal $I = (x^2 + y^2 - 1, xy)$ and suppose
        $x \prec y$. A possible characteristic set for I is $\{xy, y^3 - y\}$ whose
        zeroes are a set of higher dimension than the zeroes of I. Removing
        from it the zeroes of the initial of xy, i.e., the line of equation $y = 0$,
        one gets only two of the four original points in $\mathscr{Z}(I)$.
    (ii) The hypothesis of primality for I in the assertion (4) of Theorem 3.1
        cannot be relaxed. Consider for example the ideal $I = (xy, y^2)$. The
        given set of generators is already a characteristic set for I, assuming
        $y \prec x$, but $\mathrm{in}((xy)^2) = y^2 \in I$. □

One way to interpret the preceding theorem is to say that constructing
characteristic sets helps only in answering geometrical questions, but not with
general algebraic problems. Thus, while characteristic sets are capable of
answering the *membership problem for a radical ideal*, they are not powerful
enough to handle the more general *membership problem for an arbitrary ideal*.

4. Wu's approach to the construction of characteristic sets

In this section, we shall consider the following triangulation process, due to
J.F. Ritt and Wu Wen-Tsün, which computes a so-called *extended characteristic set* of an ideal by repeated applications of the generalized pseudodivision.
Historically, this represents the first effort to effectively construct a triangular
set corresponding to a system of differential equations. Here, we focus just
on the algebraic analog.

DEFINITION 4.1. RITT'S PRINCIPLE. Let $F = \{f_1, \ldots, f_s\} \subseteq$
$k[x_1, \ldots, x_n]$ be a finite nonempty set of polynomials, and $I = (F)$ be
the ideal generated by F. The *Wu-Ritt process* described below obtains an
ascending set $\mathscr{G}$ such that *either*

    (i) $\mathscr{G}$ consists of a polynomial in $k \cap I$, or
    (ii) $\mathscr{G} = \langle g_1, \ldots, g_r \rangle$ with $\mathrm{class}(g_1) > 0$ and such that

$$g_i \in I, \quad \text{for all } i = 1, \ldots, r,$$
$$\mathrm{prem}(f_j, \mathscr{G}) = 0, \quad \text{for all } j = 1, \ldots, s.$$

Ascending chain $\mathscr{G}$ is called an *extended characteristic set* of F. □

It is trivial to see that when the algorithm terminates, it, in fact, returns
an ascending set $\mathscr{G}$ that satisfies the conditions given in Definition 4.1. The
termination follows from the following observations:

Let $F \subseteq k[x_1, \ldots, x_n]$ be a (possibly, infinite) set of polynomials. Consider the family of all ascending sets, each of whose components is in F,

$$S_F = \{\mathscr{F} = \langle f_1, \ldots, f_r \rangle : \mathscr{F} \text{ is an ascending set and } f_i \in F, 1 \leq i \leq r\}.$$

```
Wu-Ritt Process:
Input:       F = {f₁, ... , fₛ} ⊆ k[x₁, ... , xₙ].
Output:      𝒢, an extended characteristic set of F.

1.   𝒢 := ∅;     R := ∅;
2.  loop
3.      F := F ∪ R;     F' := F;     R := ∅;
4.      while F' ≠ ∅ loop
5.          Choose a polynomial f ∈ F' of minimal rank;
6.          F' := F' \ {g: class(g) = class(f) and
7.                      g is not reduced with respect to f};
8.          𝒢 := 𝒢 ∪ {f};
9.      end{loop};
10.     for all f ∈ F \ 𝒢  loop
11.         if r := prem(f, 𝒢) ≠ 0  then
12.             R := R ∪ {r};
13.         end{if};
14.     end{loop};
15.  until R = ∅;
16.  return 𝒢;
end{Wu-Ritt Process}.  □
```

A minimal element in S_F (with respect to the $\prec$ order on ascending sets) is denoted as $\mathrm{MinASC}(F)$. The following easy proposition can be shown:

PROPOSITION 4.1. *Let F be as above. Let g be a polynomial reduced with respect to $\mathrm{MinASC}(F)$. Then*

$$\mathrm{MinASC}(F \cup \{g\}) \prec \mathrm{MinASC}(F). \quad \square$$

Now, let F_i be the set of polynomials obtained at the beginning of the *i*th iteration of the loop [lines 2–15]. Starting from the set F_i, the algorithm constructs the ascending chain $\mathcal{G}_i = \mathrm{MinASC}(F_i)$ in the loop [lines 4–9]. Now, if R_i (constructed by the loop [lines 10–14]) is nonempty, then each element of R_i is reduced with respect to $\mathcal{G}_i$. Now, since

$$F_{i+1} = F_i \cup R_i,$$

we observe that

$$\mathrm{MinASC}(F_0) \succ \mathrm{MinASC}(F_1) \succ \cdots \succ \mathrm{MinASC}(F_i) \succ \cdots.$$

Since the "$\prec$" is a well-ordering on the ascending sets, the above chain must be finite and the algorithm must terminate. However, the number of steps the algorithm may take in the worst case can be nonelementary in the parameters n (the number of variables) and d (the maximum degree of the polynomials). Assume the usual ordering on the indeterminates:

$$x_1 \prec x_2 \prec \cdots \prec x_n.$$

To understand the worst-case behavior of the algorithm, we need to consider an example which consists of s polynomials in n variables with degrees (in each variable) bounded by d and each of class $= n$ and cdeg $= d$. After each iteration of the main loops, we can show that the largest degree in any variable is bounded by $T_2(k)$, where

$$T_2(0) = d,$$
$$T_2(1) = 2d,$$
$$T_2(k) = 2\,T(k-1) + T(k-2), \qquad k > 1.$$

We observe that if after the $(i - 1)$, $2 < i \le d + 1$, iterations of the loops, we have a set of s polynomials consisting of

- f_1 of class $= n$, cdeg $= d - i + 2$ and deg $\le T_1(i - 1)$,
- f_2 of class $= n$, cdeg $= d - i + 1$ and deg $\le T_2(i - 1)$, and
- $f_3, \ldots, f_s$, each of class $= n$, cdeg $= d - i + 1$ and deg $\le T_3(i - 1)$,

then, in the worst-case, we can obtain after i iterations, a set of s polynomials consisting of

- f_1' of class $= n$, cdeg $= d - i + 1$ and deg $\le T_2(i - 1)$,
- f_2' of class $= n$, cdeg $= d - i$ and deg $\le 2T_2(i - 1) + T_1(i - 1)$, and
- $f_3', \ldots, f_s'$, each of class $= n$, cdeg $= d - i$ and deg $\le T_2(i - 1) + T_3(i - 1)$.

The rest follows from the fact that the maximal value of the degrees is given by T_2.

The recurrence equation can be easily solved to get

$$T_2(k) \le \frac{d}{2\sqrt{2}}(1 + \sqrt{2})^{(k+1)}.$$

Now observe that the above process can be repeated for at most d iterations to yield a similar instance with class reduced by 1 but class degree enlarged to $c'd(1 + \sqrt{2})^d$. Thus, a bound on the number of iterations can be given by the following recurrence equations:

$$T(0, d) = 1,$$
$$T(n, d) \le T\left(n - 1, c'(1 + \sqrt{2})^d\right) + d, \qquad n \ge 1.$$

A solution to the above equation leads to

$$T(n, d) \leq c \cdot \underbrace{d(1 + \sqrt{2})^{d(1+\sqrt{2})^{\cdot^{\cdot^{\cdot d(1+\sqrt{2})^d}}}}}_{n} \, ,$$

a nonelementary function in the input-size.

In general, an extended characteristic set of an ideal is not a characteristic set of the ideal. However, an extended characteristic set *does* satisfy the following property, in a manner similar to a characteristic set.

THEOREM 4.2. *Let* $F \subseteq k[x_1, \ldots, x_n]$ *be a basis of an ideal* I, *with an extended characteristic set* $\mathcal{G} = \langle g_1, \ldots, g_r \rangle$. *Then*

$$\mathcal{Z}(\mathcal{G}) \setminus \left(\bigcup_{i=1}^{r} \mathcal{Z}(\text{in}(g_i)) \right) \subseteq \mathcal{Z}(I) \subseteq \mathcal{Z}(\mathcal{G}).$$

PROOF. Let

$$\mathcal{M}(\mathcal{G}) = \{f : \text{prem}(f, \mathcal{G}) = 0\}.$$

denote, as before, the set of all polynomials that are *pseudo-divisible* by $\mathcal{G}$. Thus, by the properties of an extended characteristic set, $(\mathcal{G}) \subseteq I \subseteq \left(\mathcal{M}(\mathcal{G}) \right)$. Using the elementary properties of the zero sets of polynomials, we get

$$\mathcal{Z}\left(\mathcal{M}(\mathcal{G}) \right) \subseteq \mathcal{Z}(I) \subseteq \mathcal{Z}(\mathcal{G}).$$

Hence, it suffices to prove that

$$\mathcal{Z}(\mathcal{G}) \setminus \left(\bigcup_{i=1}^{r} \mathcal{Z}(\text{in}(g_i)) \right) \subseteq \mathcal{Z}(\mathcal{M}(\mathcal{G})).$$

Let $p \in k^n$ be a point in the zero set $\mathcal{Z}(\mathcal{G}) \setminus (\bigcup_{i=1}^{r} \mathcal{Z}(\text{in}(g_i)))$. Let $f \in \mathcal{M}(\mathcal{G})$. Then there exists a $g = \prod_{i=1}^{r} \text{in}(g_i)^{\alpha_i} \in \mathcal{G}^{\infty}$ such that $gf \in J$. Thus, $g(p)f(p) = 0$. But by assumption, $g(p) \neq 0$. Then $f(p) = 0$, as is to be shown. □

5. Construction and complexity of characteristic sets[3]

Let $I \subseteq k[x_1, \ldots, x_n]$ be an ideal generated by a set of s generators, $f_1, \ldots, f_s$, in the ring of polynomials in n indeterminates over the field k. Further, assume that each of the polynomial f_i in the given set of generators has its degree bounded by d:

$$(\forall \, 1 \leq i \leq s) \quad [\deg(f_i) \leq d].$$

Let $\mathcal{G} = \langle g_1, \ldots, g_r \rangle$ be a characteristic set of the ideal, and let $D = \max\{\deg(g_i) : 1 \leq i \leq r\}$ be the degree of this characteristic set. In the next

---

[3]The results in this section are largely based on the paper [18]; more details may be found therein.

two subsections, we shall give upper and lower bounds for D, as function of s, d and n, for a large class of orderings on the indeterminates.

5.1. Lower bounds

THEOREM 5.1. LOWER BOUND THEOREM. *Let* k *be an algebraically closed field, and let* I *be an ideal in* $k[x_1, \ldots, x_n]$, *generated by the following* n *polynomials,* $F = \{f_1, \ldots, f_n\}$, *each of degree* d :

$$f_1 = x_1 - x_n^d,$$
$$f_2 = x_2 - x_1^d,$$
$$\vdots$$
$$f_{n-1} = x_{n-1} - x_{n-2}^d,$$
$$f_n = x_n - x_{n-1}^d.$$

Then, independent of the ordering on the indeterminates,
  (1) *every characteristic set of* F *is of degree* $D \geq d^n$;
  (2) *every extended characteristic set of* F *is of degree* $D \geq d^n$.

PROOF. Let $\pi \in S_n$ be a permutation of $[1..n]$, and the arbitrary but fixed ordering on the indeterminates be $x_i = x_{\pi(1)} \prec x_{\pi(2)} \prec \cdots \prec x_{\pi(n)}$. Note that $I \cap (k[x_i] \setminus k)$ contains a nonzero polynomial $x_i - x_i^{d^n}$ of *minimal possible degree,* d^n.
  (1) Let $\mathscr{G} = \langle g_1, \ldots, g_r \rangle$ be a characteristic set of F, with respect to the chosen ordering. Since $\dim I = 0$,

$$|\mathscr{G}| = n \quad \text{and} \quad g_1 \in I \cap (k[x_i] \setminus k).$$

Thus, $D \geq \mathrm{cdeg}(g_1) = \min\{\deg(f) : f \in I \cap (k[x_i] \setminus k)\} = d^n$.
  (2) Let $\mathscr{G}' = \langle g_1', \ldots, g_r' \rangle$ be an extended characteristic set of F, with respect to the chosen ordering. By an examination of the Wu-Ritt process, we see that $\mathrm{in}(g_j') = 1$, for all $1 \leq j \leq r$, and by Lemma 4.2,

$$\mathscr{Z}(\mathscr{G}') = \mathscr{Z}(I) = \text{a finite set.}$$

Hence

$$|\mathscr{G}'| = n \quad \text{and} \quad g_1 \in I \cap (k[x_i] \setminus k).$$

Thus, $D \geq \mathrm{cdeg}(g_1) = \min\{\deg(f) : f \in I \cap (k[x_i] \setminus k)\} \geq d^n$. $\quad\square$

In the subsequent discussion, we shall focus only on the characteristic sets of an ideal, since it is dubious whether an extended characteristic set improves the computational power of our algorithms. We remark that if we choose the following ordering on the indeterminates, $x_1 \prec x_2 \prec \cdots \prec x_n$, then a characteristic set of I of Theorem 5.1 is given by

$$\mathcal{G} = \langle g_1 = x_1 - x_1^{d^n},$$
$$g_2 = x_2 - x_1^{d},$$
$$\vdots$$
$$g_{n-1} = x_{n-1} - x_1^{d^{n-2}},$$
$$g_n = x_n - x_1^{d^{n-1}} \rangle.$$

A slightly better bound of $\deg(f_1)\deg(f_2)\cdots\deg(f_n)$ can be obtained by a simple modification of the construction in Theorem 5.1.

We note that the ideal I of Theorem 5.1 is a zero-dimensional ideal; the previous example can be modified to show that, in general, the degree of a characteristic set is $\Omega(d^{(n-\dim I)})$. It is interesting to ask if a better lower bound can be constructed for more general cases.

5.2. Upper bounds. We start by recalling the following version of *effective Hilbert's Nullstellensatz*, originally due to D. Brownawell [6] and later sharpened by J. Kollár [26].

THEOREM 5.2. EFFECTIVE HILBERT'S NULLSTELLENSATZ. *Let* $I = (f_1,$ $\ldots, f_s)$ *be an ideal in* $k[x_1, \ldots, x_n]$, *where* k *is an arbitrary field, and* $\deg(f_i) \leq d$, $1 \leq i \leq s$. *Let* $h \in K[x_1, \ldots, x_n]$ *be a polynomial with* $\deg(h) = \delta$. *Then* $h \in \sqrt{I}$ *if and only if*

$$(\exists\, b_1, \ldots, b_s \in k[x_1, \ldots, x_n]) \quad \left[h^M = \sum_{i=1}^{s} b_i f_i, \right],$$

where $M \leq 2(d+1)^n$ *and* $\deg(b_i f_i) \leq 2(\delta+1)(d+1)^n$, $1 \leq i \leq s$. $\square$

Using this version of Nullstellensatz, we shall first derive a degree bound for a characteristic set of a zero-dimensional ideal, and then use a "lifting" procedure to obtain a bound for the more general cases.

The following proposition shows that every zero-dimensional ideal I contains a univariate polynomial $h_j(x_j)$, in each variable x_j, with the additional property that $\deg(h_j) \leq 2(d+1)^{2n}$. Since the sequence $\langle h_1, \ldots, h_n \rangle$ is clearly an ascending set in S_I, we get a similar bound on the class degrees of the polynomials in a characteristic set of I. This leads to a degree bound of $2n(d+1)^{2n}$ for a characteristic set of I.

LEMMA 5.3. *Let* $I = (f_1, \ldots, f_s)$ *be a zero-dimensional ideal in* $k[x_1, \ldots, x_n]$, *where* k *is an arbitrary field, and* $\deg(f_i) \leq d$, $1 \leq i \leq$ s. *Then for every indeterminate* x_j, *there exists a univariate polynomial*

$h_j \in I \cap k[x_j]$ *such that*

$$\left(\exists\, b_{j,1}, \dots, b_{j,s} \in k[x_1, \dots, x_n]\right) \quad \left[h_j = \sum_{i=1}^{s} b_{j,i} f_i,\right],$$

where $\deg(h_j) \le 2(d+1)^{2n}$, *and* $\deg(b_{j,i} f_i) \le 4(d+1)^{2n}$, $1 \le i \le s$.

PROOF. Suppose first that k is algebraically closed. Assume that the finite set of T zeroes of I are given by

$$\mathscr{Z}(I) = \{P_i = \langle p_{i,1}, \dots, p_{i,j}, \dots, p_{i,n} \rangle \in k^n : 1 \le i \le T\} \neq \emptyset.$$

By the Bezout's inequality of J. Heintz [20], we have $T \le (d+1)^n$. Now consider the following monic polynomial $h'_j(x_j) \in k[x_j]$,

$$h'_j(x_j) = \prod_{i=1}^{t}(x_j - p_{i,j}),$$

where $p_{i,j}$ is the jth coordinate of the ith zero of I. As h'_j vanishes at all the zeros of I, by Hilbert's Nullstellensatz, it must be in the *radical* of I,

$$h'_j \in \sqrt{I}, \quad \text{that is,} \quad h_j = (h'_j)^M \in I.$$

But by construction h_j is a monic polynomial of degree TM, in $k[x_j]$. Moreover, by Theorem 5.2, we see that $M \le 2(d+1)^n$, and

$$\deg(h_j) \le 2(d+1)^{2n}, \quad 1 \le j \le n.$$

The bound on the representation of h_j comes directly from Theorem 5.2,

$$\deg(b_{j,i} f_i) \le 2(T+1)(d+1)^n \le 4(d+1)^{2n}.$$

If k is not algebraically closed, the above construction can be done in its algebraic closure. Observe now that, since the degree of h_j is bounded, its coefficients are determined by a system of linear equations over k, and thus, $h_j \in k[x_j]$. □

COROLLARY 5.4. *Let* $I = (f_1, \dots, f_s)$ *be a zero-dimensional ideal in* $k[x_1, \dots, x_n]$, *where* k *is an arbitrary field, and* $\deg(f_i) \le d$, $1 \le i \le s$. *Let* $g \in k[x_1, \dots, x_n]$ *be a polynomial with* $\deg(g) = \Delta$. *Then* $g \in I$ *if and only if*

$$\left(\exists\, a_1, \dots, a_s \in k[x_1, \dots, x_n]\right) \quad \left[g = \sum_{i=1}^{s} a_i f_i,\right],$$

where $\deg(a_i f_i) \le \max\{d, \Delta\} + 6n(d+1)^{2n}$, $1 \le i \le s$.

PROOF. Assume that $g \in I$. Then, using the univariate, monic polynomials h_j's (obtained in Lemma 5.3) as rewriting rules modulo $J = (h_1, \ldots, h_n) \subseteq I$, g may be expressed as $g = h + \sum_{i=1}^{s} a_i' f_i$, where

$$\deg(a_i' f_i) \leq \max_{1 \leq i \leq s}\{\deg(f_i)\} + \sum_{j=1}^{n} \deg_{x_j}(h_j) \leq d + 2n(d+1)^{2n}, \qquad 1 \leq i \leq s.$$

Thus, we have

$$h = g - \sum_{i=1}^{s} a_i' f_i \in J \quad \text{and} \quad \deg(h) \leq \max\{\Delta, d + 2n(d+1)^{2n}\}.$$

But, by Lemma 2.2, $\mathcal{H} = \langle h_1, \ldots, h_n \rangle$ is a characteristic set of J. Since $h \in J$, we see that $\operatorname{prem}(h, \mathcal{H}) = 0$. But since $\operatorname{in}(h_j) = 1$, for all $1 \leq j \leq n$, we have

$$h = \sum_{j=1}^{n} q_j h_j = \sum_{i=1}^{s} \left(\sum_{j=1}^{n} q_j b_{j,i} \right) f_i = \sum_{i=1}^{s} a_i'' f_i,$$

where

$$\deg(a_i'' f_i) \leq \max_j \{\deg(q_j)\} + \max_{i,j}\{\deg(b_{j,i} f_i)\}$$

$$\leq \deg(h) + 4(d+1)^{2n} \leq \max\{d, \Delta\} + 6n(d+1)^{2n}, \qquad 1 \leq i \leq s.$$

Thus g may be expressed as $g = \sum_{i=1}^{s}(a_i'' + a_i') f_i = \sum_{i=1}^{s} a_i f_i$, where

$$\deg(a_i f_i) \leq \max_{1 \leq i \leq s}\{\deg(a_i'' f_i), \deg(a_i' f_i)\}$$

$$\leq \max\{d, \Delta\} + 6n(d+1)^{2n}. \qquad \square$$

THEOREM 5.5. ZERO-DIMENSIONAL UPPER BOUND THEOREM. *Let* $I = (f_1, \ldots, f_s)$ *be a zero-dimensional ideal in* $k[x_1, \ldots, x_n]$, *where* k *is an arbitrary field, and* $\deg(f_i) \leq d$, $1 \leq i \leq s$. *Then* I *has a characteristic set* $\mathcal{G} = \langle g_1, \ldots, g_n \rangle$ *with respect to the ordering,* $x_1 \prec x_2 \prec \cdots \prec x_n$, *where for all* $1 \leq j \leq n$,

   (i) $\operatorname{class}(g_j) = j$ *and* $\deg(g_j) \leq 2n(d+1)^{2n} = O\left(d^{O(n)}\right).$
   (ii)

$$(\exists\, a_{j,1}, \ldots, a_{j,s} \in k[x_1, \ldots, x_n]) \quad \left[g_j = \sum_{i=1}^{s} a_{j,i} f_i, \right],$$

and $\deg(a_{j,i} f_i) \leq 8n(d+1)^{2n} = O(d^{O(n)})$, $1 \leq i \leq s$.

PROOF. Note that $\mathcal{H} = \langle h_1, \ldots, h_n \rangle$ of Lemma 5.3 is an ascending set in S_I. The first condition follows from the facts that, for all $1 \leq j \leq n$,

1. $\text{class}(g_j) = \text{class}(h_j) = j$, and $\text{cdeg}(g_j) \le \text{cdeg}(h_j) \le \deg(h_j) \le 2(d+1)^{2n}$.

2. For all $1 \le i < j$, $\deg_{x_i}(g_j) < \deg_{x_i}(g_i)$, as g_j is reduced with respect to the preceding g_i's. Thus

$$\deg(g_j) \le \left(\sum_{i=1}^{j} \text{cdeg}(g_i) \right) - j + 1 \ \le \ 2j(d+1)^{2n} - j + 1 \ \le \ 2n(d+1)^{2n}.$$

Since each g_j is in I, the condition (3) is an immediate consequence of Corollary 5.4. $\square$

The results on the bounds for a characteristic set of a zero-dimensional ideal can be extended to the more general classes of ideals, by the following "lifting" process. However, in this case, we need to restrict the class of orderings on the indeterminates that may be chosen.

Let $I = (f_1, \ldots, f_s)$ be an ideal in the ring $k[x_1, \ldots, x_n]$ where k is an arbitrary field, $\deg(f_i) \le d$. Assume that the indeterminates, $x_{\pi(1)} = u_1$, $\ldots$, $x_{\pi(l)} = u_l$, are the *independent* variables with respect to I. That is, these independent variables form the largest subset of $\{x_1, \ldots, x_n\}$ such that $I \cap k[u_1, \ldots, u_l] = (0)$. Thus, $\dim(I) = l > 0$. The remaining $r = (n - l)$ indeterminates, $x_{\pi(l+1)} = v_1, \ldots, x_{\pi(n)} = v_r$, form the set of *dependent* variables. We write $k[u_1, \ldots, u_l, v_1, \ldots, v_r]$ for the ring of polynomials, $k[x_1, \ldots, x_n]$.

The ordering on the indeterminates is assumed to be so chosen that

$$(\forall\, 1 \le i \le l) \quad (\forall\, 1 \le j \le r) \quad [u_i \prec v_j].$$

For a polynomial $f \in k[u_1, \ldots, u_l, v_1, \ldots, v_r]$, we shall use the notation, $\deg_U(f)$ and $\deg_V(f)$, to imply

$$\deg_U(f) = \sum_{i=1}^{l} \deg_{u_i}(f) \le \deg(f) \quad \text{and} \quad \deg_V(f) = \sum_{j=1}^{r} \deg_{v_j}(f) \le \deg(f).$$

Let I' be the *extended ideal* of I in the ring $k(u_1, \ldots, u_l)[v_1, \ldots, v_r]$, i.e., $I' = I\{k(u_1, \ldots, u_l)[v_1, \ldots, v_r]\}$. Then I' is a zero-dimensional ideal in $k(u_1, \ldots, u_l)[v_1, \ldots, v_r]$, and by the results of the previous subsection, I' has a characteristic set $\mathscr{G}' = \langle g_1', \ldots, g_r' \rangle$, where $\text{class}(g_j') = \pi(l + j)$ and $\deg_V(g_j') \le 2r(d + 1)^{2r}$. Additionally, by the results of the previous section, each g_j' can be expressed as

$$(4) \qquad g_j' = \sum_{i=1}^{s} a_{j,i}' f_i \quad \text{where } \deg_V(a_{j,i}' f_i) \le 8r(d + 1)^{2r}.$$

Thus the system of linear equations defined by (3) consists of at most

$$\Gamma = \binom{8r(d+1)^{2r} + r}{r}$$

equations in at most $(s+1)\Gamma$ unknowns, where the known coefficients are polynomials of degree at most d in $k[u_1, \ldots, u_l]$. Note that there is one equation per each power-product in r variables of degree $\leq 8r(d+1)^{2r}$, and one unknown per coefficient of the polynomials g_j' and $a_{j,i}'$'s. Thus the coefficients of g_j' and $a_{i,j}'$'s are rational functions in $k(u_1, \ldots, u_d)$, with the numerators and denominators being determined by determinants of polynomial matrices of order $\leq \Gamma$ and with entries of degree $\leq d$ in $k[u_1, \ldots, u_l]$. Hence, these numerators and denominators, themselves, have their degrees bounded by $d\Gamma$.

Now, the left- and right-hand sides of equation (4) can be multiplied by an appropriate polynomial, $m \in k[u_1, \ldots, u_l]$, to clear out the denominators, and yield the equation

(5)
$$g_j = \sum_{i=1}^{s} a_{j,i} f_i,$$

where all the coefficients are in $k[u_1, \ldots, u_l]$. Note that m can be chosen to be the *least common multiplier* of all the denominators, and since there are at most $(s+1)\Gamma$ such denominators, we have

$$\deg_U(m) \leq d(s+1)\Gamma^2,$$

$$\deg_U(g_j) \leq d\Gamma + d(s+1)\Gamma^2,$$

$$\deg_U(a_{j,i} f_i) \leq d + d\Gamma + d(s+1)\Gamma^2.$$

In the following theorem we show that $\mathscr{G} = \langle g_1, \ldots, g_r \rangle$ is, in fact, a characteristic set of the ideal I.

THEOREM 5.6. GENERAL UPPER BOUND THEOREM. *Let $I = (f_1, \ldots, f_s)$ be an ideal in $k[x_1, \ldots, x_n]$, where k is an arbitrary field, and $\deg(f_i) \leq d$, $1 \leq i \leq s$. Assume that $x_1, \ldots, x_l$ are the independent variables with respect to I. Let $r = n - \dim I = n - l$. Then I has a characteristic set $\mathscr{G} = \langle g_1, \ldots, g_r \rangle$ with respect to the ordering*

$$x_1 \prec x_2 \prec \cdots \prec x_n,$$

where for all $1 \leq j \leq r$,

   (i) $\operatorname{class}(g_j) = j + l$ *and*

$$\deg(g_j) \leq 4(s+1)(9r)^{2r} d(d+1)^{4r^2} = O(s \, d^{O(r^2)}).$$

(ii) $(\exists\ a_{j,1}, \ldots, a_{j,s} \in k[x_1, \ldots, x_n])\ [g_j = \sum_{i=1}^{s} a_{j,i} f_i,]$, and

$$\deg(a_{j,i} f_i) \leq 11(s+1)(9r)^{2r} d(d+1)^{4r^2} = (Os\ d^{O(r^2)}),\ 1 \leq i \leq s.$$

PROOF. As before, let us assume that the independent variables with respect to I are $u_1 = x_1$, ..., and $u_l = x_l$, and we write $k[u_1, \ldots, u_l, v_1, \ldots, v_r]$ for the ring $k[x_1, \ldots, x_n]$. Let $\mathscr{G} = \langle g_1, \ldots, g_r \rangle$ be the sequence of polynomials defined in the preceding paragraph. As g_j's were derived from an ascending set $\mathscr{G}'$ of $S_{I'}$ in $k(u_1, \ldots, u_l)[v_1, \ldots, v_r]$, by multiplication with a polynomial in $k[u_1, \ldots, u_l]$, and since $g_j \in I$ (by construction), clearly $\mathscr{G}$ is an ascending set in S_I.

Furthermore, $\mathscr{G}$ is a minimal set in S_I, since, if there were another ascending set $\mathscr{H} \prec \mathscr{G}$ in S_I, then $\mathscr{H}$ itself would be an ascending set in $S_{I'}$ and of smaller rank than $\mathscr{G}'$, when the polynomials in $\mathscr{H}$ are treated as polynomials in $k(u_1, \ldots, u_l)[v_1, \ldots, v_r]$; this, however, would contradict the assumption that $\mathscr{G}'$ is a characteristic set of the extended ideal I'.

By construction, $\mathrm{class}(g_j) = j + l$. The rest involves some simple calculations. □

5.3. Algorithms. Now, we can describe how to compute a characteristic set of an ideal, by using the degree bounds of the *General Upper Bound Theorem* and fairly simple ideas from linear algebra. In particular, we need, as subroutines, fast sequential and parallel algorithms for computing the *rank* and *determinants* of a matrices of order N over arbitrary field. For this purpose, we assume that the sequential algorithms have time complexity $O(N^{2.376})$ [1, 11] and that the parallel algorithms have time complexity $O(\log^2 N)$, (and polynomial size) [5, 12, 33]. Our complexity model assumes an $O(1)$-complexity cost for all field operations over k.

Let I be an ideal given by a set of generators $\{f_1, \ldots, f_s\} \in k[x_1, \ldots, x_n]$, where k is an arbitrary field, $\deg(f_i) \leq d$. Assume that after some reordering of the indeterminates, the indeterminates $x_1, \ldots, x_n$ are so arranged that the first l of them are *independent* with respect to I, and the remaining $r = (n - l)$ variables, *dependent*. The ordering on the indeterminates are so chosen that

$$x_1 \prec x_2 \prec \cdots \prec x_n.$$

These conditions can be verified in $O(s^7 d^{O(n^2)})$ sequential time, or equivalently, in $O(n^4 \log^2(sd))$ parallel time by an algorithm of Dickenstein et al. [14].

Assume, inductively, that the first $(j-1)$ elements $g_1, \ldots, g_{j-1}$, of a characteristic set, $\mathscr{G}$, of I have been computed, and we wish to compute the jth element g_j of $\mathscr{G}$. By the Theorem 5.6, we know that $\mathrm{class}(g_1) =$

$(l+1), \ldots,$ class$(g_{j-1}) = (l+j-1)$ and class$(g_j) = (l+j)$. Let

$$\text{cdeg}(g_1) = d_{l+1}, \ldots, \text{cdeg}(g_{j-1}) = d_{l+j-1}.$$

Thus, the polynomial g_j, sought, must be a nonzero polynomial of least degree in x_{l+j}, in $I \cap k[x_1, \ldots, x_{l+j}]$ such that

$$\deg_{x_{l+1}}(g_1) < d_{l+1}, \ldots, \deg_{x_{l+j-1}} < d_{l+j-1}.$$

Furthermore, we know, from the *General Upper Bound Theorem*, that

$$(6) \qquad (\exists\, a_{j,1}, \ldots, a_{j,s} \in k[x_1, \ldots, x_n]) \quad \left[g_j = \sum_{i=1}^{s} a_{j,i} f_i, \right],$$

and $\deg(g_j), \deg(a_{j,i} f_i) \leq 11(s+1)(9r)^{2r} d(d+1)^{4r^2}$, $1 \leq i \leq s$. Thus g_j satisfying all the properties can be determined by solving an appropriate system of linear equations, involving the power products of degree less than $11(s+1)(9r)^{2r} d(d+1)^{4r^2}$, i.e., $O(s^{O(n)}(d+1)^{O(n^3)})$-many linear equations.

In particular, the sequential and parallel complexity of the algorithm to compute the jth element of a characteristic set are given, respectively, by $O(s^{O(n)}(d+1)^{O(n^3)})$ and $O(n^6 \log^2(s+d+1))$. Thus, a characteristic set of the given ideal I, can be computed by at most n-many iterations in $O(s^{O(n)}(d+1)^{O(n^3)})$ sequential time and $O(n^7 \log^2(s+d+1))$ parallel time.

THEOREM 5.7. COMPLEXITY OF CHARACTERISTIC SETS. *Let $I = (f_1, \ldots, f_s)$ be an ideal in $k[x_1, \ldots, x_n]$, where k is an arbitrary field, and $\deg(f_i) \leq d$, $1 \leq i \leq s$. Then with respect to any ordering on the indeterminates $x_1 \prec x_2 \prec \cdots \prec x_n$, where the first $\dim(I)$-many variables are independent, one can compute a characteristic set of I, in $O(s^{O(n)}(d+1)^{O(n^3)})$ sequential time or $O(n^7 \log^2(s+d+1))$ parallel time. The polynomials in the computed characteristic set are of degree $O(s(d+1)^{O(n^2)})$.* □

If the ideal is zero-dimensional, then the arguments in this section can be specialized to yield better bounds (see [18]).

6. Applications: Mechanical geometric theorem proving

Let us consider how the concepts of characteristic sets (or extended characteristic sets) can be used in a simple geometric theorem prover. The first such mechanical theorem prover, devised by Wu Wen-Tsün [35, 36, 37] in China, has come to be known as the *China Prover* and has been successfully used to prove many classical and some new theorems in plane analytic geometry. Some further improvements in this direction has been achieved by Shang-Ching Chou [10].

Other approaches, based on different constructive methods in computer algebra, have also been proposed for this problem: For instance, B. Kutzler and S. Stifter [27] and D. Kapur [24] have proposed methods based on Gröbner bases; G. Carrá and G. Gallo [8] have devised a method using the dimension of the underlying algebraic variety; Jia-Wei Hong [22] has introduced a seminumerical algorithm using an interesting gap theorem and "proof-by-example" techniques. However, none of these other methods have yet achieved the kind of wide-spread acceptance that Wu's method has had.

The method based on the Wu-Ritt characteristic sets is not, however, a real theorem prover, in the sense that it does not follow any logical proof theoretic techniques; it simply takes an algebraic translation of a set of geometric statements and tries to verify its validity in the manner to be made more precise. For most of the geometric statements, however, the translation from geometry to algebra is far from being completely automated. The fundamental principle that the translation method relies on is fairly simple, i.e., the classical coordinate method of Descartes introduced in the seventeenth century.

DEFINITION 6.1. By an *elementary geometry statement*, we mean a formula of the following kind:

$$(7) \qquad (f_1 = 0 \, \wedge \, f_2 = 0 \, \wedge \, \cdots \, \wedge \, f_s = 0) \quad \Rightarrow \quad (g = 0),$$

where the f_i's and g are polynomials in $k[x_1, \ldots, x_n]$, the variables x_i's are assumed to be bound by universal quantification and their ranges are assumed to be the field k, the base field of the underlying geometry. We further assume that the base field is algebraically closed. □

The conjunct $\bigwedge_i(f_i = 0)$ is called the *premise* of the geometry statement, and will be assumed to be nontrivial, i.e., the set of points in k^n satisfying the premise is nonempty. The statement $g = 0$ is its *conclusion*.

To prove a statement, then, is to show that a geometric formula is valid. For a complete analysis of the classes of geometric propositions included in Definition 6.1, refer to [10].

However, one problem with the above algebraic statement is that it does not mention certain *geometric degeneracies* that are implicitly excluded: For instance, when a geometric statement mentions a *triangle*, it is conventionally assumed to mean those nondegenerate triangles whose vertices are noncollinear. But, on the other hand, spelling out all the conceivable geometric degeneracies makes the process unappealingly cumbersome. Wu's approach permits one to circumvent this problem as it produces these nondegeneracy conditions as a natural by-product. In fact, Wu's approach proves an elementary geometry statement, in the sense that it shows that the *conclusion is true whenever the premises are generically true*. In geometric terms, it shows that the conclusion polynomial vanishes on *some open subset* (under the Zariski topology) of the zero set of the system of polynomials, $f_1, f_2, \ldots, f_s$.

Following Wu, in this case, we will say that the corresponding geometric statement is *generically true*.

The Wu's algorithm is as follows: Assume that the input to the algorithm is an elementary geometry formula which is obtained after the theorem has been translated algebraically.

```
Wu's Algorithm:
Input:      Premises = F = {f₁ , ... , f_s} ⊆ k[x₁ , ... , x_n],
            Conclusion = g ∈ k[x₁ , ... , x_n].
Output:     Trivial, if the premises are contradictory;
            True, if the geometry statement is generically true.

1. Compute 𝒢 , a characteristic set of I ;
2. if 𝒢 = ⟨1⟩  then
3.    return Trivial;
2. elsif prem(g, 𝒢) = 0  then
3.    return True;
4. else
5.    return Unknown;
end{Wu's Algorithm}. □
```

In order to understand Wu's Algorithm we need to make the following observations:

- If $\mathscr{G} = \langle 1 \rangle$ then, since every $g_i \in \mathscr{G}$ also belongs to the ideal (F), we see that $(F) = (1)$ and that $\mathscr{Z}(F) = \emptyset$. In this case, the system of premises are inconsistent and the geometry statement is trivially true.

  Also observe that if the ideal $(F) = (1)$ then, by definition, its characteristic set must be $\langle 1 \rangle$. Thus, the algorithm always correctly detects a trivial geometry statement.

- Suppose $\mathscr{G} = \langle g_1, \ldots, g_r \rangle \neq \langle 1 \rangle$. If $r = \text{prem}(g, \mathscr{G}) = 0$ then

$$(8) \quad \text{in}(g_r)^{\alpha_r} \text{in}(g_{r-1})^{\alpha_{r-1}} \cdots \text{in}(g_1)^{\alpha_1} g = \sum_{i=1}^{r} q_i g_i \in (f_1, \ldots, f_s).$$

Thus at every point $p \in k^n$, at which f_i's (hence, g_i's) vanish, but not $\text{in}(g_i)$'s, we note that the conclusion polynomial g must also vanish. The clause

$$\text{in}(g_1) = 0 \vee \cdots \vee \text{in}(g_r) = 0$$

has been interpreted by Wu as the associated degeneracy condition for the original geometry statement. Thus, when Wu's Algorithm returns the value "True," the following holds:

$$(\forall p \in k^n) \quad [(f_1(p) = 0 \wedge \cdots \wedge f_s(p) = 0) \wedge$$
$$\neg(\text{in}(g_1)(p) = 0 \vee \cdots \vee \text{in}(g_r)(p) = 0) \quad \Rightarrow \quad g(p) = 0] .$$

Geometrically, we have

$$\mathcal{Z}(g) \supseteq \mathcal{Z}(f_1, \ldots, f_s) \setminus \bigcup_{i=1}^{r} \mathcal{Z}(\text{in}(g_i)),$$

that is, g vanishes on *some open set* (in the sense of Zariski topology) of $\mathcal{Z}(f_1, \ldots, f_s)$.

We also note that, in line 1 of Wu's Algorithm, one could have used an extended characteristic set instead of characteristic set and the argument would still hold with appropriate changes.

- One problem with Wu's Algorithm, as presented here, is that it is not complete, and in fact, it is only the first half of the algorithm developed by Wu. Nevertheless, just this portion of the algorithm has found many applications, as one can argue heuristically that the algorithm succeeds in vast majority of the cases [10].

  Wu's Algorithm, as presented here, guarantees that whenever the algorithm claims that a statement is generically true, it is indeed so; however, the converse does not hold. In fact, we may have a conclusion g that vanishes on some open set of the zero set of the premises, but Wu's Algorithm may fail to detect this without further *decomposition* of the characteristic set. We will say more about this in §7.

- Even the complete Wu's Algorithm has several hard-to-eliminate drawbacks. First it is unable to work with arbitrary fields (not algebraically closed, e.g., $\mathbb{R}$). If propositions about real geometry are investigated in this way, a false proposition will be rejected (because it is false also in the complex field), but sometimes a true theorem (over real closed field) may unfortunately be rejected. For example, consider the hypothesis $x^2 + y^2 = 0$ and the conclusion $y = 0$. This is of course true on $\mathbb{R}^2$ but false on $\mathbb{C}^2$.

- Lastly, the method is unable to handle inequalities. So geometric propositions involving "*internal*," "*external*," or "*between*" are not in the range of such a prover. In this sense, the so called "Wu geometry" is smaller in scope than the more general "Tarski geometry" which includes all the propositions that can be proved by real quantifier elimination algorithms [21].

However, in spite of many obvious shortcomings, Wu's Algorithm remains a powerful and efficient tool for a large class of geometry statements.

6.1. Complexity of Wu's Algorithm. The techniques of the previous section can be adapted to develop good sequential and parallel procedures for computing the pseudoremainder of a given polynomial f and hence, efficient implementation of Wu's Algorithm. We state the relevant lemmas and omit their proofs; the easy proofs can be found in [18].

LEMMA 6.1. QUANTITATIVE PSEUDO-DIVISION LEMMA.   *Consider two polynomials f and $g \in k[x_1, \ldots, x_n]$, where k is an arbitrary field, $\deg(f) \le d$ and $\deg(g) \le \delta$. If $r = \text{prem}(g, f)$ is the pseudoremainder of g with respect to f then $\deg(r) \le (d+1)(\delta+1) - 1$. Also, the pseudoremainder r can be computed in sequential time $O((d\delta + 1)^{O(n)})$ and parallel time $O(n^2 \log^2(d\delta + 1))$.* □

LEMMA 6.2. QUANTITATIVE GENERALIZED PSEUDO-DIVISION LEMMA.   *Consider an ascending set $\mathscr{F} = \langle f_1, \ldots, f_r \rangle \subseteq k[x_1, \ldots, x_n]$ and a polynomial $g \in k[x_1, \ldots, x_n]$, where k is an arbitrary field, $\deg(f_i) \le d$, for all $1 \le i \le s$, and $\deg(g) \le \delta$. If $g_0 = \text{prem}(g, \mathscr{F})$ is the generalized pseudoremainder of g with respect to $\mathscr{F}$ then $\deg(g_0) \le (d+1)^r(\delta + 1)$. Also, the generalized pseudoremainder g_0 can be computed in sequential time $O(\delta^{O(n)}(d+1)^{O(nr)})$ and parallel time $O(n^2 r^3 \log^2(d\delta + 1))$.* □

Combining all of the above results, we have

THEOREM 6.3. COMPLEXITY OF WU'S ALGORITHM.   *Consider a set of premises $F = \{f_1, \ldots, f_s\} \subseteq k[x_1, \ldots, x_n]$ and a conclusion $g \in k[x_1, \ldots, x_n]$, where k is an arbitrary field, $\deg(f_i) \le d$, for all $1 \le i \le s$, and $\deg(g) \le d$.*

Then the Wu's Algorithm *for geometric theorem proving takes $O(s^{O(n^2)}(d+1)^{O(n^4)})$ sequential time and $O(n^7 \log^2(s+d+1))$ parallel time.*

□

As an immediate corollary of the above results, we see that a geometric theorem (involving no inequalities) can be verified in $O(2^{O(C(P+C)^3)})$ sequential time and $O(C^2(P+C)^7)$ parallel time, where it is assumed that the geometric theorem is presented by P points in the plane, and C constructions involving straight lines and circles.

7. Concluding remarks

Although the concept characteristic set has proven to be quite useful in the realm of geometry, several extensions are necessary to improve its power: one such idea is that of an irreducible characteristic set.

DEFINITION 7.1. IRREDUCIBLE CHARACTERISTIC SET.   Let $I = (f_1, \ldots, f_s)$ be an l-dimensional ideal in $k[x_1, \ldots, x_n]$. Further assume that the first l variables, $x_1, \ldots, x_l$, are the independent variables, the last $r = (n - l)$ variables, $x_{l+1}, \ldots, x_n$, the dependent variables, and the ordering on the variables is

$$x_1 \prec \cdots \prec x_l \prec x_{l+1} \prec \cdots \prec x_n.$$

A characteristic set of I, $\mathscr{G} = \langle g_1, \ldots, g_r \rangle$, is said to be an *irreducible characteristic set* of I, if

$$g_1 = \text{ irreducible over } k_1 = k(x_1, \ldots, x_l),$$
$$g_2 = \text{ irreducible over } k_2 = QF(k_1[x_{l+1}]/(g_1)),$$
$$\vdots$$
$$g_j = \text{ irreducible over } k_j = QF(k_{j-1}[x_{l+j-1}]/(g_{j-1})),$$
$$\vdots$$
$$g_r = \text{ irreducible over } k_r = QF(k_{r-1}[x_{r-1}]/(g_{r-1})),$$

where QF denotes the quotient field of fractions over a domain.

The definition above is constructive, in the sense that the irreducibility of an ascending set can be tested algorithmically, since there are factorization algorithms over a field, and over the successive algebraic extensions of a given field. However, these algorithms have a very high intrinsic time complexity [13].

The significance of the notion of an irreducible characteristic set becomes clear from the following proposition whose proof, using the concept of a *generic point* of an irreducible variety, can be found in [34] or [36].

PROPOSITION 7.1. *Let I be an ideal, and $\mathscr{G}$, a characteristic set of I. Then*

$$\mathscr{G} \text{ is irreducible } \quad \Leftrightarrow \quad I \text{ is prime. } \quad \square$$

The idea of irreducible characteristic set extends the domain of applicability to a larger class of algebraic problems.

(i) **Test for deciding primality of an ideal:** This can be achieved by using the characteristic set algorithm together with the test for irreducibility of a univariate polynomial over an arbitrary field. The validity of the algorithm is a simple consequence of the preceding proposition.

(ii) **Test for deciding membership in a prime ideal:** Recall that if $I =$ prime and its characteristic set is $\mathscr{G}$, then

$$g \in I \quad \Leftrightarrow \quad \text{prem}(g, \mathscr{G}) = 0.$$

Thus membership problem for a prime ideal can be solved by simply computing a characteristic set and then using the pseudodivision algorithm.

(iii) For a given ideal I, by using the factorization algorithm, one can construct a sequence of *irreducible characteristic sets* $\mathscr{G}_1, \ldots, \mathscr{G}_k$

such that $\mathscr{Z}(I) = \mathscr{Z}(\mathscr{G}_1) \cup \cdots \cup \mathscr{Z}(\mathscr{G}_k)$. Thus,

$$g \in \sqrt{I} \quad \Leftrightarrow \quad (\forall\, i)\, [\mathrm{prem}(g, \mathscr{G}_i) = 0].$$

This provides a test for checking the membership in the radical of an ideal.

(iv) Lastly, using above *geometric decomposition*, one can construct a *complete scheme* for geometric theorem proving, where one decides if a formula of the kind

$$(\forall\, p \in k^n) \quad [\, (f_1(p) = 0 \wedge \cdots \wedge f_s(p) = 0) \wedge$$
$$\neg(\mathrm{in}(g_1)(p) = 0 \vee \cdots \vee \mathrm{in}(g_r)(p) = 0) \quad \Rightarrow \quad g(p) = 0\,],$$

is "generically true."

REFERENCES

1. Alfred V. Aho, John E. Hopcroft, and Jeffery D. Ullman, *The design and analysis of computer algorithms*, Addison-Wesley, Reading, MA, 1974.

2. D. Bayer and M. Stillman, *A criterion for detecting m-regularity*, Invent. Math. **87** (1987), 1–11.

3. _____, *On the complexity of computing syzygies*, J. Symbolic Comput. **6** (1988), 135–147.

4. David Bayer, *The division algorithm and the Hilbert scheme*, PhD thesis, Harvard University, Cambridge, MA, 1982.

5. Stuart J. Berkowitz, *On computing the determinant in small parallel time using a small number of processors*, Inform. Process. Lett. **18** (1984), 147–150.

6. D. Brownawell, *Bounds for the degree in the nullstellensatz*, Ann. Math. (2) **126** (1987), 577–591.

7. B. Buchberger, *Gröbner bases: An algorithmic method in polynomial ideal theory*, Recent Trends in Multidimensional System Theory, Reidel, Dordrecht, 1985.

8. G. Carrá and G. Gallo, *A procedure to prove geometrical statements*, Lecture Notes in Comput. Sci., vol. 365, 1986, pp. 141–150.

9. S. C. Chou, *Proving and discovering theorems in elementary geometries using Wu's method*, PhD thesis, Department of Math., Univ. of Texas at Austin, Austin, TX, 1985.

10. _____, *Mechanical geometry theorem proving*, Reidel, Kluwer Academic Publishers Group, Dordrecht, Boston, 1988.

11. Don Coppersmith and Shmuel Winograd, *Matrix multiplication via arithmetic progressions*, Proc. of the Nineteenth Annual ACM Sympos. on Theory of Computing, vols. 25–27, New York City, NY, 1987, pp. 1–6.

12. L. Csanky, *Fast parallel matrix inversion algorithms*, SIAM J. of Comput. **5** (1976) 618–623.

13. J. H. Davenport, Y. Siret, and E. Tournier, *Computer algebra*, Academic Press, NY, 1988.

14. A. Dickenstein, N. Fitchas, M. Giusti, and C. Sessa, *The membership problem for unmixed polynomial ideals is solvable in subexponential time*, preprint, 1989.

15. T. Dubé, B. Mishra, and C. K. Yap, *Complexity of Buchberger's algorithm for Gröbner bases*, Extended Abstract, 1986.

16. Thomas William Dubé, *Quantitative analysis problems in computer algebra: Gröbner bases and the nullstellensatz*, PhD thesis, Courant Institute of Mathematical Sciences, New York University, NY, January 1989.

17. G. Gallo, *La Dimostrazione Automatica in Geometria e Questioni di Complessita' Correlate*, Tesi di Dottorato, Catania, 1989.

18. G. Gallo and B. Mishra, *Efficient algorithms and bounds for Wu-Ritt characteristic sets*, Proceedings of MEGA 90: Meeting on Effective Methods in Algebraic Geometry, Castiglioncello, Livorno, Italy, April 1990.

19. M. Giusti, *Some effectivity problems in polynomial ideal theory*, Lecture Notes in Comput. Sci., vol. 174, Springer-Verlag, Berlin and New York, 1984, pp. 159–171.

20. J. Heintz, *Definability and fast quantifier elimination over algebraically closed fields*, Theoretical Comput. Sci. **24** (1983), 239–277.

21. J. Heintz, T. Recio, and M.-F. Roy, *Algorithms in real algebraic geometry and applications to computational geometry*, 1991 (in this volume).

22. Jiawei Hong, *Proving by example and gap theorem*, 27th Annual Sympos. on Foundations of Comput. Sci., IEEE Computer Society Press, Toronto, Ontario, Canada, 1986, pp. 107–116.

23. D. T. Huynh, *A superexponential lower bound for Gröbner bases and Church-Rosser commutative Thue systems*, Informat. and Control **86** (1986), 196–206.

24. D. Kapur, *Geometry theorem proving using Gröbner bases*, J. Symbolic Comput. **2** (1986), 399–412.

25. E. R. Kolchin, *Differential algebra and algebraic groups*, Academic Press, NY, 1973.

26. János Kollár, *Sharp effective nullstellensatz*, preprint, 1988.

27. B. Kutzler and S. Stifter, *Automated geometry theorem proving using Buchberger's algorithm*, Proc. of the 1986 Sympos. on Symbolic and Algebraic Comput., ACM, Waterloo, Canada, 1986, pp. 209–214.

28. D. Lazard, *Gröbner bases, Gaussian elimination and resolution of systems of algebraic equations*, Lecture Notes in Comput. Sci., vol. 162, Springer-Verlag, Berlin and New York, 1983, pp. 146–157.

29. E. W. Mayr and A. R. Meyer, *The complexity of the word problems for commutative semigroups and polynomial ideals*, Adv. in Math. **46** (1982), 305–329.

30. Bhubaneswar Mishra, *Algorithmic algebra*, Texts & Monographs Comput. Sci., Springer-Verlag, Berlin and New York (to appear).

31. Bud Mishra and Chee Yap, *Notes on Gröbner bases*, Informat. Sci. **48** (1989), 219–252.

32. H. M. Möller and F. Mora, *Upper and lower bounds for the degree of Gröbner bases*, Lecture Notes in Comput. Sci., vol. 174, Springer-Verlag, Berlin and New York, 1984, pp. 172–183.

33. K. Mulmuley, *A fast parallel algorithm to compute the rank of a matrix over an arbitrary field*, Combinatorica **7** (1987), 101–104.

34. J. F. Ritt, *Differential algebra*, Amer. Math. Soc., Providence, RI, 1950.

35. Wu Wen-Tsün, *On the decision problem and the mechanization of theorem proving in elementary geometry*, Scientia Sinica **21** (1978), 157–179.

36. _____, *Basic principles of mechanical theorem proving in geometries*, J. Systems. Sci. Math. Sci. **4** (1984), 207–235; also in J. of Automated Reasoning **2** (1986), 221–252.

37. ___, *Some recent advances in mechanical theorem-proving of geometries*, Automated Theorem Proving: After 25 Years, Contemp. Math., vol. 29, Amer. Math. Soc., Providence, RI, 1984, pp. 235–242.

38. Chee-keng Yap, *A double-exponential lower bound for degree-compatible Gröbner bases*, Report 181, NYU-Courant Robotics Laboratory, NY, 1988.

COURANT INSTITUTE, NEW YORK UNIVERSITY, NEW YORK, NEW YORK 10012

DIMACS Series in Discrete Mathematics
and Theoretical Computer Science
Volume **6**, 1991

Algorithms in Real Algebraic Geometry
and Applications to Computational Geometry

JOOS HEINTZ, TOMAS RECIO, AND MARIE-FRANÇOISE ROY

Introduction

Real algebraic geometry has had important theoretical developments in the last ten years. In the same period, its connections with various algorithmic issues has become progressively more evident.

We think that enough significant results have been obtained, and enough general ideas developed, that it is time to attempt a first general overview of these connections.

In the first section of this survey we will consider a list of important basic results in real algebraic geometry, which are connected in various ways to algebraic and geometric algorithms. We shall consider successively univariate polynomials, semi-algebraic sets, differential geometry, and real algebra. For each result we shall indicate briefly algorithmic issues and applications. We shall develop two particular examples: how to solve the piano mover's problem and how to obtain some lower bounds in computational geometry.

In the second section, we concentrate on the effectiveness and complexity of classical problems, from quantifier elimination to the piano mover's problem. We shall use a number of results quoted in §1. These problems have been solved with doubly exponential complexity using Collins' cylindrical algebraical decomposition method. A number of singly exponential results (with good parallelization) have been obtained recently. We shall present a theorem which is the key to a unified treatment of these results and then explain how to use it for a list of various problems, including the complexity of constructing Whitney stratifications. Important differences between algebraic and numerical complexity for these problems are explained at the end of this section. We conclude with a philosophical discussion of singly exponential

1980 *Mathematics Subject Classification* (1985 *Revision*). Primary 14P10.

versus doubly exponential complexity.

1. Basic results in real algebraic geometry

We shall consider the following subjects: basic results on the real roots of univariate polynomials, stability and finiteness properties of semi-algebraic sets, fundamental concepts in differential geometry, and new developments in real algebra; and we shall try to indicate very briefly how each subject is related to various algorithmic issues.

The field of real numbers is denoted by $\mathbf{R}$.

1.1. Polynomials in one variable

Sturm-Sylvester theorem. Sturm's theorem is the basic algorithm for computing exactly the number of real roots of a polynomial [St].

The following generalization of Sturm's theorem due to Sylvester [Sy] has been forgotten and rediscovered a number of times in the last 150 years.

DEFINITIONS. Let P and Q be two univariate polynomials with coefficients in $\mathbf{R}$.

The *Sturm sequence* of P and Q is defined as follows:

$$P_0 = P,$$

$P_1 = R$, the remainder of the euclidean division of $P'Q$ by P,

P_{i+1} is the negative of the remainder of P_{i-1} divided by P_i.

We denote by $V(P, Q, -\infty)$ (resp. $V(P, Q, +\infty)$) the number of sign variations in the Sturm sequence of P and Q at $-\infty$ (resp. $+\infty$), and we define

$$V(P, Q) = V(P, Q, -\infty) - V(P, Q, +\infty).$$

We denote by

$\quad r(P)$ the number of real roots of P,

$\quad r_{>0}(P, Q)$ the number of real roots of P with $Q > 0$,

$\quad r_{<0}(P, Q)$ the number of real roots of P with $Q < 0$,

$\quad r_{=0}(P, Q)$ the number of real roots of P with $Q = 0$.

THEOREM 1. $r_{>0}(P, Q) - r_{<0}(P, Q) = V(P, Q)$. *In particular* $r(P) = V(P, 1)$.

COROLLARY.

$$\begin{bmatrix} 1 & 1 & 1 \\ 0 & 1 & -1 \\ 0 & 1 & 1 \end{bmatrix} \cdot \begin{bmatrix} r_{=0}(P, Q) \\ r_{>0}(P, Q) \\ r_{<0}(P, Q) \end{bmatrix} = \begin{bmatrix} V(P, 1) \\ V(P, Q) \\ V(P, Q^2) \end{bmatrix}.$$

This allows one to compute $r_{>0}(P, Q)$, $r_{<0}(P, Q)$ *and* $r_{=0}(P, Q)$.

Other methods for computing the number $r_{>0}(P, Q) - r_{<0}(P, Q)$ use the signature of a quadratic form [H1]. For a discussion and comparison of these different methods, see [GLomRR3].

Hermite's method was generalized to the bivariate case (by Hermite himself [H2]) and quite recently to the multivariate case [P].

Algorithmic issues and applications. The Sturm-Sylvester theorem is a basis for formal computations with inequalities [BKR, RS]. It is a key tool for obtaining quantifier elimination methods valid in any real closed field as indicated in §2.

Thom's lemma. We refer to any one of the strict order relations > 0, $= 0$, < 0 as *sign conditions*, and we refer to the superset of these which includes the relations ≥ 0 and $\leq$ as *generalized sign conditions*.

If $\epsilon = (\epsilon(0), \dots, \epsilon(d-1))$ is a d-tuple of generalized sign conditions, we denote by $\bar{\epsilon}$ the d-tuple obtained by relaxing the strict inequalities of ϵ, that is by replacing > 0 (resp. < 0) by ≥ 0 (resp. ≤ 0).

In general, a set defined by a list of univariate polynomial inequalities is not connected. Its closure for the euclidean topology is not obtained by relaxing strict inequalites in its definition: isolated points may appear as in the following example. The closure of the set $\{x \in \mathbf{R} \mid x^3 - x^2 > 0\}$ is not $\{x \in \mathbf{R} \mid x^3 - x^2 \geq 0\}$ but $\{x \in \mathbf{R} \mid x^3 - x^2 \geq 0\}\backslash\{0\}$. This is no longer the case when one considers the family of a polynomial and all its derivatives.

THEOREM 2. *Let P be a polynomial of degree d with real coefficients, $P^0, P^1, \dots, P^{(d-1)}$ its derivatives (with the notation $P^0 = P$, $P^1 = P'$) and $\epsilon = (\epsilon(0), \dots, \epsilon(d-1))$ an d-tuple of generalized sign conditions.*

Let

$$A_\epsilon = \{x \in \mathbf{R} \mid \forall i \in \{0, \dots, d-1\}, \ P^{(i)}(x)\epsilon(i))\}.$$

Then A_ϵ is either empty or connected. If A_ϵ is nonempty, its closure is $A_{\bar{\epsilon}}$.

COROLLARY. *Let P be a polynomial of degree d with integer coefficients. Let x and x' be two elements of $\mathbf{R}$. Suppose the signs $\epsilon(i)$ and $\epsilon'(i)$ of $P^{(i)}(x)$ and $P^{(i)}(x')$, $i = 0, \dots, d-1$, are given. Then we have the following properties.*

If $\epsilon = \epsilon'$ and both $\epsilon(0)$ and $\epsilon'(0)$ are the relation $= 0$, then $x = x'$. That is to say, if all the derivatives at two roots satisfy the same sign conditions, then they must be equal.

Otherwise, let k be the smallest integer such that $\epsilon(d-k)$ and $\epsilon'(d-k)$ are different. Then $\epsilon(d-k+1) = \epsilon'(d-k+1)$ is different from $= 0$, and if $\epsilon(d-k+1) = \epsilon'(d-k+1)$ are > 0, then

$$x > x' \iff P^{(d-k)}(x) > P^{(d-k)}(x'),$$

whereas if $\epsilon(d-k+1) = \epsilon'(d-k+1)$ are < 0, then

$$x > x' \iff P^{(d-k)}(x) < P^{(d-k)}(x').$$

The proofs of these two results are very easy [BCR].

Algorithmic issues and applications. Many geometric algorithms would be more *robust*, i.e. they would not suffer the failures due to round-off errors inherent in floating point arithmetic, if the arithmetic were exact. To the extent that one can remain within the realm of algebraic numbers, Theorem 2 and its corollary provide the basis for doing exact NC computations with real algebraic numbers [**CR, CuLMPR**].

Thom's lemma is also used for the computation of the topology of real algebraic curves [**R**] and in automatic theorem proving involving inequalities [**Gu**].

Moreover, multivariate generalizations of Thom's lemma give algorithmical methods for determining the closure of particular semi-algebraic sets, and thus an explicit description of connected components of semi-algebraic sets ([**CR**]; see also [**SS**]).

Fewnomials. Descarte's rule [**D**] gives a bound on the number of real positive roots of a polynomial.

THEOREM 3. *For any polynomial P the sign-variations in the coefficients of P overcounts the number of positive real roots of P, including multiplicities.*

In particular, the sign-variations in the coefficients gives the exact number of positive real roots (with multiplicities) in the case when all the roots of the polynomial are real.

An important consequence of Descartes's rule is the following: fewnomials (that is polynomials with few monomials) have few real roots [**BR**].

The *additive complexity* of a polynomial is the minimal number of +- signs needed to write the polynomial. It is clearly bounded by the number of monomials, but is a more invariant quantity. The number of real roots is also bounded in terms of the additive complexity [**Kh1, Kh2, Ri2**].

Algorithmic issues and applications. Since fewnomials have few real roots, it would be nice to have algorithms for computing the exact number of real roots sensitive to the number of monomials, or the additive complexity. Unfortunately all known methods depend on the degree.

Descartes's rule can be also used to study the theory of pfaffian manifolds [**Kh1, Kh2, Ri2**]. In the multivariate case too, the number of connected components of semi-algebraic sets can be bounded in terms of the number of monomials or of the additive complexity [**BR, Ri3**].

Random polynomials. It seems to be a well known experimental result that real polynomials have few real roots. A fundamental probabilistic result making precise this idea is due to Kac [**K**].

THEOREM 4. *Let P be a random polynomial whose coefficients are independent random variables with standard normal distribution. Let N_n be the random variable that associates to P the number N_n of different real roots of P. The expected value of N_n, EN_n is assymptotically $(2/\pi)\log n$.*

. This theorem was later generalized in several directions for many natural random variables (see [**BS**]).

Algorithmic issues and applications. Using these results, it has been possible to start the study of the average complexity of basic univariate algorithms (see [**CuR**]).

The worst case complexity of many algorithms in real algebraic geometry is frequently exponential or even doubly exponential. This would appear to prohibit their usefulness in practice. However, because of the preceding results, there is hope than one may design algorithms that "on average" (under various assumptions about the distribution) may perform much better.

1.2. Semi-algebraic sets. An *algebraic set* V of $\mathbf{R}^n$ is a set defined by a polynomial equation. A *semi-algebraic set* S of $\mathbf{R}^n$ is a set defined by a boolean combination of polynomial inequalities.

Many important geometric constructions can be described by semi-algebraic sets, and these sets enjoy many remarkable finiteness properties.

Projection. It is clear that the projection of an algebraic set is in general not an algebraic set, and that in the description of the projection inequalities are needed (think of the projection of a circle on a line). The following important result means that polynomial inequalities are sufficient to describe the projections of (semi-)algebraic sets.

THEOREM 5. *The projection of a semi-algebraic $S \subset \mathbf{R}^{n+1}$ on $\mathbf{R}^n$ is a semi-algebraic set.*

A very elementary proof of this result, due to Hörmander [**Hö**], uses only basic results about univariate polynomials (see [**BCR**]).

A major consequence of the projection theorem is the *Tarski-Seidenberg principle*. Before stating it, it is necessary to make precise the definition of formulae.

A *formula of the language of ordered fields with parameters in a ring* **A** is one obtained from polynomial inequalities with coefficients in **A** using the following logical symbols : $\vee$ (disjunction), $\wedge$ (conjunction), $\neg$ (negation), $\exists$, $\forall$ (existential and universal quantifications) over variables appearing in the polynomials.

For example, a semi-algebraic set is described by a quantifier-free formula, whereas the closure of a semi-algebraic set is described by a formula. It is possible to quantify over polynomials of fixed degree d, since they are given by $(d+1)$-tuples of coefficients. Quantification over integers, polynomials of arbitrary degrees, and semi-algebraic sets are not allowed.

We have then, as a corollary of the projection theorem, the famous quantifier elimination theorem, called the Tarski-Seidenberg principle:

THEOREM 6 (Tarski-Seidenberg principle, or quantifier elimination). *A subset of $\mathbf{R}^n$ defined by a first-order formula of the language of ordered fields with coefficients in $\mathbf{R}$ is semi-algebraic.*

The proof is immediate from the projection theorem, by induction on the number of quantifiers. An existential quantifier corresponds to a projection, and $\forall$ is equivalent to $\neg \exists \neg$.

Algorithmic issues and applications. The closure, interior, and boundary of a semi-algebraic set are semi-algebraic. The distance function between two semi-algebraic sets is a semi-algebraic function (that is, its graph is semi-algebraic). Many important geometric constructions—i.e., critical points of semi-algebraic functions, limits of tangent planes, singular locus, etc.—can be described by formulas of the language of ordered fields.

As soon as results on the complexity of a quantifier elimination method are known, results on the complexity of these various geometric constructions are easy to deduce.

Connected components. In general, irreducible real algebraic sets may have several connected components, which marks a significant difference from the algebraically closed case. These connected components are no longer algebraic sets, but rather semi-algebraic sets, according to the following theorem.

THEOREM 7. *The connected components of a semi-algebraic set are finite in number and are semi-algebraic sets.*

For a proof, see [**BCR**].

Algorithmic issues and applications. The fact that semi-algebraic sets have more than one connected component is the reason why the negative answer to the isotopy conjecture for oriented configurations of points follows from Mnëv's universality theorem (see [**Mn**] or [**GP2**]).

The most famous application of the preceding theorem is the *"piano mover's problem."*

Let us consider the following problem: Suppose you are given a room with doors and other openings and an object (a robot, often called "the piano"); you are to determine if there is a permissible rigid motion of the piano from its initial position p_0 inside the room to a desired final position p_1 outside the room, while the piano avoids the obstacles, defined by the walls of the room [**SS**] (cf. Figure 1).

The set O of obstacles and the piano P are presumed to be semi-algebraic subsets of $\mathbf{R}^n$.

Let C denote the configuration space and F the free space, that is, the subset of configurations such that the piano does not run across the walls.

It is not difficult to see that the set C can be represented as a semi-algebraic subset of some $\mathbf{R}^N$. If $n = 2$, that is, if the problem is planar, a frame is fixed to the piano and the positions of the piano are given by the image of this frame. Two coordinates are needed to determine the origin of the frame and it is possible to add two more coordinates (for example a rotation matrix) to determine the position completely. The group of planar isometries is of dimension 3, and is a semi-algebraic subset of $\mathbf{R}^4$. So one can take $N = 4$.

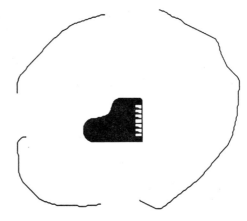

FIGURE 1

If p is a position, that is, an element of C, we shall denote by P_p the subset of $\mathbf{R}^n$ corresponding to the piano in the position p. Since P is semi-algebraic, P_p is semi-algebraic too. We have $F = \{p \in C \mid P_p \cap O = \varnothing\}$.

This set, defined by a first-order formula of the language of ordered fields, is therefore semi-algebraic, and a quantifier elimination algorithm will give an explicit description of it.

Given two allowed positions p_0 and p_1, the problem we started with reduces to the following: Do p_0 and p_1 belong to the same connected component of F? If they do, the piano can be moved from p_0 to p_1, and if not, such a motion is impossible.

So we are led to consider the following general problem: Given a semi-algebraic set S explicitly described, decide whether two elements x_0 and x_1 of S belong to the same connected component of S. If they do, explicitly construct a path between them lying in S.

Semi-algebraic triviality theorem. The following theorem is a fundamental tool for obtaining various finiteness results.

THEOREM 8. *Let F be a semi-algebraic family of semi-algebraic sets indexed by S (that is, F is a semi-algebraic subset of $\mathbf{R}^{n+k}$, with $\pi(F) \subset S$ where π is the projection of $\mathbf{R}^{n+k}$ on $\mathbf{R}^n$). Then there is a finite semi-algebraic partition (S_i) of S such that the fibers F_t of F (that is $\{x \in F \mid \pi(x) = t\}$) are homeomorphic for all t in S_i.*

The proof of this result can be found in [**BCR**]. It is a consequence of another important finiteness result: every semi-algebraic set admits a finite triangulation (see also [**BCR**]).

Algorithmic issues and applications. As a consequence of this theorem, several objects of interest have only a finite number of topological types. For example, given a semi-algebraic family F indexed by S, the number of topological types of the Voronoi diagrams of the fibers of F is finite and the partition of S in sets where the topological type is fixed is semi-algebraic.

The algebraic sets of fixed degree form a semi-algebraic family and thus have a finite number of topological types.

Bounding the number of these types, counting them, or even more interesting, finding algorithms to describe them, are natural questions.

Hilbert's sixteenth problem asks what are the topological types of a curves of fixed degree in the real projective plane. The complete answer is known only up to degree 6 (see [**Gu, Wi, Ri1**]).

1.3. Differential geometry

Thom-Milnor's bounds on the number of connected components. We shall denote by $B_0(S)$ the *number of connected components* of a semi-algebraic set S.

Thom-Milnor results (and their various generalizations) state that the sum of all Betti numbers of a semi-algebraic set is polynomial in the number and the degree of the polynomials defining S, singly exponential in the number n of variables [**Mil2, T**].

We are interested here only in the induced bound on the number $B_0(S)$ (which is the zeroth Betti number). More precisely we have the following result:

THEOREM 9. *Let S be a semi-algebraic set of $\mathbf{R}^n$ defined by polynomials of degree at most d.*

(a) *If S is the zero set of one equation, then $B_0(S) \leq d^n$.*

(b) *If S is defined by one inequality, then $B_0(S) \leq d^n$.*

(c) *If S is the zero set of several equations, then $B_0(S) \leq (2d-1)^{n-1}d$.*

(d) *If S is defined by a conjunction of several equations and s strict inequalities, then $B_0(S) \leq (2sd+1)^n(sd+1)$.*

(e) *If S is defined by a conjunction of several equations and s strict or non-strict inequalities, then $B_0(S) \leq (2(s+1)d+1)^n((s+1)d+1)$.*

(f) *The total number of connected components of semi-algebraic sets defined by a conjunction of sign conditions on all polynomials of a given family F of s polynomials of degree d in n variables is bounded by $(\mathrm{O}(sd))^{n+1}$.*

The proofs of the results (a), (b), (c), (d), (e) can be found in [**BR, MPR, or BCR**]. The proof of (f) is a small variation on ideas in [**GV1**]. The basic idea of the proof of Theorem 9 is simple in the case of a compact smooth hypersurface. A well-chosen height function has a finite number of critical points, and at least two critical points on every connected component of the set. Hence, the number of connected components is bounded by the number of critical points divided by two. The number of critical points (which are defined by equations of degree bounded by d) can be bounded using Bezout's theorem. Various tricks (using for example small deformations) permit one to reduce the general cases to this case.

Algorithmic issues and applications. Thom-Milnor bounds are essential ingredients in the new complexity bounds for quantifier elimination and other related problems (see §2).

Thom-Milnor bounds have natural applications in computational geometry, such as the counting of polytopes, order types, realizable oriented matroids, arrangements of hyperplanes [**Al, GP1**].

Another important application of Thom-Milnor bounds is *lower bounds* in computational geometry [**SY, B, MPR**]. Now, let us consider this last subject in more detail.

The idea is to translate a problem in computational geometry into a "*membership problem for a semi-algebraic set* S," (i.e., does a point belongs to some semi-algebraic set S?), and then to make precise the following intuitive idea: a semi-algebraic set which is geometrically complicated, for example with many connected components, cannot be described by a simple algorithm. The algorithms considered here for solving a given semi-algebraic membership problem are described by algebraic computation trees (for a definition, see [**B**]).

THEOREM 10. *Any algebraic computation tree tree solving the "membership problem for the semi-algebraic set S" of* $\mathbf{R}^n$ *has a depth which is in the class* $\Omega(\log_2 B_0(S) - n)$.

The proof of this theorem can be found in [**B**]. The ideas are very similar to more classical results of [**Str1, Str2**].

For example, let us consider the problem called "*element uniqueness*": given n reals, decide whether two of them are equal. The associated semi-algebraic set is

$$\bigcup_{\sigma \in S_n} \{(x_1, \dots, x_n) \in \mathbf{R}^n \mid x_{\sigma(1)} < \dots < x_{\sigma(n)}\}.$$

It has $n!$ connected components. Hence, using the preceding theorem, we get an $\Omega(n\log(n))$ lower bound for the depth of any algebraic computation tree solving "*element uniqueness*." Here we count all arithmetic and comparisons operations. Naturally, upper bounds of the same complexity are classical for this problem (sorting algorithms).

This method has recently been generalized to other problems [**MPR**] where the associated semi-algebraic set has few connected components but nevertheless is still geometrically complicated, because, for example, its intersection with a polynomial of low degree has many connected components. In order to adapt to this new situation it is necessary to consider the nonscalar complexity, i.e. to count only the number of multiplications and of comparisons, and to re-prove Theorem 10 with this new notion of complexity. As a corollary we get lower bounds for the depth of algebraic computation trees solving the membership problem for sets having a large number of connected components when intersected with the zero set of an equation of low degree.

New lower bounds can be obtained by this method [**MPR**]. For example, a lower bound for "*largest empty circle*" is in $\Omega(n\log(n))$.

Let us indicate what the "*largest empty circle*" problem is and how to prove this lower bound.

The "*largest empty circle*" problem is the following: Given n points in the plane, find the largest circle whose center is in the convex hull of the points and which does not contain any of the points in its interior.

We shall consider two other related problems:

"*Maximum gap*": Given n points on the real line line, find the maximum gap between two consecutive points.

"*Even distribution*": Given n points on the real line decide if they are evenly distributed with spacing 1, that is, if there exists a permutation σ of the points such that for all $i = 1, \ldots, n-1$, $0 < x_{\sigma(i+1)} - x_{\sigma(i)} < 1$.

It is easy to see that "*even distribution*" $\Leftarrow$ "*maximum gap*" $\Leftarrow$ "*largest empty circle*" (where $A \Leftarrow B$ means that problem A is easy to solve when a solution to problem B is known; so a lower bound for A is a lower bound for B).

The semi-algebraic set associated to "*even distribution*" is

$$S = \bigcup_{\sigma \in S_n} \{x_1, \ldots, x_n \in \mathbf{R}^n \mid \forall i = 1, \ldots, n-1, 0 < x_{\sigma(i+1)} - x_{\sigma(i)} < 1\}$$

which is connected, so that Theorem 10 does not apply directly. But the intersection of this semi-algebraic set with some ellipsoid does give many connected components, and we can apply the modified Theorem 10 and its corollary.

Morse theory. This theory permits the reconstruction of the topology of a compact hypersurface by the consideration of a quadratic form in the neighbourhood of a finite number of critical points of a well-chosen function [**Mil1**].

Algorithmic issues and applications. The existence and construction of Morse functions is a tool in the recent proofs of quantifier elimination with optimal complexity. It is hoped that a semi-algebraic Morse theory with singly exponential complexity will be developed in the future. For the moment, only the local aspects of Morse theory are known with good computational complexity. The global aspects involve vector fields and are not of semi-algebraic nature. Such a theory should lead to effective triangulation results for smooth hypersurfaces with a singly exponential number of simplices, that is, to a good algorithmic description of the topology of a smooth object.

Stratifications. It is not the case that all algebraic or semi-algebraic sets are smooth. This is the reason why one considers stratified sets.

A *stratification* of a semi-algebraic set is a finite decomposition of the set into semi-algebraic differential varieties, such that the closure of a stratum is obtained by taking the union of strata of smaller dimensions.

A typical constructions of a stratification for an algebraic set consists in taking the singular locus, then the singular locus of the singular locus, etc., all of which are algebraic sets. The strata are then the various connected components of the subsets of regular points of these algebraic sets.

A Whitney stratification of a semi-algebraic set is a stratification with nice differential control (limits of tangent planes ...) when passing from one stratum to another stratum of smaller dimension contained in its closure (see for example [BCR]).

More generally, several variations on the notions of stratification have been studied (see for example [Bek]).

One interesting feature of such a construction is that in the neigbourhood of a small stratum the topological structure of bigger strata is a product (topological triviality).

In the semi-algebraic setting all these notions of stratification lead to semi-algebraic strata. The essential reason for this is that strata are described by first-order formulas.

ALGORITHMIC ISSUES AND APPLICATIONS. The problem of the complexity of the construction of such stratifications is interesting in itself (see next section).

Whitney stratifications were first used by Canny for his solution to the piano mover's problem [Ca1]. In fact, Whitney stratifications are not at all essential in this problem and more general results were obtained later without any use of stratifications [GV2, HRS3, HRS4, GHRSV, Go].

One may hope that the construction of Whitney stratifications is a first step towards good bounds on the computation of the topology of general semi-algebraic sets (for example on the number of simplices of a triangulation).

1.4. Real algebra

Real closed fields. A *real closed field* $\mathbf{R}$ is a real field, i.e. where -1 is not a sum of squares, admitting a unique ordering with positive cone the squares of $\mathbf{R}$, and such that every polynomial in $\mathbf{R}[X]$ of odd degree has a root in $\mathbf{R}$.

Every ordered field admits a unique *real closure*, that is an algebraic extension which is real closed.

For example, the field of real numbers is a real closed field, as is the field of *real algebraic numbers* (real numbers satisfying an equation with integer coefficients). The field of real algebraic numbers is the real closure of $\mathbf{Q}$.

Real closed fields are not necessarily archimedean, as we shall see in the next section.

A major goal of modern real algebraic geometry has been to extend most geometric results to the general context of arbitrary real closed fields [BCR], but some care is required to achieve this goal. A real closed field admits a unique ordering, and semi-algebraic sets over a real closed field are equipped with the generalization of the euclidean topology. In order to get a good notion of connected component it is necessary to consider semi-algebraically connected sets, that is semi-algebraic sets which are not the disjoint union of open semi-algebraic subsets. With this definition the results described in the beginning of §1 are still valid (see [**BCR**]) (with the exception of Morse theory, which needs further care, since integration of vector fields is not available over an arbitrary real closed field).

An important result in the theory of real closed fields is the Transfer Principle which is a corollary of Theorem 5 generalized to arbitrary real closed fields.

THEOREM 11. *Let* **R** *and* **R**′ *be two real closed fields with* **R** ⊂ **R**′. *A formula* Φ *of the language of ordered fields with parameters in* **R** *is true in* **R**′ *if and and only if it is true in* **R**.

Particular cases of real closed fields with a geometrical meaning are real (algebraic) numbers and non-archimedean real closed fields which contain fields of rational functions. In the next section we develop the important case of Puiseux series.

Algorithmic issues and applications. Algorithms which work for arbitrary real closed fields should use very basic algebraic subroutines. This allows the algorithmic study of rings embedded in arbitrary real closed fields, with oracles to give arithmetic operations and signs of elements (see [**LomR, HRS2**]).

Puiseux series. Let us consider the following fundamental example of a non-archimedean real closed field which plays an important role in geometry and computer algebra.

Let **R** be a real closed field. If X is a variable, one denotes by $\mathbf{R}(X)\widehat{}$ the *field of Puiseux series in* X *with coefficients in* **R**. It consists of the element 0 and the series

$$\sum_{i \geq i_0, i \in \mathbf{Z}} a_i X^{i/q}$$

with $i_0 \in \mathbf{Z}$, $a_i \in \mathbf{R}$, $a_{i_0} \neq 0$ and $q \in \mathbf{N}$ [**Wa**].

The motivation behind this definition is the following. In order to construct a real closed field containing $\mathbf{R}(X)$ it is necessary to consider rational exponents. For example, in order to solve the equation $Y^3 - X^2 = 0$ (a cusp) one needs $X^{2/3}$. The following result says that the consideration of rational exponents is enough to ensure that the field be real closed.

THEOREM 12. *The field* $\mathbf{R}(X)\widehat{}$ *is real closed. The element* X *is infinitely small and positive (positive and smaller than any positive elements of* **R***). Positive elements are elements with positive coefficient on their lowest degree term.*

For a proof see [**Wa**] or [**LiR**].

The field of Puiseux series $\mathbf{R}(X)\widehat{}$ admits as a subfield the field of rational functions $\mathbf{R}(X)$ equipped with the order 0_+ (the sign of a nonzero rational function is given by its sign to the right of 0).

The real closure of $\mathbf{R}(X)$ equipped with the order 0_+ is the field of *algebraic Puiseux series*, that is, the subfield of elements of $\mathbf{R}(X)\widehat{}$ satisfying an algebraic equation with coefficients in $\mathbf{R}[X]$. They can be interpreted geometrically as the half-branches of algebraic curves above a small open interval to the right of 0.

Algorithmic issues and applications. Puiseux series will provide in §2 the required tool for exact formal computations when small deformations are used.

Puiseux series are a natural ingredient in the computation of analytic branches of real curves [**CuPRRR**].

Bröcker-Scheiderer results. A *basic open semi-algebraic set* is a semi-algebraic set defined by a conjunction of strict polynomial inequalities. It is intuitively clear (and not too difficult to prove) that in general, basic open semi-algebraic sets in $\mathbf{R}^n$ cannot be defined by less than n inequalities. (Take the example of the generalized octant $\{(x_1, \ldots, x_n) \in \mathbf{R}^n \mid x_1 > 0, \ldots, x_n > 0\}$.)

It is a remarkable fact that n inequalities are always sufficient.

THEOREM 13. *A basic open semi-algebraic set in $\mathbf{R}^n$ can be defined by n strict inequalities.*

This means, for example, that the interior of a polygon in the plane with an arbitrary number of vertices can be defined by two equations (not linear ones).

The statement of the result is quite geometrical, but its proof in fact relies on deep algebraic results (including Pfister's theory of multiplicative forms) [**Br, Sc, Ma**].

For the moment, this result is only partially algorithmic (see [**Bu**].

Algorithmic issues and application. Some lower bounds in computational geometry have been obtained using Bröcker-Scheiderer's results [**PR**].

Real effective Nullstellensatz. The problem is the following: how to construct an algebraic characterization of the fact that a polynomial f vanishes at the common (real) zeros of a set of polynomials $f_1, \ldots, f_s$?

For complex zeroes, the Hilbert's Nullstellensatz states that then there exists an integer k and polynomials $A_1, \ldots, A_s$ such that

$$f^k = A_1 f_1 + \cdots + A_s f_s.$$

In the complex case, it has been known for a long time [**He**] that the result is constructive. Several recent papers [**Bro, CGH, Ko**] have proved that f^k and $A_1, \ldots, A_s$ can be chosen with degrees singly exponential in the number of variables.

The answer in the real case is given by the following real Nullstellensatz.

THEOREM 14. *There exists an integer k and polynomials $A_1, \ldots, A_s$, $g_1, \ldots, g_t$ such that*

$$f^{2k} + g_1^{\,2} + \cdots + g_t^{\,2} = A_1 f_1 + \cdots + A_s f_s.$$

This theorem leads naturally to the notion of *real radical* of an ideal I, that is, polynomials f such that there exists an integer k and a sum of squares of polynomials s such that $f^{2k} + s \in I$.

The classical proof of this result involves Zorn's lemma.

It is only recently that an algorithmic proof of a general result implying the real Nullstellensatz and various Positivstellensätze has been obtained [**Lom1**], leading to hyperexponential complexity bounds [**Lom2**].

No clear idea of the complexity bounds that can be hoped for exists at present.

Algorithmic issues and applications. The main issue is to understand the algorithmic complexity of real algebra versus the algorithmic complexity of real geometry.

The real Nullstellensatz gives a decision method for the membership problem to the real radical. A more difficult related problem is the calculation of generators of the real radical.

2. Algorithms in real algebraic geometry

2.1. Preliminaries. The first important result is the Tarski-Seidenberg principle [**Ta, Se**], which says that the elimination of quantifiers is computable, with primitive recursive, hyper-exponential complexity. Moreover, Tarski's method is valid for any real closed field.

Collins defined his CAD (Cylindrical Algebraical Decomposition) method and, using modern computer algebra (particularly the theory of subresultants), showed that the elimination of quantifiers is polynomial in most parameters (degree, size of coefficients, and number of equations), and doubly exponential in the number of variables. Results on lower bounds [**We, DH**] prove that this is the best that can be hoped for when considering these complexity parameters. The same results hold for topological problems ([**SS**] or [**CR**]).

It is only recently that singly exponential complexity results have been obtained, first in the algebraically closed case (starting with [**CG**], based on [**H**]), then in the real case (detailed references will be given in §2.5). In the real closed case, the first step was to obtain a singly exponential algorithm to test whether a semi-algebraic set is empty or not (the formula for this problem is particularly simple and involves only existential quantifiers). The second step was to introduce a new complexity parameter, the number of alternating blocks of quantifiers in the prefix of a prenex formula, and to prove that quantifier elimination is doubly exponential in the number of blocks. In the mathematically significant situations, this number of alternations is fixed and small, and the complexity of these quantifier elimination methods implies a number of singly exponential results. In the general situation, the number of alternations of quantifiers can be close to the number of variables, and the lower bounds known for CAD are still valid.

Another interesting aspect of the new methods is the parallelization they allow. We shall say that an algorithm which runs in sequential time $f(n)$ is *well-parallelizable* if it has parallel complexity $(\log(f(n)))^{O(1)}$ and uses $(f(n))^{O(1)}$ processors. (For a definition of parallel complexity, see [**Ga**].) As soon as all the subroutines used in the algorithms are based on linear algebra, good parallelization is available [**Be**].

2.2. Several algorithmic problems. Throughout the sequel, let us denote by S a semi-algebraic subset of $\mathbf{R}^n$ defined by a boolean combination of polynomial inequalities with coefficients in a ring $\mathbf{A}$ contained in the real closed field $\mathbf{R}$ (so that $\mathbf{A}$ is ordered by the order induced by $\mathbf{R}$). Typically $\mathbf{A}$ is the ring of integers,and $\mathbf{R}$ is the field of real numbers. We denote by n the number of variables of these polynomials and by D the sum of their degrees.

We shall study the complexity of the algorithms considered, that is, the number of arithmetic operations over $\mathbf{A}$ and the comparisons of elements of $\mathbf{A}$, as a function of the two parameters n and D.

When we consider a first-order formula of the language of real closed fields we shall introduce the parameter m counting the number of alternation of quantifiers in the formula.

Here are the problems that we want to study:

(1) (test of emptyness) Decide whether a semi-algebraic set S is empty or not.

(2) (construction of representatives) Construct at least one point in every semi-algebraically connected component of S.

(3) (projection) Compute the projection of a semi-algebraic subset of $\mathbf{R}^n$ on $\mathbf{R}^{n-k}$.

(4) (decision) Decide whether a statement of the language of ordered fields is true or not.

(5) (quantifier elimination) Eliminate quantifiers in a formula of the language of ordered fields.

(6) (decision for connectivity) Decide whether two points of S belong to the same semi-algebraically connected component.

(7) (path construction) If possible, construct a connecting path between the two points.

(8) (description of connected components) Give an explicit semi-algebraic description of the connected components of S.

(9) (construction of Whitney stratifications) Describe explicitly a Whitney stratification of a semi-algebraic set.

(10) (triangulation) Give an explicit description of a semi-algebraic homeomorphim between a semi-algebraic set and a simplicial complex.

These problems have many relations, without being equivalent, e.g. (5) implies (1), (3), and (4); (1) is a consequence of (2), (3), and (4); (8) implies (6); (9) uses (5); etc.

We shall try to give an idea of the methods used to solve these problems and the complexity of these methods.

Let us remark that since n polynomials of degree d in n variables define, using Bezout's theorem, as many as d^n points, and since it is impossible to describe a single exponential number of objects in less that a single exponential time, the best we can hope for general methods solving these problems is a complexity singly exponential in n.

2.3. The univariate case. The first point to make is that the Sturm-Sylvester theory (see subsection 1.1) can be obtained using the theory of subresultants [**Lo, GLomRR2**], suitably modified to maintain sign information [**GLomRR1**] or [**GLomRR3**]. The subresultant algorithms use arithmetic operations in the ring of coefficients only (no fractions); run in time polynomial in the degree of polynomials, are well parallelizable, and have nice bounds on coefficent size in case the coefficient ring is the integers.

Starting from this modification of Sturm-Sylvester theory and using variants of [**BKR**], one can obtain the following result:

THEOREM 15. *Let $P, Q_1, \ldots, Q_s$ be polynomials in one variable with degrees less than or equal to d and coefficients in $\mathbf{A}$. There exists an algorithm over $\mathbf{A}$ with sequential complexity polynomial in d and s, well parallelizable, solving the problem*

$(*)$ "*determine at the roots of P in $\mathbf{R}$*
         the signs of the polynomials $Q_1, \ldots, Q_s$."

For a proof, see [**BKR, RS**].

Using Thom's lemma, its corollary, and the preceding Theorem 15 applied to the derivatives of P and to Q, one can prove that exact computations with real algebraic numbers are in NC.

THEOREM 16. *Let P and Q be polynomials in one variable with degrees less than or equal to d and coefficients in $\mathbf{A}$. There exists an algorithm over $\mathbf{A}$ with sequential complexity polynomial in d and s, well parallelizable, allowing one*

- *to characterize the roots of P in $\mathbf{R}$ by the signs they give to the derivatives of P,*
- *to order these roots,*
- *to decide for each of these roots the sign of Q.*

For a proof see [**RS, CuLMPR**].

The essential point to remember is that there are nonnumerical methods valid in a very general situation (in any real closed field) which solve basic univariate problems in polynomial sequential time and polylog parallel time.

2.4. Cylindrical algebraic decomposition. We shall now explain the classical method for solving problems (1)–(10) (enumerate under "Several algorithmic problems" above) and give an idea of its complexity.

The technique is the following [**Co**]: starting from a family of polynomials $P_1, \dots, P_s$ in n variables, compute a family of polynomials $Q_1, \dots, Q_{s'}$ in $n-1$ variables which define a semi-algebraic partition (T_i) of $\mathbf{R}^{n-1}$ such that on every connected component of T_i there is a cylindrical situation, i.e. for every i there is a finite number l_i of semi-algebraic continuous functions $(\xi_{i,j})_{j=1,\dots,l_i}$ with

$$\xi_{i,1}(x_1, \dots, x_{n-1}) < \cdots < \xi_{i,l_i}(x_1, \dots, x_{n-1}),$$

such that:

- For all $(x_1, \dots, x_{n-1})$ of T_i the set of the zeros of the nonidentically null P_j's over T_i is

$$\{\xi_{i,1}(x_1, \dots, x_{n-1}), \dots, \xi_{i,l_i}(x_1, \dots, x_{n-1})\}.$$

- For all $(x_1, \dots, x_{n-1}) \in T_i$, the sign of each $P_k(x_1, \dots, x_{n-1}, X_n)$ between two roots $\xi_{i,j}(x_1, \dots, x_{n-1})$ and $\xi_{i,j+1}(x_1, \dots, x_{n-1})$ is fixed.

The functions $\xi_{i,j}$ slice up the cylinder above T_i.

The procedure then iterates, starting from the new polynomials in $n-1$ variables and applying the same method, reducing by one dimension on each iteration.

EXAMPLES. (1) For the surface defined by the equation $Y = Z^3 - XZ$, there are three functions $\xi_{i,j}$ above T_1 and one function above T_2 (cf. Figure 2).

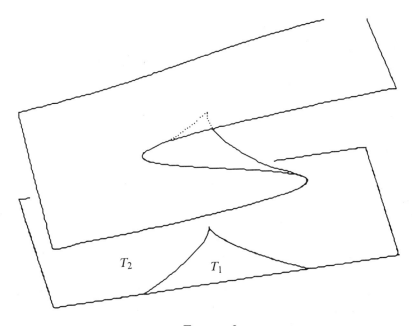

FIGURE 2

(2) This is the situation for the cubic with equation $Y^2 = X^3 - X$:

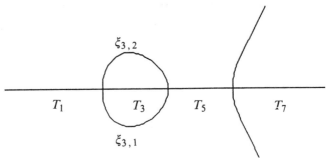

FIGURE 3

For a *plane curve* given by a polynomial $P(X, Y) = 0$, the construction eliminates the variable Y and gets as polynomial in one variable the discriminant $D(X)$. The T_i are then the zero set of D, and the set where D is different from 0. The connected components of the T_i are the roots of D and the intervals between two real roots of D. On each connected component of the T_i it is clear that it is possible to continuously follow the real roots of $P(x, Y)$ as functions of x, and this defines the $\xi_{i,j}$.

The *cylindrical algebraic decomposition method* consists of two phases

(a) a projection phase, where the variables are eliminated one-by-one,

(b) a going-up phase, where the sign conditions and cylinders are reconstructed progressively, going from the line to the plane, then at the end to $\mathbf{R}^n$.

In phase (a) elimination theory is used: resultants, subresultants, etc.

In phase (b) one starts from the univariate polynomials obtained at the end of phase (a) and characterizes their roots and the intervals between their roots using, for example (for the reals), a dichotomy algorithm based on Sturm's theorem. Sturm's theorem is then used in the fibers to go from dimension k to dimension $k + 1$.

Using several variants of cylindrical algebraic decomposition, it is possible to solve problems (1)–(10) ([**Co**], [**SS**], or [**C**]).

The complexity of calculating a cylindrical algebraic decomposition is polynomial in the degree d and the number s of input equations, and doubly exponential in n, the dimension. This complexity appears to be intrinsically related to the iterative method of eliminating variables.

Indeed, if one eliminates one variable between polynomials of degree d in n variables, one gets polynomials of degree d^2 in $n - 1$ variables. Eliminating a second time gives polynomials of degree d^4 in $n - 2$ variables, and at the end of the process one has polynomials in one variable of degree d^{2^n}. This doubly exponential behaviour is unavoidable, and it is possible to give doubly exponential lower bound result for quantifier elimination [**We, DH**].

2.5. The new techniques. Recent results based on the ideas appearing in Grigor'ev's and Vorobjov's papers [**G, GV1**], show that it is possible to obtain algorithms with complexity polynomial in d and s, and singly exponential in n for problems (1), (2), (3), (6), (7), and (8). Introducing a new complexity parameter, the number m of blocks of quantifiers, one can obtain a complexity doubly exponential in m for (4) (decision) and (5) (quantifier elimination). (See [**Ca1, Ca2, CaGV, G, GHRSV, GV1, GV2, HRS1, HRS2, HRS3, HRS4, HRS5, Re1, Re2**].)

The key geometric idea is the following: avoid cascading projections by working directly on the object using Morse functions with a finite number of critical points (that is, using the Thom-Milnor ideas sketched in subsection 1.3 in an algorithmic context), and then as soon as one has reduced to a one-dimensional problem to project directly on a line.

Let us take an example to illustrate this situation. If we consider a Morse function on the torus, we shall have to deal with only four critical points (Figure 4a). On the other hand, if we project it, we have a lot of extra points

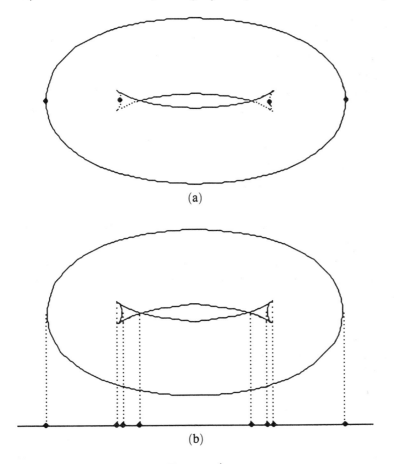

(a)

(b)

FIGURE 4

to study that have nothing to do with the original geometric situation and are created by the chosen projection (Figure 4b). In the first case, the number of points to consider will be single exponential, and in the second case it will be doubly exponential. The equation of the torus is

$$4X^2 + 2Y^2 + 2Z^2 - 4YZ - (X^2 + Y^2 + Z^2 + 3).$$

The Morse function we consider is the Y coordinate.

In order to make this nice and simple idea work efficiently it is necessary to develop some quite sophisticated algorithmic and methodological equipment, and the proofs become rather technical.

In particular, in order to apply Morse theory one should be in the smooth case, hence it is useful to use small deformations to remove singularities.

Since we are working in a symbolic context (the situation depends on parameters) these small deformations cannot be performed numerically, and it is necessary to use infinitesimal deformations, that is, to work in extensions of the ground field which are fields of Puiseux series. In fact, since all the computations are based on linear algebra subroutines and are made in the ring of coefficients, no implementation of Puiseux series is needed, and all the computations take place in various polynomial ground rings.

We indicate now how it is possible to obtain all these new results as a consequence of a unifying technical but fundamental result.

The first thing is to prove the following result solving problems (1) (emptyness test), and (2) (construction of representatives) in single exponential time.

THEOREM 17. *Let* **A** *be a domain contained in a real closed field* **R**. *Let* S *be a semi-algebraic subset of* $\mathbf{R}^n$ *defined by a boolean formula of length* L *over a family* F *of* s *polynomials* $F_1, \dots, F_s$ *with coefficients in* **A** *of total degrees* $\deg(F_1), \dots, \deg(F_s)$, *and* $D = \sum_{i=1, \dots, s} \deg(F_i)$. *Then there exists an algorithm over* **A** *of sequential complexity* $O(L)D^{n^{O(1)}}$, *well parallelizable, which decides whether* S *is empty or not. Moreover, if* S *is not empty, the algorithm constructs at least one point in every (semi-algebraically) connected component of* S, *and the coordinates of these points satisfy polynomial equations of degree* $D^{O(n)}$.

The steps of the proof [GV1, HRS2] are the following:

• Consider first the case of a smooth bounded hypersurface S. For such a hypersurface the following properties are used: given a Morse function on a smooth bounded hypersurface S, the set of critical points of S is finite and intersects every connected component of S. On the other hand, the degrees of the equations defining these critical points are bounded by the degree of the polynomial defining the hypersurface.

Hence, for a well-chosen Morse function the set of critical points is the zero set of n polynomials of degree at most d defining a 0-dimensional ideal. The coordinate ring of this 0-dimensional ideal is a finitely generated vector space of dimension singly exponential in n. Checking the existence of real points in the zero set of this ideal uses the algorithms in §2.3.

All the computations are reduced to linear algebra. The complexity is singly exponential in n, essentially because we are dealing with a vector space of dimension singly exponential in n.

Let us remark that the points obtained on each connected component are described through Thom's lemma by a polynomial and a list of signs on its derivatives. They are not obtained by numerical methods (which would not be available anyway in a general real closed field).

• Then the general case is reduced to the case of a smooth bounded hypersurface. This is done by a technique similar to the one used to prove Thom-Milnor bounds. An extension of the ground field is necessary.

This can be done by adding an infinitely large element Ω in order to reduce to the bounded case, then a first infinitesimal ϵ in order to study the hypersurfaces close to the boundary of the set, and lastly a second infinitesimal δ to reduce to the smooth case.

Now instead of computing in the ring $\mathbf{A}$ one has to calculate in the ring $\mathbf{A}[\Omega, \epsilon, \delta]$, and the associated real closed field is the composite of Puiseux series fields.

All the algorithms use only linear algebra subroutines and can be easily parametrized.

When one considers a parametric situation, where the sets to consider are the fibers S_t of a semi-algebraic set contained in $\mathbf{R}^n$, with parameters varying in $\mathbf{R}^k$, one is led to the following result:

THEOREM 18. *Let S be a semi-algebraic subset of* $\mathbf{R}^n$ *defined by s polynomials in n variables with coefficients in* $\mathbf{A}$ *such that the sum of their degrees is less than or equal to* D.

Then there exists an algorithm over $\mathbf{A}$ *with sequential complexity* $D^{n^{O(1)}}$, *well parallelizable, which constructs*

• *a partition of* $\mathbf{R}^k$ *into semi-algebraic subsets* T_i,

• *for all m, a finite family of continuous semi-algebraic functions from* T_i *into* $\mathbf{R}^{n-k}$ $(\xi_{i,j})_{j=1,\dots,l_i}$ *such that*

(a) *the graph of* $\xi_{i,j}$ *is contained in S*

(b) *for all* $y \in T_m$ *the graphs of the* $\xi_{i,j}$ *intersect every (semi-algebraically) connnected component of* $S \cap (\{y\} \times \mathbf{R}^{n-k})$.

The principle of the proof of the theorem [HRS2] is straightforward. The sets T_i's are those subsets of the parameter space for which the computations needed to test S_t for emptiness are identical.

This technical theorem will be the key for all the applications we have in mind.

Let us point out two important particular cases of this result:

• The case where $k = n - 1$. This is the classical cylindrical algebraic decomposition of S, since the fibers are then finite.

• The case where $k = 1$. The theorem then gives parametrized curves, allowing one to continuously follow points in the connected components of the fibers. This case will be used for the piano mover's problem (problem (7)).

Starting from the preceding theorem, a solution to problem (4) (quantifier elimination) with a complexity doubly exponential in the number of blocks of quantifiers is obtained by induction on the number of blocks of quantifiers [HRS2]. For a more precise statement, see [Re2].

Problem (7) (construction of semi-algebraic path) is first solved in the case of a smooth hypersurface using Theorem 18 in the case $k = 1$, constructing connecting curves in one direction, and then linking these curves through particular fibers, working by induction on the dimension of fibers [HRS3]. An important point is that the dimension of the fibers is smaller than the dimension of the set we started with, but the fibers are defined by polynomials of the same degree.

The general case uses the refined complexity bounds for quantifier elimination mentioned before, and small deformations (i.e. Puiseux series) techniques [HRS4]. See also another approach in [Go].

With this method it is possible to construct semi-algebraic paths described by polynomials with degrees singly exponential in n.

Using this solution to problem (7) it is possible to count the exact number of connected components by connecting the representative points obtained before (see Theorem 17) in each component.

The solution to problem (8) (description of connected components of semi-algebraic sets) is a parametric version of the solution to problem (7). Indeed, as soon as a point in every connected component has been constructed, a connected component will be the set of points connected to this given point by algorithm (7).

To get singly exponential complexity in this last problem requires some subtlety, in particular it is useful to preprocess some calculations [HRS5].

The solution to problem (9) (construction of a Whitney stratification) is now relatively easy. What is needed is to follow the classical geometric construction of stratifications, which is done by constructing strata of progressively smaller dimensions, and then proving that in each step of the construction the strata constructed are semi-algebraic and that the construction can be performed with single exponential complexity (mainly because all objects to construct are described by first-order formulas admitting a bounded number of quantifier alternations). Hence the construction is doubly exponential in the depth of the skeleton of the stratification (see [MR] for more details).

There is still no satisfactory result for problem (10), since no algorithmic triangulation method with less than a doubly exponential number of simplices is known for the moment.

2.6. Some remarks on numerical aspects. Consider the following problem: $(7')$ decide whether two points of S belong to the same connected component, and if the answer is yes, construct a piecewise linear path connecting them.

As opposed to problem (7), for which solutions polynomial in d and s (and singly exponential in n) were obtained, problem $(7')$ is intrinsically exponential in d, as shown by the following counterexample [**HKSS**].

One starts from the construction due to M. Mignotte [**Mi**] of a polynomial $P(X) = X^d - (3X - 1)^2$ which, for large d, has two real roots x_1 and x_2 very close to $1/3$, at distance less than $2 \times 3^{-(d+2/2)}$.

Consider now the equation $P(X^2 + Y^2) = 0$, whose zero set is the union of two circles with radii equal to the square roots of the real roots of P. One can easily find in the region between these two circles points such that the number of line segments linking them inside the region is necessarily exponential in d.

Hence it is essential, in order to achieve the desired complexity in problem (7), to use semi-algebraic paths and global algebraic nonnumerical methods.

2.7. Singly exponential versus doubly exponential complexity. All the preceding results lead to a general question: where is the boundary between singly and doubly exponential problems?

The general concept proposed by Michel Demazure for describing the situation is the following :

Algebra is doubly exponential;
geometry is singly exponential.

In the algebraically closed case, several results confirm this philosophy. The general membership problem to an ideal, a fundamental algebraic problem, is doubly exponential [**MM**], but membership to the radical of the ideal, which expresses a geometric condition on zero sets, is singly exponential [**Bro, CGH, Ko**]. The complexity of the equidimensional set theoretical decomposition is singly exponential [**GH**], etc. The complexity of quantifier elimination in the algebraically closed case is doubly exponential in the number of blocks of quantifiers [**FGM**], that is singly exponential as soon as the number of blocks of quantifiers is fixed , which is always the case in geometric situations that have been considered. General quantifier elimination in the algebraically closed case is doubly exponential.

In the real case, the natural geometric problems we have considered are singly exponential, or doubly exponential in the intrinsic geometrical difficulty of the problem (number of block of quantifiers or depth of the skeleton of the stratification). Again, general quantifier elimination is doubly exponential. But very little is known for the moment on the complexity of real algebra (real Nullstellensatz and related problems such as the real radical computation).

BIBLIOGRAPHY

[Al] N. Allon, *The number of polytopes, configurations and real matroids*, Mathematika **33** (1986), 62–71.

[Bek] K. Bekka, *Regular stratifications of subanalytic sets*, J. London Math. Soc. (to appear).

[BR] R. Benedetti and J.-J. Risler, *Real algebraic and semi-algebraic sets*, Hermann, Paris, 1990.

[B] M. Ben-Or, *Lower bounds for algebraic computation trees*, Proc. 15th ACM Annual Sympos. on Theory of Comp., 1982, pp. 80–86.

[BKR] M. Ben-Or, D. Kozen, and J. Reif, *The complexity of elementary algebra and geometry,*[1] J. Comput. Sys. Sci. **32** (1986), 251–264.

[Be] S. J. Berkowitz, *On computing the determinant in small parallel time using a small number of processors*, Inform. Process. Lett. **18** (1984), 147–150.

[BS] A. T. Bharucha-Reid and M. Sambandham, *Random polynomials*, Academic Press, New York, 1986.

[BCR] J. Bochnak, M. Coste, and M.-F. Roy, *Géométrie algébrique réelle*, Springer-Verlag, 1987.

[Br] L. Bröcker, *Characterizaion of basic semi-algebraic sets*, Münster University, Germany, 1987.

[Bro] D. Brownawell, *Bounds for the degrees in the Nullstellensatz*, Ann. of Math. (2) **126** (1987), 577–591.

[Bu] J. Buresi, *Calcul effectif d'une solution générique du théorème de Bröcker-Scheiderer*, Université de Rennes, France, preprint, 1990.

[Ca1] J. Canny, *Some algebraic and geometric computations in PSPACE*, ACM Sympos. on the Theory of Computation, 1988, pp. 460–467.

[Ca2] ——, *The complexity of robot motion planning*, MIT Press, Cambridge, MA, 1989.

[CGH] L. Caniglia, J. Heintz, and A. Galligo, *Some new effectivity bounds in computational geometry*, Lecture Notes in Comp. Sci., vol. 357, Springer-Verlag, Berlin, 1989, pp. 131–151.

[CaGV] J. Canny, D. Grigor'ev, and N. Vorobjov, *Describing connected components of semi-algebraic sets in subexponential time*, Preprint, Leningrad.

[CG] A. L. Chistov and D. Grigor'ev, *Complexity of quantifier elimination in the theory of algebraically closed fields*, Lecture Notes in Comp. Sci., vol. 199, Springer-Verlag, Berlin, 1984, pp. 63–69.

[Co] G. Collins, *Quantifier elimination for real closed fields by cylindric algebraic decomposition*, Lecture Notes in Comp. Sci., vol. 33, Springer-Verlag, Berlin, 1975, pp. 134–183.

[C] M. Coste, *Effective semi-algebraic geometry*, Lecture Notes in Comp. Sci., vol. 391, Springer-Verlag, Berlin, 1989, 1–27.

[CR] M. Coste and M.-F. Roy, *Thom's lemma, the coding of real algebraic numbers and the topology of semi-algebraic sets*, J. Symbolic Computation **5** (1988), 121–129.

[CuLMPR] F. Cucker, H. Lanneau, B. Mishra, P. Pedersen, and M.-F. Roy, *Real algebraic numbers are in NC*, Applicable Algebra in Engineering, Communication and Computing (to appear).

[CuPRRR] F. Cucker, L. M. Pardo, M. Raimodo, T Recio, and M.-F. Roy, *On local and global analytic branches of a real algebraic curve*, Lecture Notes in Comp. Sci., vol. 356, Springer-Verlag, Berlin, 1989, pp. 161–182.

[CuR] F. Cucker and M.-F. Roy, *Kac's theorem on the number of real roots and its consequences in average complexity*, J. Symbolic Comput. **5** (1990), 10.

[DH] J. Davenport and J. Heintz, *Real quantifier elimination is doubly exponential*, J. Symbolic **5** (1988), 29–35.

[D] R. Descartes, *Géométrie*, 1636, A Source Book in Mathematics, Harvard University Press, Cambridge, MA, 1969, pp. 90–93.

[FGM] N. Fitchas, A. Galligo, and J. Morgenstern, *Precise sequential and parallel complexity bounds for quantifier elimination over algebraically closed fields*, J. Pure Appl. Algebra **67** (1990), 1–14.

[Ga] J. von zur Gathen, *Parallel arithmetic computations: a survey*, Lecture Notes in Comp. Sci., vol. 233, Springer-Verlag, Berlin, 1986, pp. 93–112.

[GH] M. Giusti and J. Heintz, *Algorithmes—disons rapides—pour la décomposition d'une variété algébrique en composantes irréductibels et équidimensionnelles*, MEGA 1990 Proceedings, Birkhäuser, Bonn, 1991, pp. 169–196.

[GLomRR1] L. Gonzalez, H. Lombardi, T. Recio, and M.-F. Roy, *Sturm-Habicht sequences*, Proceedings ISSAC 1989, pp. 136–146.

[GLomRR2] ____, *Sous-résultants et spécialisation de la suite de Sturm* I, Informatique Théorique et Applications **24** (1990), 561–588.

[GLomRR3] ____, *Sous-résultants et spécialisation de la suite de Sturm* II, Informatique Théorique et Applications (to appear).

[GP1] J. E. Goodman and R. Pollack, *Upper bound for configuration of polytopes in* $\mathbf{R}^d$, Discrete and Comp. Geom. **1** (1986), 219–227.

[GP2] ____, *Allowable sequences and order types in discrete and computational geometry*, Recent Advances in Discrete and Computational Geometry (J. Pach ed.), Springer-Verlag, Berlin (to appear).

[Go] L. Gournay, *Construction of road-maps in semi-algebraic sets*, AAECC 1991 (to appear).

[G] D. Grigor'ev, *Complexity of deciding Tarski algebra*, J. Symbolic Comput. **5** (1988), 65–108.

[GHRSV] D. Grigor'ev, J. Heintz. M.-F. Roy, P. Solerno, and N. Vorobjov, *Comptage des composantes connexes des ensembles semi-algébriques en temps simplement exponentiel* C. R. Acad. Sci. Paris **313** (1990), 879–882.

[GV1] D. Grigor'ev and N. Vorobjov, *Solving systems of polynomial inequalities in subexponential time*, J. Symbolic Comput. **5** (1988), 37–64.

[GV2] ____, *Counting connected components of semi-algebraic sets in subexponential time*, Leningrad, preprint.

[Gud] D. A. Gudkov, *The topology of real algebraic varieties*, Russian Math. Surv. **29** (1974), 1–79.

[Gu] A. Guergueb, *Exemples de démonstration automatique en géométrie réelle*, Thèse, Université de Rennes, 1990.

[H] J. Heintz, *Definability and fast quantifier elimination over algebraically closed fields*, Theoret. Comput. Sci. **24** (1983), 239–277.

[HKSS] J. Heintz, T. Krick, A. Slissenko, and P. Solernó, *Le problème de la visibilité indirecte est intrinsèquement exponentiel dans le plan*, manuscript, Univ. of Buenos Aires, 1990.

[HRS1] J. Heintz, M.-F. Roy, and P. Solernó, *On the complexity of semialgebraic sets*, Proc. IFIP 89 San Francisco, North-Holland, Amsterdam, 1989, pp. 293–298.

[HRS2] ____, *Sur la complexité du principe de Tarski-Seidenberg*, Bull. Soc. Math. France **118** (1990), 101–126.

[HRS3] ____, *Single exponential path finding in semi-algebraic sets*. I, *The case of smooth compact hypersurface*, Proc. AAECC-8 1990, Tokyo (to appear).

[HRS4] ____, *Single exponential path finding in semi-algebraic sets*. II, *The general case*, Abhyankar's Conference Proceedings 1990 (to appear).

[HRS5] ____, *Description des composantes connexes d'un ensemble semi-algébrique en temps simplement exponentiel* C. R. Acad. Sci. Paris **313** (1991), 169–190.

[He] G. Hermann, *Die Frage der endlich vielen Schritten in der Theorie des Polynomideale*, Math. Ann. **96** (1926), 736–788.

[H1] C. Hermite, *Remarques sur le théorème de Sturm*, C. R. Acad. Sci. Paris **36** (1853), 52–54.

[H2] ____, *Sur l'extension du théorème de M. Sturm a un système d'équations simultanées*, Oeuvres de Charles Hermite, vol. III, pp. 1–34.

[Hö] L. Hörmander, *The analysis of partial differential operators*, vol. 2, Springer-Verlag, Berlin and New York, 1983.

[K] M. Kac, *On the average number of real roots of a random algebraic equation*, Bull. Amer. Math. Soc. **49** (1943).

[Kh1] A. G. Kovansky, *On a class of systems of transcendental equations*, Soviet Math. Dokl. **22** (1980), 762–765.

[Kh2] ____, *Real analytic varieties with the finiteness property and complex Abelian integrals*, Funct. Anal. **18** (1984), 119–127.

[Ko] J. Kollar, *Sharp effective Nullstellensatz*, J. Amer. Math. Soc. **1** (1988), 963–975.

[LiR] Z. Ligatsikas and M.-F. Roy, *Séries de Puiseux sur un corps réel clos*, C. R. Acad. Sci. Paris **311** (1990), 625–628.

[Lom1] H. Lombardi, *Effective real Nullstellensatz and variants*, MEGA 1990 Proceedings, Birkhäuser, Bonn, 1991, pp. 263–288.

[Lom2] ____, *Mathématiques constructives et complexité en temps polynomial, deux exemples: l'algèbre réelle, les différentes représentations des nombres réels*, Thèse d'habilitation à diriger des recherches, Université de Nice, 1990.

[LomR] H. Lombardi and M.-F. Roy, *Constructive elementary theory of real closed fields*, MEGA 1990 Procedings, Birkhäuser, Bonn, 1991, pp. 249–262.

[Lo] R. Loos, *Generalized poynomial reaminder sequences*, Computer Algebra, Symbolic and Algebraic Computation (Buchberger, Collins, and Loos, eds.), Springer-Verlag, 1982, pp. 115–138.

[Ma] L. Mahé, *Une démonstration élémentaire du théorème de Bröcker -Scheiderer*, C. R. Acad. Sci. Paris **309** (1989), 613–616.

[MM] E. Mayr and M. Meyer, *The complexity of the word problems for commmutative semigroups and polynomial ideals*, Adv. in Math. **46** (1982), 305–329.

[Mi] M. Mignotte, *Some useful bounds*, Computer Algebra, Symbolic and Algebraic Computation (Buchberger, Collins, and Loos, eds.), Springer-Verlag, Berlin and New York, 1982, pp. 259–263.

[Mil1] M. Milnor, *Morse theory* Princeton Univ. Press, Princeton, NJ, 1963.

[Mil2] ____, *On the Betti numbers of real algebraic varieties*, Proc. Amer. Math. Soc. **15** (1964), 275–280.

[Mn] N. E. Mnëv, *The universality theorems on the classification problem of configuration varieties and convex polytopes varieties*, Lecture Notes in Math., vol. 1346, Springer-Verlag, Berlin, 1988, pp. 527–544.

[MPR] J. L. Montana, L. M. Pardo, and T. Recio, *The non scalar model of complexity in computational geometry*, MEGA 1990 Procedings, Birkhäuser, Bonn, 1991, pp. 347–362.

[MR] T. Mostowski and E. Rannou, *On the complexity of construction of Whitney stratifications* (submitted to AAECC 1991).

[PR] L. M. Pardo and T. Recio, *Rabin's width of a complete proof and the width of a semialgebraic set*, Lecture Notes in Comp. Sci., vol. 378, Springer-Verlag, Berlin, 1989.

[P] P. Pedersen, *Counting real zeroes* Ph. D. Thesis, Courant Institute, New York Univ., NY, 1991.

[Re1] J. Renegar, *On the computational complexity and geometry of the first order theory of the reals*, Technical Report 856, Cornell University, Ithaca, NY, 1989.

[Re2] ____, *Recent progress on the complexity of the decision problem for the reals*, This volume.

[Ri1] J.-J. Risler, *Sur le 16-ème problème de Hilbert: un résumé et quelques questions*, Séminaire sur la Géométrie Algébrique Réelle, Publ. Math. Univ. Paris VII **9** (1980), 11–25.

[Ri2] ____, *Complexité et géométrie réelle (d'après Khovansky)*, Séminaire Bourbaki **637** (1984).

[Ri3] ____, *Some aspects of complexity in real algebraic geometry*, J. Symbolic Comput. **5** (1988), 109–119.

[R] M.-F. Roy, *Computation of the topology of a real algebraic curve*, Astérisque **192** (1990), 17–33.

[RS] M.-F. Roy and A. Szpirglas, *Complexity of computations with real algebraic numbers*, J. Symbolic Comput. **10** (1990), 39–51.

[Sc] C. Scheiderer, *Stability index of real varieties*, Invent. Math. **97** (1989), 467–483.

[SS] J. Schwartz and M. Scharir, *On the "piano movers" problem*. II, *General techniques for computing topological properties of real algebraic manifolds*, Adv. Appl. Math. **4** (1983), 298–351.

[Se] A. Seidenberg, *A new decision method for elementary algebra*, Ann. of Math. **60** (1954), 365–374.

[SY] J. M. Steele and A. C. Yao, *Lower bounds for algebraic decision trees* J. Algorithms **3** (1982), 1–8.

[Str1] V. Strassen, *Die Berechungskomplexität von elementarsymmetrischen Funktionen und Interpolationskoeffizienten*, Numer. Math. **20** (1973), 238–251.

[Str2] ____, *The complexity of continued fractions*, SIAM J. Comp. **12/1** (1983), 1–27.

[St] C. Sturm, *Mémoire sur la résolution des équations numériques*, Ins. France Sc. Math. Phys. **6** (1835).

[Sy] J. T. Sylvester, *On a theory of syzygetic relations of two rational integral functions, comprising an application to the theory of Sturm's function*, Trans. Roy. Soc. London (1853); Reprinted in Collected math papers, vol. 1, Chelsea, NY, 1983, pp. 429–586.

[Ta] A. Tarski, *A decision problem for elementary algebra and geometry*, Berkeley, 1951.

[T] R. Thom, *Sur l'homologie des variétés réelles*, Differential and Combinatorial Topology, Princeton Univ. Press, Princeton, NJ, 1965, pp. 255–265.

[Wa] R. Walker, *Algebraic curves*, Princeton Univ. Press, Princeton, NJ, 1950.

[We] V. Weispfenning, *The complexity of linear problems in fields*, J. Symbolic Comput. **5** (1988), 3–27.

[Wi] G. Wilson, *Hilbert sixteenth problem*, Topology **17** (1978), 53–73.

UNIVERSITY OF BUENOS AIRES, ARGENTINA

UNIVERSITY OF SANTANDER, SPAIN

UNIVERSITY OF RENNES, FRANCE

DIMACS Series in Discrete Mathematics
and Theoretical Computer Science
Volume 6, 1991

Ehrhart Polynomials of Convex Polytopes, h-Vectors of Simplicial Complexes, and Nonsingular Projective Toric Varieties

TAKAYUKI HIBI

ABSTRACT. We develop the theory of combinatorics on Ehrhart polynomials of convex polytopes by means of some fundamental results on Cohen-Macaulay rings and nonsingular projective toric varieties.

Introduction

Let $\mathscr{P} \subset \mathbb{R}^N$ be a *rational* convex polytope, i.e., a convex polytope, any of whose vertices has rational coordinates. Given a positive integer n we write $i(\mathscr{P}, n)$ for the number of those rational points $(\alpha_1, \alpha_2, \ldots, \alpha_N)$ in $\mathscr{P}$ such that each $n\alpha_i$ is an integer. In other words,

$$i(\mathscr{P}, n) := \#(n\mathscr{P} \cap \mathbb{Z}^N).$$

Here $n\mathscr{P} := \{n\alpha; \alpha \in \mathscr{P}\}$ and $\#(X)$ is the cardinality of a finite set X.

Even though the history of the research on enumeration of certain rational points in convex polytopes goes back to the nineteenth century, the systematic study of $i(\mathscr{P}, n)$ originated in the work of Ehrhart (who was a teacher in a *lycée*) beginning around 1955. The monograph [Ehr1] is an exposition of Ehrhart's research over a period of many years. Since Ehrhart built up the foundation on $i(\mathscr{P}, n)$, this interesting topic has been studied by, e.g., Macdonald [Mac1, Mac2], McMullen [Mc1, Mc2], and Stanley [Sta4].

Nowadays, the technique of commutative algebra and algebraic geometry is recognized as one of the basic and powerful tools for the study of combinatorics. Consult, e.g., [Hoc2, Rei, Sta2, Sta5, Sta9, Sta11, Sta13, Bil, B-R, H1]. Such algebraic technique can be also applied to the investigation of $i(\mathscr{P}, n)$. In particular, the theory of canonical modules [Sta3] of Cohen-Macaulay rings generated by monomials [Hoc1] plays an important role in our study of $i(\mathscr{P}, n)$.

1991 *Mathematics Subject Classification.* Primary 05E25; Secondary 13H10.

The purpose of this paper is to invite the reader to a short tour for a survey of recent development on $i(\mathscr{P}, n)$ with some concrete problems that might stimulate further study of the topic. Our treatment will be rather sketchy; in a subsequent paper, a more comprehensive account will be given.

Our main research object is the functions $i(\mathscr{P}, n)$ of *integral* convex polytopes $\mathscr{P}$ (i.e., convex polytopes $\mathscr{P}$ such that each vertex of $\mathscr{P}$ has integer coordinates). Ehrhart established that, when $\mathscr{P}$ is integral, the function $i(\mathscr{P}, n)$ possesses the following fundamental properties:

(0.1) $i(\mathscr{P}, n)$ is a polynomial in n of degree $d(= \dim \mathscr{P})$. (Thus $i(\mathscr{P}, n)$ can be defined for *every* integer n.)

(0.2) $i(\mathscr{P}, 0) = 1$.

(0.3) ("loi de réciprocité" [Ehr2]) $(-1)^d i(\mathscr{P}, -n) = \#(n(\mathscr{P} - \partial\mathscr{P}) \cap \mathbb{Z}^N)$ for every integer $n > 0$.

We say that $i(\mathscr{P}, n)$ is the *Ehrhart polynomial* of $\mathscr{P}$. Consult, e.g., [Sta8, pp. 235–241] for an introduction to Ehrhart polynomials.

We organize this paper as follows. First, in §1, we define a certain combinatorial sequence $\delta(\mathscr{P}) = (\delta_0, \delta_1, \ldots, \delta_d) \in \mathbb{Z}^{d+1}$, called the δ-*vector* of $\mathscr{P}$, arising from the generating function for $i(\mathscr{P}, n)$ of an integral convex polytope $\mathscr{P}$ of dimension d (see equation (1)). We consider what can be said about the δ-vector of an arbitrary integral convex polytope (cf. Theorem (1.3)) and then turn to the problem of finding integral convex polytopes that possess symmetric δ-vectors (cf. Theorem (1.4)). On the other hand, the purpose of §2 is to discuss a relation between $\delta(\mathscr{P})$ and the h-vector $h(\Delta)$ of some triangulation Δ of the boundary $\partial\mathscr{P}$ of $\mathscr{P}$. Finally, in §3, we study a class of integral convex polytopes that are related with finite partially ordered sets (cf. equation (7)). Via the theory of nonsingular projective toric varieties (e.g., [Sta5, Tei]), we prove that certain combinatorial sequences arising from enumeration on linear extensions of finite partially ordered sets are unimodal (Corollary (3.4)).

The author would like to thank Professor Richard P. Stanley for exciting discussions on $i(\mathscr{P}, n)$ and some related topics while the author was staying at Massachusetts Institute of Technology during the 1988–89 academic year.

1. Ehrhart polynomial

Let $\mathscr{P} \subset \mathbb{R}^N$ be an integral convex polytope of dimension d and $\partial\mathscr{P}$ the boundary of $\mathscr{P}$. We define the sequence $\delta_0, \delta_1, \delta_2, \ldots$ of integers by the formula

$$(1) \qquad (1 - \lambda)^{d+1}\left[1 + \sum_{n=1}^{\infty} i(\mathscr{P}, n)\lambda^n\right] = \sum_{i=0}^{\infty} \delta_i\lambda^i.$$

Then, thanks to the basic facts (0.1) and (0.2) on $i(\mathscr{P}, n)$, a fundamental result on generating functions, e.g., [Sta8, Corollary 4.3.1] guarantees that $\delta_i = 0$ for every $i > d$. When $\mathscr{P} \subset \mathbb{R}^N$ is an integral convex polytope of dimension d, we say that the sequence $\delta(\mathscr{P}) := (\delta_0, \delta_1, \ldots, \delta_d)$, which appears in equation (1), is the *δ-vector* of $\mathcal{P}$. Thus, in particular,

(2) $$\delta_0 = 1 \quad \text{and} \quad \delta_1 = \#(\mathcal{P} \cap \mathbb{Z}^N) - (d + 1).$$

One of the fundamental results on δ-vectors of integral convex polytopes obtained earlier is the following

(1.1) PROPOSITION ([**Sta4, Theorem 2.1**]). *The δ-vector $\delta(\mathcal{P}) = (\delta_0, \delta_1, \ldots, \delta_d)$ of an integral convex polytope $\mathcal{P} \subset \mathbb{R}^N$ of dimension d is nonnegative, i.e., $\delta_i \geq 0$ for every $0 \leq i \leq d$.*

We refer the reader to [**Sta4, p. 337**] for a historical comment on the above Proposition (1.1). Also, see [**B-M, p. 254**].

On the other hand, it follows easily from (0.3) that

(3) $$\delta_d = \#((\mathcal{P} - \partial\mathcal{P}) \cap \mathbb{Z}^N).$$

Moreover, when $N = d$, the leading coefficient of $i(\mathcal{P}, n)$ coincides with the volume (= Lebesgue measure) $\text{vol}(\mathcal{P})$ of $\mathcal{P}$ (cf. [**Sta8, Proposition 4.6.30**]), i.e.,

(4) $$(\delta_0 + \delta_1 + \cdots + \delta_d)/d! = \text{vol}(\mathcal{P}).$$

Our final goal for the study of Ehrhart polynomials is to find a complete (combinatorial) characterization of the δ-vectors of integral convex polytopes.

(1.2) EXAMPLE. Let $N = d = 3$ and $q > 0$ an integer. Also, let $\mathcal{P} \subset \mathbb{R}^3$ be the tetrahedron with the vertices $(0, 0, 0)$, $(q, 1, 0)$, $(1, 0, 1)$, and $(0, 1, 1)$. Then $\#(\mathcal{P} \cap \mathbb{Z}^3) = 4$ and $(\mathcal{P} - \partial\mathcal{P}) \cap \mathbb{Z}^3$ is empty. Also, the volume of $\mathcal{P}$ is $(q + 1)/3!$. Hence, by (2), (3), and (4), we have $\delta(\mathcal{P}) = (1, 0, q, 0)$; thus, $i(\mathcal{P}, n) = ((q + 1)n^3 + 6n^2 + (11 - q)n + 6)/3!$.

When $d = 2$, thanks to [**Sco**], we can give a complete characterization of the δ-vectors of integral convex polytopes. In fact, the δ-vectors arising from integral convex polytopes of dimension 2 are the following: (i) $(1, n, 0)$, $n \geq 0$, (ii) $(1, n, 1)$, $1 \leq n \leq 7$, and (iii) $(1, n, m)$, $2 \leq m \leq n \leq 3m + 3$.

Now, what can be said about the δ-vector of an arbitrary integral convex polytope?

(1.3) THEOREM. *Suppose that $\mathcal{P} \subset \mathbb{R}^N$ is an integral convex polytope of dimension d with the δ-vector $\delta(\mathcal{P}) = (\delta_0, \delta_1, \ldots, \delta_d)$.*
   (a) ([**H5, Theorem A**]) *We have the linear inequality*

(5) $$\delta_d + \delta_{d-1} + \cdots + \delta_{d-i} \leq \delta_0 + \delta_1 + \cdots + \delta_i + \delta_{i+1}$$

for every $0 \leq i \leq [(d - 1)/2]$.
   (b) ([**Sta14, Proposition 4.1**]) *Assume that $\delta_j \neq 0$ and $\delta_{j+1} = \delta_{j+2} = \cdots = \delta_d = 0$. Then the inequality*

(6) $$\delta_0 + \delta_1 + \cdots + \delta_i \le \delta_j + \delta_{j-1} + \cdots + \delta_{j-i}$$

holds for every $0 \le i \le [j/2]$.

The above inequalities (5) and (6) in Theorem (1.3) are typical examples of the combinatorial consequences of some recent algebraic works toward the problem of finding a combinatorial characterization of the Hilbert functions of Cohen-Macaulay integral domains. See also [H2].

We give here a sketch of proof of Theorem (1.3) for the reader who is familiar with the theory of canonical modules of Cohen-Macaulay rings. Consult [Sta14] and [H9] for the detailed information. First, let X_1, X_2, ..., X_N and T be indeterminates over a field k. We write $A(\mathscr{P})_n$ for the vector space over k spanned by those monomials $X_1^{\alpha_1} \cdots X_N^{\alpha_N} T^n$ such that $(\alpha_1, \ldots, \alpha_N) \in n\mathscr{P} \cap \mathbb{Z}^N$. Also, set $A(\mathscr{P})_0 = k$. Then the direct sum $A(\mathscr{P}) := \bigoplus_{n \ge 0} A(\mathscr{P})_n$ of $\{A(\mathscr{P})_n\}_{n=0,1,2\ldots}$ as a vector space over k turns out to be a noetherian graded ring of Krull dimension $d + 1$ with the *Hilbert function* $H(A(\mathscr{P}), n) := \dim_k A(\mathscr{P})_n = i(\mathscr{P}, n)$. Now, Hochster [Hoc1] guarantees that $A(\mathscr{P})$ is a Cohen-Macaulay ring. Moreover, by virtue of [Sta3, equation (21), p. 82], we can describe, explicitly, a graded ideal $I = \bigoplus_{n \ge 1} (I \cap A(\mathscr{P})_n) \subset A(\mathscr{P})_+ := \bigoplus_{n \ge 1} A(\mathscr{P})_n$ of $A(\mathscr{P})$ with $I \cong K_{A(\mathscr{P})}$. Here $K_{A(\mathscr{P})}$ is the canonical module of the Cohen-Macaulay ring $A(\mathscr{P})$. Let $\rho := \min\{n ; I \cap A(\mathscr{P})_n \ne (0)\} (> 0)$ and $0 \ne a \in I \cap A(\mathscr{P})_\rho$. Then we have the exact sequence (**) $0 \to A(\mathscr{P}) \to I \to I/aA(\mathscr{P}) \to 0$ since $A(\mathscr{P})$ is an integral domain. A fundamental fact [H-K] (see also [H1, Lemma (1.7)]) in the theory of canonical modules of Cohen-Macaulay rings implies that $A(\mathscr{P})/I$ is a Cohen-Macaulay ring of dimension d and that $I/aA(\mathscr{P})$ is a Cohen-Macaulay module of dimension d over $A(\mathscr{P})$. On the other hand, thanks to [Sta3, equation (12), p. 71], we can compute the Hilbert functions of $A(\mathscr{P})/I$ and $I/aA(\mathscr{P})$ by means of the δ-vector of $\mathscr{P}$. Then the required inequalities (5) and (6) follow immediately from a standard (and well-known) fact, e.g., [Sta3, Corollary 3.11] on Hilbert functions of Cohen-Macaulay rings and modules.   Q.E.D.

Note that, in the above proof, the noetherian graded ring $A(\mathscr{P})$ is not necessarily generated by $A(\mathscr{P})_1$. However, there exists a system of parameters for $A(\mathscr{P})$ consisting of elements of $A(\mathscr{P})_1$ (if k is infinite). It might be of interest to ask when $A(\mathscr{P})$ is generated by $A(\mathscr{P})_1$.

At present, there is little hope of giving a purely combinatorial proof for Theorem (1.3).

We now study an analogue of the Dehn-Sommerville equations of the h-vectors of simplicial convex polytopes (see, e.g., [B-L, Sta5]) for δ-vectors of integral convex polytopes.

In general, we say that a convex polytope $\mathscr{P}$ of dimension d is *of standard type* if $\mathscr{P} \subset \mathbb{R}^d$ and the origin of $\mathbb{R}^d$ is contained in the interior $\mathscr{P} - \partial\mathscr{P}$ of $\mathscr{P}$. When $\mathscr{P} \subset \mathbb{R}^d$ is of standard type, the *polar set* (or *dual polytope*)

$\mathcal{P}^*$ of $\mathcal{P}$ is defined to be

$$\mathcal{P}^* := \{(\alpha_1, \alpha_2, \ldots, \alpha_d) \in \mathbb{R}^d ; \alpha_1\beta_1 + \alpha_2\beta_2 + \cdots + \alpha_d\beta_d \leq 1$$
$$\text{for every } (\beta_1, \beta_2, \ldots, \beta_d) \in \mathcal{P}\}.$$

Note that $\mathcal{P}^* \subset \mathbb{R}^d$ is also a convex polytope of standard type and $(\mathcal{P}^*)^* = \mathcal{P}$. Moreover, if $\mathcal{P}$ is rational, then $\mathcal{P}^*$ is also rational (cf. [**Grü, p. 47**]).

Fix an integer $d > 1$ and let $\mathscr{C}_0(d)$ be the set of integral convex polytopes $\mathcal{P} \subset \mathbb{R}^d$ of standard type. Also, we write $\mathscr{C}^*(d)$ for the set of those $\mathcal{P} \in \mathscr{C}_0(d)$ such that the polar set $\mathcal{P}^*$ of $\mathcal{P}$ is an integral convex polytope.

Even though the following Theorem (1.4) is easy to prove (based on the Ehrhart law of reciprocity (0.3)), this result plays an important role in our theory of combinatorics on δ-vectors.

(1.4) THEOREM ([**H6**]). *The δ-vector $\delta(\mathcal{P}) = (\delta_0, \delta_1, \ldots, \delta_d)$ of $\mathcal{P} \in \mathscr{C}_0(d)$ is symmetric, i.e., $\delta_i = \delta_{d-i}$ for every $0 \leq i \leq d$, if and only if $\mathcal{P} \in \mathscr{C}^*(d)$.*

The above Theorem (1.4) can be generalized for the functions $i(\mathcal{P}, n)$ of rational convex polytopes $\mathcal{P}$ (see [**H7**]). Also, it would be of interest to compare our Theorem (1.4) and [**H7**]) with [**Ish, Theorem 7.7**].

If $\mathcal{P} \subset \mathbb{R}^N$ is an integral convex polytope of dimension d, then there exists an integral convex polytope $\mathcal{Q} \subset \mathbb{R}^d$ of dimension d with $\delta(\mathcal{P}) = \delta(\mathcal{Q})$. See, e.g., [**Sta8, pp. 238–239**]. Hence, for the combinatorial study of δ-vectors $\delta(\mathcal{P}) = (\delta_0, \delta_1, \ldots, \delta_d)$ of integral convex polytopes $\mathcal{P}$ of dimension d with $\delta_d > 0$, thanks to equation (3), we have only to consider the integral convex polytopes $\mathcal{P} \subset \mathbb{R}^d$ of standard type.

Since $\mathcal{P} \in \mathscr{C}^*(d)$ implies $\mathcal{P}^* \in \mathscr{C}^*(d)$, it is quite reasonable to ask if we can compute $\delta(\mathcal{P}^*)$ in terms of $\delta(\mathcal{P})$. However, the following Example (1.5) falls short of our expectation.

(1.5) EXAMPLE. Let $d = 3$. First, we consider $\mathcal{P} \in \mathscr{C}^*(3)$ with the vertices $(1, 1, 1), (-1, 0, 0), (0, -1, 0), (0, 0, -1), (0, 1, 1), (1, 0, 1)$, and $(1, 1, 0)$. Then $\delta(\mathcal{P}) = (1, 4, 4, 1)$ and $\delta(\mathcal{P}^*) = (1, 19, 19, 1)$. On the other hand, let $\mathcal{Q} \in \mathscr{C}^*(3)$ be the bipyramid, which is the convex hull of $\{(1, 0, 0), (0, 1, 0), (1, 1, 0), (-1, 0, 0), (0, -1, 0), (0, 0, 1), (0, 0, -1)\}$ in $\mathbb{R}^3$, with $\delta(\mathcal{Q}) = (1, 4, 4, 1)$. Then $\mathcal{Q}^* \subset \mathbb{R}^3$ is the prism with $\delta(\mathcal{Q}^*) = (1, 20, 20, 1)$. Hence $\delta(\mathcal{P}^*) \neq \delta(\mathcal{Q}^*)$ even though $\delta(\mathcal{P}) = \delta(\mathcal{Q})$.

A somewhat interesting question related with Example (1.5) is the problem when convex polytopes $\mathcal{P} \in \mathscr{C}^*(d)$ satisfy $\delta(\mathcal{P}) = \delta(\mathcal{P}^*)$. We remark that the equality $\delta(\mathcal{P}) = \delta(\mathcal{P}^*)$ implies $\text{vol}(\mathcal{P}) = \text{vol}(\mathcal{P}^*)$ by equation (4).

On the other hand, during the DIMACS workshop on Polytopes and Convex Sets (Rutgers University, January 8—12, 1990), Stanley and the author discussed the following question: Let $\mathcal{P}$ and $\mathcal{Q}$ be integral convex polytopes in $\mathbb{R}^N$ of dimension d and suppose that each vertex of $\mathscr{P}$ is a vertex of $\mathscr{Q}$ (thus, in particular, $\mathscr{P} \subset \mathscr{Q}$). Then is $\delta(\mathscr{Q}) \geq \delta(\mathscr{P})$? (Namley, is the ith component of $\delta(\mathscr{Q})$ greater than or equal to the ith component of $\delta(\mathscr{P})$ for every $0 \leq i \leq d$?)[1]

2. Simplicial complexes

Let us review the definition of f-vectors and h-vectors of simplicial complexes. Let Δ be a *simplicial complex* on the *vertex set* $V = \{x_0, x_1, \ldots, x_v\}$. Thus, Δ is a collection of subsets of V such that (i) $\{x_i\} \in \Delta$ for every $0 \leq i \leq v$ and (ii) if $\sigma \in \Delta$ and $\tau \subset \sigma$ then $\tau \in \Delta$. An element $\sigma \in \Delta$ is called an *i-face* of Δ if $\#(\sigma) = i + 1$. A *facet* of Δ is a maximal face (with respect to inclusion) of Δ. Let $d := \max\{\#(\sigma) ; \sigma \in \Delta\}$. Then the *dimension* of Δ is $\dim(\Delta) := d - 1$. We write $f_i = f_i(\Delta)$, $0 \leq i < d$, for the number of i-faces of Δ. The vector $f(\Delta) = (f_0, f_1, \ldots, f_{d-1})$ is called the *f-vector* of Δ. Define the *h-vector* $h(\Delta) = (h_0, h_1, \ldots, h_d)$ of Δ by the formula

$$\sum_{i=0}^{d} f_{i-1}(\lambda - 1)^{d-i} = \sum_{i=0}^{d} h_i \lambda^{d-i}$$

with $f_{-1} = 1$. In particular, $h_0 = 1$, $h_1 = \#(V) - d$, and

$$h_0 + h_1 + \cdots + h_d = f_{d-1}.$$

We now study a certain triangulation Δ of the boundary $\partial\mathscr{P}$ of a convex polytope $\mathscr{P} \in \mathscr{C}^*(d)$ and discuss a relation between the δ-vector of $\mathscr{P}$ and the h-vector of Δ.

Let $\mathscr{P} \in \mathscr{C}^*(d)$ and set $V := \partial\mathscr{P} \cap \mathbb{Z}^d$. We write $\mathscr{T}$ for the set of simplices $\sigma \subset \mathbb{R}^d$ such that each vertex of σ is contained in V. Thus in particular $\{x\} \in \mathscr{T}$ for each $x \in V$. We say that a subset Δ of $\mathscr{T}$ is a *triangulation of $\partial\mathscr{P}$ with the vertex set V* if the following conditions are satisfied:

(i) $\{x\} \in \Delta$ for each $x \in V$,
(ii) if $\sigma \in \Delta$ and τ is a face of σ, then $\tau \in \Delta$,
(iii) if $\sigma, \tau \in \Delta$, then $\sigma \cap \tau$ is a common face of both σ and τ,
(iv) $\bigcup_{\sigma \in \Delta} \sigma = \partial\mathscr{P}$.

(2.1) LEMMA. *The boundary $\partial\mathscr{P}$ of every $\mathscr{P} \in \mathscr{C}^*(d)$ possesses a triangulation with the vertex set $V = \partial\mathscr{P} \cap \mathbb{Z}^d$.*

A triangulation Δ of the boundary $\partial\mathscr{P}$ of $\mathscr{P} \in \mathscr{C}^*(d)$ with the vertex set $V = \partial\mathscr{P} \cap \mathbb{Z}^d$ might be regarded as a simplicial complex on V of dimension $d - 1$ whose geometric realization is $\partial\mathscr{P}$. Since $\partial\mathscr{P}$ is homeomorphic to the $(d-1)$-sphere, the h-vector $h(\Delta) = (h_0, h_1, \ldots, h_d)$ of the simplicial

---
[1]Stanley [Sta15] answered this question affirmatively by the use of a modification of the Cohen-Macaulay ring $A(\mathscr{P})$, which appears in the sketch of the proof of Theorem (1.3).

complex Δ on V satisfies the *Dehn-Sommerville equation* $h_i = h_{d-i}$ for every $0 \leq i \leq d$. Consult, e.g., [**Hoc2**] and [**Sta6**].

On the other hand, a triangulation Δ of the boundary $\partial \mathcal{P}$ of $\mathcal{P} \in \mathscr{C}^*(d)$ with the vertex set $V = \partial \mathcal{P} \cap \mathbb{Z}^d$ is called *compressed* (cf. [**Sta4, p. 337**]) if, for each facet σ of Δ, the determinant of the matrix $(\mathbf{x}_{i1}, \mathbf{x}_{i2}, \dots, \mathbf{x}_{id})$ is equal to ± 1, where $\{\mathbf{x}_{ij}\}_{1 \leq j \leq d}$ is the set of vertices of σ. (Note that if $\mathscr{Q} \subset \mathbb{R}^d$ is the simplex which is the convex hull of $\{(0, 0, \dots, 0)\} \cup \sigma$, then $d! \operatorname{vol}(\mathscr{Q})$ coincides with the absolute value of the determinant of the matrix $(\mathbf{x}_{i1}, \mathbf{x}_{i2}, \dots, \mathbf{x}_{id})$.)

When $d \leq 3$ and $\mathcal{P} \in \mathscr{C}^*(d)$, every triangulation of $\partial \mathcal{P}$ with the vertex set $V = \partial \mathcal{P} \cap \mathbb{Z}^d$ is compressed, because the volume of an integral convex polytope $\mathscr{Q} \subset \mathbb{R}^2$ of dimension 2 with $\#(\mathscr{Q} \cap \mathbb{Z}^2) = 3$ is equal to $1/2$.

(2.2) PROPOSITION (cf. [**Sta4**] and [**B-M**]). *Suppose that Δ is a triangulation of the boundary $\partial \mathcal{P}$ of $\mathcal{P} \in \mathscr{C}^*(d)$ with the vertex set $V = \partial \mathcal{P} \cap \mathbb{Z}^d$. Let $h(\Delta) = (h_0, h_1, \dots, h_d)$ be the h-vector of Δ and $\delta(\mathcal{P}) = (\delta_0, \delta_1, \dots, \delta_d)$ the δ-vector of $\mathcal{P}$. Then $\delta(\mathcal{P}) \geq h(\Delta)$, i.e., $\delta_i \geq h_i$ for every $0 \leq i \leq d$. Moreover, $h(\Delta) = \delta(\mathcal{P})$ if and only if Δ is compressed.*

(2.3) EXAMPLE. Let $d = 4$ and $\mathcal{P} \in \mathscr{C}^*(4)$ the convex polytope with the vertex set $\{(0, 0, 0, 1), (1, 1, 0, 1), (1, 0, 1, 1), (0, 1, 1, 1), -(0, 0, 0, 1), -(1, 1, 0, 1), -(1, 0, 1, 1), -(0, 1, 1, 1)\}$. Then $\delta(\mathcal{P}) = (1, 4, 22, 4, 1)$, which is not an "0-sequence" (see, e.g., [**Sta3, H2**]), thus there exists no compressed triangulation of $\partial \mathcal{P}$ with the vertex set $V = \partial \mathcal{P} \cap \mathbb{Z}^4$. In fact, $\mathcal{P}$ is a simplicial convex polytope with the h-vector $h(\mathcal{P}) = (1, 4, 6, 4, 1)$ and the set of faces of $\mathcal{P}$ is the unique triangulation of $\partial \mathcal{P}$ with the vertex set V.

Now, let $\delta(\mathcal{P}) = (\delta_0, \delta_1, \dots, \delta_d)$ be the δ-vector of $\mathcal{P} \in \mathscr{C}^*(d)$ and suppose that $h(\Delta) = (h_0, h_1, \dots, h_d)$ is the h-vector of a triangulation Δ of $\partial \mathcal{P}$ with the vertex set $V = \partial \mathcal{P} \cap \mathbb{Z}^d$, whose existence is guaranteed by Lemma (2.1). Then the Lower Bound Theorem by Barnette [**Bar1, Bar2**] implies the inequality $h_i \geq h_1$ for every $1 \leq i < d$. On the other hand, since we have $\delta_1 (= \#(\mathcal{P} \cap \mathbb{Z}^d) - (d+1)) = h_1 (= \#(V) - d)$, we obtain from Proposition (2.2) the following result.

(2.4) COROLLARY. *The δ-vector $\delta(\mathcal{P}) = (\delta_0, \delta_1, \dots, \delta_d)$ of an arbitrary convex polytope $\mathcal{P} \in \mathscr{C}^*(d)$ satisfies the inequality $\delta_i \geq \delta_1$ for every $1 \leq i < d$.*

In [**H10**], the above Corollary (2.4) is generalized as follows:

(2.5) THEOREM ([**H10**]). *Let $\mathcal{P} \subset \mathbb{R}^N$ be an integral convex polytope of dimension d with the δ-vector $\delta(\mathcal{P}) = (\delta_0, \delta_1, \dots, \delta_d)$ and suppose that $(\mathcal{P} - \partial \mathcal{P}) \cap \mathbb{Z}^N$ is nonempty, i.e., $\delta_d \neq 0$. Then we have the inequality $\delta_1 \leq \delta_i$ for every $1 \leq i < d$.*

Fix an integer $d > 1$ and let $\mathcal{Z}^*(d)$ be the set of δ-vectors $\delta(\mathcal{P})$ of $\mathcal{P} \in \mathcal{C}^*(d)$. Also, we write $\mathcal{Z}_t^*(d)$ for the subset of $\mathcal{Z}^*(d)$, which consists of δ-vectors $\delta(\mathcal{P})$ of $\mathcal{P} \in \mathcal{C}^*(d)$ such that $\partial\mathcal{P}$ possesses a compressed triangulation with the vertex set $V = \partial\mathcal{P} \cap \mathbb{Z}^d$. We remark that $\mathcal{Z}^*(d) = \mathcal{Z}_t^*(d)$ if $d \leq 3$; however, $\mathcal{Z}_t^*(d) \neq \mathcal{Z}^*(d)$ for every $d \geq 4$.

It would be of interest to find a combinatorial characterization of the sequences in $\mathcal{Z}^*(d)$ (or $\mathcal{Z}_t^*(d)$).

On the other hand, by virtue of [**Hen, Theorem 3.6**], the supremum of volumes $\text{vol}(\mathcal{P})$ of $\mathcal{P} \in \mathcal{C}^*(d)$ is bounded. Hence, Lemma (1.1) and equation (4), together, imply the following:

(2.6) PROPOSITION. *The set $\mathcal{Z}^*(d)$ is finite for every $d > 1$.*

We do not know the exact values of $\nu(d) := \max\{\text{vol}(\mathcal{P}) ; \mathcal{P} \in \mathcal{C}^*(d)\}$ and $\mu(d) := \#(\mathcal{Z}^*(d))$ when $d \geq 3$. Note that $\nu(2) = 9/2$ [**Sco**] and $\mu(2) = 7$; however, $\nu(d) > (d + 1)\,d/d!$ if $d \geq 3$ (cf. [**Z-P-W**]). Also, see [**Hen, §4**].

3. Toric varieties

We are now in the position to study toric varieties arising from triangulations of the boundaries of convex polytopes. We refer the reader to, e.g., [**Oda**] and [**Dan**] for basic information on toric varieties.

Suppose that Δ is a triangulation of the boundary $\partial\mathcal{P}$ of $\mathcal{P} \in \mathcal{C}^*(d)$ with the vertex set $V = \partial\mathcal{P} \cap \mathbb{Z}^d$. Then we can construct a simplicial complete fan $\mathcal{F}(\Delta)$ (cf. [**Dan, §5**]) in the $\mathbb{Q}$-vector space $\mathbb{Q}^d$ associated with Δ in the obvious way ([**Sta10, p. 218; Oda, Proposition 2.19**]). In fact, for each face σ of Δ, we define a simplicial convex polyhedral cone $C(\sigma)$ (with apex at the origin) to be the union of all rays whose vertex is the origin and which pass through σ. Then the set $\mathcal{F}(\Delta)$ of all such cones $C(\sigma)$ forms a complete fan. We write $\mathcal{X}(\Delta)$ for the complete toric variety associated with $\mathcal{F}(\Delta)$. Here we should remark that $\mathcal{X}(\Delta)$ is nonsingular [**Oda, Theorem 1.10**] if and only if Δ is compressed (cf. Proposition (2.2)). In general, the toric variety $\mathcal{X}(\Delta)$ is not necessarily projective (even though $\mathcal{X}(\Delta)$ is nonsingular, see [**Oda, p. 84**]). On the other hand, if the toric variety $\mathcal{X}(\Delta)$ is nonsingular and projective, then the δ-vector $\delta(\mathcal{P})$ of $\mathcal{P}$ coincides with the h-vector of some simplicial convex polytope of dimension d (cf. [**Sta10, p. 219**]); thus, in particular, $\delta(\mathcal{P})$ is unimodal, i.e,. $\delta_0 \leq \delta_1 \leq \cdots \leq \delta_{[d/2]}$ ([**Sta5**] and [**Sta12, Theorem 20**]). Consult, e.g., [**Sta11**] for further results related with toric varieties and unimodal sequences.

It would be of interest to ask if there exists a natural class of convex polytopes $\mathcal{P} \in \mathcal{C}^*(d)$ such that the boundary $\partial\mathcal{P}$ of $\mathcal{P}$ possesses a triangulation Δ with the vertex set $V = \partial\mathcal{P} \cap \mathbb{Z}^d$ for which the corresponding toric variety $\mathcal{X}(\Delta)$ is projective (or nonsingular).

On the other hand, there exists a convex polytope $\mathcal{P} \in \mathcal{C}^*(d)$ and triangulations Δ, Δ' of the boundary $\partial\mathcal{P}$ of $\mathcal{P}$ with the vertex set $V = \partial\mathcal{P} \cap \mathbb{Z}^d$

such that $\mathscr{X}(\Delta)$ is nonsingular (resp. projective), but $\mathscr{X}(\Delta')$ is not nonsingular (resp. projective) if $d \geq 4$ (resp. $d \geq 3$).

Throughout the remainder of this section, we suppose that X is a finite partially ordered set (*poset* for short) with elements $y_1, y_2, \ldots, y_d$ labeled so that if $y_i < y_j$ in X, then $i < j$ in $\mathbb{Z}$. A totally ordered subset of X is called a *chain* of X. Set $l := \max\{\#(C); C \text{ is a chain of } X\}$. We say that X is *pure* if every maximal chain of X has the cardinality l. If $\alpha \in X$, then $r(\alpha)$ denotes the greatest integer $m > 0$ for which there exists a chain $\alpha = \beta_1 < \beta_2 < \cdots < \beta_m$ in X.

Now, we write $\mathscr{Q}(X)$ for the subset of $\mathbb{R}^d$, which consists of those points $(\alpha_1, \alpha_2, \ldots, \alpha_d)$, such that

   (i) $0 \leq \alpha_i + r(\alpha_i) \leq l + 1$ for each $1 \leq i \leq d$, and
   (ii) $\alpha_i + r(\alpha_i) \leq \alpha_j + r(\alpha_j)$ if $y_i \geq y_j$ in X.

Let $\mathscr{O}(X)$ be the *order polytope* [Sta7] associated with X. Then

$$(7) \qquad \mathscr{Q}(X) = (l+1)\mathscr{O}(X) - (r(\alpha_1), r(\alpha_2), \ldots, r(\alpha_d)).$$

Thus, $\mathscr{Q}(X) \subset \mathbb{R}^d$ is an integral convex polytope of dimension d. Moreover, $\mathscr{Q}(X)$ is of standard type, i.e., $\mathscr{Q}(X) \in \mathscr{C}_0(d)$. We say that $\mathscr{Q}(X)$ is the *fat order polytope* associated with X.

(3.1) LEMMA. *The fat order polytope* $\mathscr{Q}(X) \in \mathscr{C}_0(d)$ *associated with* X *is contained in* $\mathscr{C}^*(d)$ *if and only if* X *is pure.*

SKETCH OF PROOF. Thanks to [Sta7] and equation (7), we know the equations of supporting hyperplanes $\mathscr{H} \subset \mathbb{R}^d$ of $\mathscr{Q}(X)$ such that $\mathscr{H} \cap \mathscr{Q}(X)$ is a facet of $\mathscr{Q}(X)$. In other words, we have information on the vertices of the dual polytope $\mathscr{Q}(X)^*$ of $\mathscr{Q}(X)$ (cf. [Grü, p. 47]). Thus, the required result follows easily from Theorem (1.4). Q.E.D.

We now state a combinatorial result on triangulations of the boundary $\partial\mathscr{Q}(X)$ of the fat order polytope $\mathscr{Q}(X)$ when X is pure.

In [H8], we prove that certain (complete and simplicial) toric varieties arising from canonical triangulations [Sta7] of order polytopes are projective. Thus, combining [H8] with [T-E, Chap. III, §2], we obtain the following:

(3.2) THEOREM. *When* X *is pure, there exists a triangulation* Δ *of the boundary* $\partial\mathscr{Q}(X)$ *of* $\mathscr{Q}(X)$ *with the vertex set* $V = \partial\mathscr{Q}(X) \cap \mathbb{Z}^d$ *such that the complete toric variety* $\mathscr{X}(\Delta)$ *is nonsingular and projective.*

Our next work is to compute the δ-vector $\delta(\mathscr{Q}(X))$ of $\mathscr{Q}(X)$. Let $w_i = w_i(X)$, $0 \leq i < d$, be the number of permutations $\pi = c_1 c_2 \cdots c_d$ of $1, 2, \ldots, d$ with the properties that (i) if $y_{c_p} < y_{c_q}$ in X, then $p < q$ (i.e., π is a *linear extension* of X) and (ii) $\#(\{r; c_r > c_{r+1}\})$, the number of *descents* of π, is equal to i. Thus, in particular, $w_0 = 1$. Set $s := \max\{i; w_i \neq 0\}$. Then we easily see the equality $s = d - l$. We say that the sequence $w(X) := (w_0, w_1, \ldots, w_s)$ is the w-*vector* of X. Consult, e.g., [**Sta1, Sta7; Sta8, Chap. 4, §5; H3, H4**] for the combinatorial background of w-vectors of finite posets.

In general, if $\varphi(\lambda) = \sum_{i \geq 0} u_i \lambda^i \in \mathbb{R}[\lambda]$ and $\xi > 0$ is an integer, then we write $[\varphi(\lambda)]^{(\xi)}$ for $\sum_{i \geq 0} u_{i\xi} \lambda^i$.

(3.3) PROPOSITION. *Let* $w(X) = (w_0, w_1, \ldots, w_s)$ *be the* w-*vector of* X *and* $\delta(\mathscr{Q}(X)) = (\delta_0, \delta_1, \ldots, \delta_d)$ *the* δ-*vector of* $\mathscr{Q}(X) \in \mathscr{C}_0(d)$. *Then we have the equality*

$$(8) \qquad \sum_{i=0}^{d} \delta_i \lambda^i = \left[(1 + \lambda + \lambda^2 + \cdots + \lambda^l)^{d+1} \sum_{j=0}^{s} w_j \lambda^j \right]^{(l+1)}.$$

We refer the reader to, e.g., [**H2, §4**] for some information on equation (8). As an immediate consequence of Theorem (3.2) with Proposition (3.3), we obtain a class of unimodal sequences in our theory of δ-vectors.

(3.4) COROLLARY. *Suppose that* X *is pure with the* w-*vector* $w(X) = (w_0, w_1, \ldots, w_s)$. *Then the combinatorial sequence* $(\delta_0, \delta_1, \ldots \delta_d)$ *defined by equation* (8) *is* (*symmetric and*) *unimodal.*

It is conjectured that the w-vector $(w_0, w_1, \ldots, w_s)$ of an arbitrary finite poset is unimodal, i.e., $w_0 \leq w_1 \leq \cdots \leq w_j \geq \cdots \geq w_s$ for some $0 \leq j \leq s$. Consult [**Sta12, pp. 505–506**] for further information. On the other hand, thanks to [**H8**] and [**Sta15, Lemma 2.2**], we easily prove the inequalities $w_{[(d+1)/2]} \geq w_{[(d+1)/2]+1} \geq \cdots \geq w_s$.

(3.5) EXAMPLE. Let $d = 4$ and suppose that $X = \{y_1, y_2, y_3, y_4\}$ is the pure poset with the partial order $y_1 < y_3$, $y_2 < y_3$, and $y_2 < y_4$. Then $l = 2$, $s = 2$, and $w(X) = (1, 3, 1)$. Hence, thanks to equation (8), the δ-vector of the fat order polytope $\mathscr{Q}(X) \in \mathscr{C}^*(4)$ is $\delta(\mathscr{Q}(X)) = (1, 80, 245, 80, 1)$. On the other hand, the vertices of $\mathscr{Q}(X)$ are $(-2, -2, -1, -1)$, $(1, -2, -1, -1)$, $(-2, 1, -1, -1)$, $(1, 1, -1, -1)$, $(-2, 1, -1, 2)$, $(1, 1, 2, -1)$, $(1, 1, -1, 2)$, and $(1, 1, 2, 2)$. Moreover, the vertices of the dual polytope $\mathscr{Q}(X)^*$ of $\mathscr{Q}(X)$ are $\beta_1 = (1, 0, 0, 0)$, $\beta_2 = (0, 1, 0, 0)$, $\beta_3 = (0, 0, -1, 0)$, $\beta_4 = (0, 0, 0, -1)$, $\beta_5 = (-1, 0, 1, 0)$, $\beta_6 = (0, -1, 1, 0)$, and $\beta_7 = (0, -1, 0, 1)$. Since we know the vertices and the facets of $\mathscr{Q}(X)^*$, it is possible by routine computation to determine if there exists a compressed triangulation of $\partial(\mathscr{Q}(X)^*)$ with the vertex set $V = \partial(\mathscr{Q}(X)^*) \cap \mathbb{Z}^4$. In fact, the triangulation Δ of $\partial(\mathscr{Q}(X)^*)$ whose facets are 2345, 1234, 2357, 1237, 3457, 4567, 1347, 1467, 1245, 1456, 1257, and 1567 is a compressed triangulation of $\partial(\mathscr{Q}(X)^*)$ with the vertex set $V = \partial(\mathscr{Q}(X)^*) \cap \mathbb{Z}^4$. Here, for example, the notation 2345 means the simplex in $\mathbb{R}^4$ with the vertex set $\{\beta_2, \beta_3, \beta_4, \beta_5\}$. Note that $\mathrm{vol}(\mathscr{Q}(X)^*) = 12/4!$ and $\delta(\mathscr{Q}(X)^*) = (1, 3, 4, 3, 1)$.

It is not difficult to see that there exists a compressed triangulation of $\partial(\mathscr{Q}(X)^*)$ with the vertex set $V = \partial(\mathscr{Q}(X)^*) \cap \mathbb{Z}^d$ for *every* pure poset X with $\#(X) = d$. Also, when X is pure, does there exist a nice formula like equation (8) to compute $\delta(\mathscr{Q}(X)^*)$?

We conclude this section with another example of combinatorial sequences contained in $\mathscr{Z}_t^*(d)$ which is related with nonsingular Fano toric varieties.

(3.6) EXAMPLE. Suppose that a convex polytope $\mathscr{P} \in \mathscr{C}^*(d)$ is simplicial. Then, thanks to Proposition (2.2), we easily see that $h(\mathscr{P}) = \delta(\mathscr{P})$ if and only if $\mathscr{P}$ is a Fano polyhedron (polytope) in the sense of [V-K, p. 223]. In [V-K] Voskresenskij and Klyachko give a complete classification of centrally symmetric Fano polytopes. (Especially, see [V-K, p. 234] on the description of the Fano polytope for the del Pezzo variety associated with the root system of type A of an even rank.) Thus, in particular, we should say that the δ-vectors $\delta(\mathscr{P})$ arising from centrally symmetric simplicial convex polytopes $\mathscr{P} \in \mathscr{C}^*(d)$ with $h(\mathscr{P}) = \delta(\mathscr{P})$ are already known. Is it possible to find a combinatorial characterization of the δ-vectors $\delta(\mathscr{P})$ of simplicial convex polytopes $\mathscr{P} \in \mathscr{C}^*(d)$ with $h(\mathscr{P}) = \delta(\mathscr{P})$?

REFERENCES

[Bar1] D. W. Barnette, *The minimal number of vertices of a simple polytope*, Israel J. Math. **10** (1971), 121–125.

[Bar2] ____, *A proof of the lower bound conjecture for convex polytopes*, Pacific J. Math. **46** (1973), 349–354.

[B-M] U. Betke and P. McMullen, *Lattice points in lattice polytopes*, Monatsh. Math. **99** (1985), 253–265.

[Bil] L. J. Billera, *The algebra of continuous piecewise polynomials*, Adv. in Math. **76** (1989), 170–183.

[B-L] L. J. Billera and C. Lee, *A proof of the sufficiency of McMullen's conditions for f-vectors of simplicial convex polytopes*, J. Combin. Theory Ser A. **31** (1981), 237–255.

[B-R] L. J. Billera and L. L. Rose, *Modules of piecewise polynomials and their freeness*, Math. Z. (to appear).

[Dan] V. I. Danilov, *The geometry of toric varieties*, Russian Math. Surveys **33** (1978), 97–154.

[Ehr1] E. Ehrhart, *Polynômes arithmétiques et Méthode des Polyèdres en Combinatoire*, Birkhaüser, Basel and Stuttgart, 1977.

[Ehr2] ____, *Démonstration de la loi de réciprocité du polyedre rationnel*, C. R. Acad. Sci. Paris Ser. A **265** (1967), 91–94.

[Grü] B. Grünbaum, *Convex polytopes*, John Wiley and Sons, Inc. (Interscience), London, New York, and Sydney, 1967.

[Hen] D. Hensley, *Lattice vertex polytopes with interior lattice points*, Pacific J. Math. **105** (1983), 183–191.

[H1] T. Hibi, *Canonical ideals of Cohen-Macaulay partially ordered sets*, Nagoya Math. J. **112** (1988), 1–24.

[H2] ____, *Flawless 0-sequences and Hilbert functions of Cohen-Macaulay integral domains*, J. Pure Appl. Algebra **60** (1989), 245–251.

[H3] ____, *Linear and non-linear inequalities concerning a certain combinatorial sequence which arises from counting the number of chains of a finite distributive lattice*, Discrete Appl. Math. (to appear).

[H4] ____, *Hilbert functions of Cohen-Macaulay integral domains and chain conditions of finite partially ordered sets*. J. Pure Appl. Algebra **72** (1991), 265–273.

[H5] ___, *Some results on Ehrhart polynomials of convex polytopes*, Discrete Math. **83** (1990), 119–121.

[H6] ___, *Dual polytopes of rational convex polytopes*, Combinatorica (to appear).

[H7] ___, *A combinatorial self-reciprocity theorem for Ehrhart quasi-polynomials of rational convex polytopes*, European J. Combin. (to appear).

[H8] ___, *Toric varieties arising from canonical triangulations of poset polytopes are projective* (submitted).

[H9] ___, *The Ehrhart polynomial of a convex polytope* (in preparation).

[H10] ___, *A lower bound theorem for Ehrhart polynomials of convex polytopes*, Adv. in Math. (to appear).

[Hoc1] M. Hochster, *Rings of invariants of tori, Cohen-Macaulay rings generated by monomials, and polytopes*, Ann. of Math. **96** (1972), 318–337.

[Hoc2] ___, *Cohen-Macaulay rings combinatorics, and simplicial complexes*, Ring Theory II (B. R. McDonald and R. Morris, eds.), Lecture Notes in Pure and Appl. Math., no. 26, Dekker, New York, 1977, pp. 171–223.

[H-K] J. Herzog and E. Kunz, *Der kanonische Modul eines Cohen-Macaulay-Rings*, Lecture Notes in Math., vol. 238, Springer-Verlag, Berlin, Heidelberg, and New York, 1971.

[Ish] M.-N. Ishida, *Torus embeddings and dualizing complexes*, Tôhoku Math. J. **32** (1980), 111–146.

[Mac1] I. G. Macdonald, *The volume of a lattice polyhedron*, Proc. Cambridge Philos. Soc. **59** (1963), 719–726.

[Mac2] ___, *Polynomials associated with finite cell-complexes*, J. London Math. Soc. (2) **4** (1971), 181–192.

[Mc1] P. McMullen, *Valuations and Euler-type relations on certain classes of convex polytopes*, Proc. London Math. Soc. (3) **35** (1977), 113–135.

[Mc2] ___, *Lattice invariant valuations on rational polytopes*, Arch. Math. **31** (1978), 509–516.

[Oda] T. Oda, *Convex bodies and algebraic geometry: An introduction to the theory of toric varieties*, Springer-Verlag, Berlin, Heidelberg, and New York, 1988.

[Rei] G. Reisner, *Cohen-Macaulay quotients of polynomial rings*, Adv. in Math. **21** (1976), 30–49.

[Sco] P. R. Scott, *On convex lattice polygons*, Bull. Austral. Math. Soc. **15** (1976), 395–399.

[Sta1] R. P. Stanley, *Ordered structures and partitions*, Mem. Amer. Math. Soc. **119** (1972).

[Sta2] ___, *The upper bound conjecture and Cohen-Macaulay rings*, Stud. Appl. Math. **54** (1975), 135–142.

[Sta3] ___, *Hilbert functions of graded algebras*, Adv. in Math. **28** (1978), 57–83.

[Sta4] ___, *Decompositions of rational convex polytopes*, Annals of Discrete Math. **6** (1980), 333–342.

[Sta5] ___, *The number of faces of a simplicial convex polytope*, Adv. in Math. **35** (1980), 236–238.

[Sta6] ___, *Combinatorics and Commutative Algebra*, Birkhaüser, Boston, Basel, and Stuttgart, 1983.

[Sta7] ___, *Two poset polytopes*, Discrete Comput. Geom. **1** (1986), 9–23.

[Sta8] ___, *Enumerative Combinatorics*, vol. I, Wadsworth & Brooks/Cole, Monterey, CA, 1986.

[Sta9] ___, *On the number of faces of centrally-symmetric simplicial polytopes*, Graphs Combin. **3** (1987), 55–66.

[Sta10] ___, *The number of faces of simplicial polytopes and spheres*, Discrete Geometry and Convexity, Ann. of New York Acad. Sci., vol. 440, 1985, pp. 212–223.

[Sta11] ___, *Generalized h-vectors, intersection cohomology of toric varieties, and related results*, Commutative Algebra and Combinatorics (M. Nagata and H. Matsumura, eds.), Adv. Stud. Pure Math., vol 11, North-Holland, Amsterdam, 1987, pp. 187–213.

[Sta12] ___, *Log-concave and unimodal sequences in algebra, combinatorics, and geometry*, Graph Theory and its Applications: East and West, Ann. of New York Acad. Sci. vol. 576, 1989, pp. 500–535.

[Sta13] ___, *F-vectors and h-vectors of simplicial posets*, J. Pure Appl. Algebra **71** (1991), 319–331. (in press).

[Sta14] ____, *On the Hilbert function of a graded Cohen-Macaulay domain*, J. Pure Appl. Algebra **73** (1991), 307–314.

[Sta15] ____, *A monotonicity property of h-vectors and h^*-vectors*, preprint, 1991.

[Tei] B. Teissier, *Variétés toriques et polytopes, Séminaire Bourbaki* 1980/81, *Exp.* 565, Lecture Notes in Math, vol. 901, Springer-Verlag, Berlin, Heidelberg, and New York, 1981, pp. 71–84.

[T-E] G. Kempf, F. Knudsen, D. Mumford, and B. Saint-Donat, *Toroidal Embeddings* I, Lecture Notes in Math., vol. 339, Springer-Verlag, Berlin, Heidelberg, and New York, 1973.

[V-K] V. E. Voskresenskij and A. A. Klyachko, *Toroidal Fano varieties and root systems*, Math. USSR Izv. **24** (1985), 221–244.

[Z-P-W] J. Zaks, M. A. Perles, and J. M. Wills, *On lattice polytopes having interior lattice points*, Elem. Math. **37** (1982), 44–46.

DEPARTMENT OF MATHEMATICS, FACULTY OF SCIENCE, HOKKAIDO UNIVERSITY, KITA-KU, SAPPORO 060, JAPAN

DIMACS Series in Discrete Mathematics
and Theoretical Computer Science
Volume 6, 1991

Unimodular Fans, Linear Codes, and Toric Manifolds

PETER KLEINSCHMIDT, NIELS SCHWARTZ,
AND BERND STURMFELS

ABSTRACT. Using methods from coding theory, we prove a linear upper bound (in the dimension) for the number of vertices of a neighborly unimodular fan. As a consequence we obtain a new class of bounds for the rational Betti numbers of a compact toric manifold.

1. Introduction

Unimodular fans in real d-space provide a combinatorial model for toric manifolds (= smooth toric varieties). In this model the face numbers of a unimodular fan Σ, collected in its f-vector, are in direct correspondence with the ranks of the even-dimensional rational cohomology of the toric manifold, collected in the h-vector of Σ; see [2, 8]. It is an open problem to find an intrinsic characterization for h-vectors of compact toric manifolds, or equivalently, for f-vectors of (complete) unimodular fans. In this article we use methods from coding theory to derive a new class of necessary conditions for f-vectors of unimodular fans.

A *unimodular fan* is a finite system Σ of cones in $\mathbf{R}^d$, pointed at the origin, which satisfies the conditions (i)–(v) below. Here (i)–(iii) state that Σ is a polyhedral cell decomposition of $\mathbf{R}^d$, thus we refer to the elements of Σ as *faces*.

(i) The union of the faces of Σ is $\mathbf{R}^d$;

(ii) the intersection of any two faces of Σ is again a face of Σ;

(iii) if σ is a face of Σ and σ' is a face of the cone σ, then σ' is a face of Σ;

1991 *Mathematics Subject Classification*. Primary 14M25; Secondary 94B05, 52B11.

Supported by the German-Israeli Foundation for Scientific Research and Development, Grant I-84-095.6/88.

Supported in part by the Center for Discrete Mathematics and Theoretical Computer Science.

Supported in part by the National Science Foundation, Grant DMS-9002056.

(iv) every r-dimensional face of Σ is the positive hull of r primitive lattice points in $\mathbf{Z}^d$,

(v) if $x_1, \ldots, x_d$ are the primitive lattice points whose positive hull is a d-dimensional face of Σ, then $|\det(x_1, \ldots, x_d)| = 1$.

If Σ satisfies only (i)–(iv), then it will be called a *simplicial fan*. The d-dimensional faces of Σ are called *facets* and the 1-dimensional faces are called *vertices*. The set of vertices of Σ can be identified with the set $\{x_1, \ldots, x_n\}$ of all primitive lattice points needed to generate the facets of Σ as in (iv). For $0 \le i \le d$, we write $f_i = f_i(\Sigma)$ for the number of i-dimensional faces of Σ. So $f_0 = 1$, $f_1 = n$ is the number of vertices, and f_d is the number of facets. The vector $f(\Sigma) = (f_0, f_1, f_2, \ldots, f_d)$ is the *f-vector* of Σ.

A simplicial fan Σ is said to be *r-neighborly* if every set of r vertices of Σ spans a face of Σ. In other words, a simplicial fan Σ with n vertices is r-neighborly if and only if $f_i(\Sigma) = \binom{n}{i}$ for $i = 0, 1, \ldots, r$. An easy geometric argument shows that there are no r-neighborly fans for $n > d + 1$ and $r > \lfloor d/2 \rfloor$. A simplicial fan is called *neighborly* if it is $\lfloor d/2 \rfloor$-neighborly. Examples of neighborly fans for all d and n are the interior point fans of *cyclic polytopes* [4].

In [5] it was shown that for $n \ge d + 3 \ge 7$ there exists no *unimodular* fan with the combinatorial type of the cyclic polytope, that is, the determinant condition (v) leads to a contradiction for these "cyclic fans." Moreover, it was shown that no unimodular fan has the same f-vector as the cyclic polytope for $d = 4$ and $n = 7, 8$. The same conclusion holds for $n \ge 2^d$ by the following result of Davis and Januszkiewicz [3, Corollary 1.22].

PROPOSITION 1 (Davis and Januszkiewicz). *Let Σ be a 2-neighborly, d-dimensional unimodular fan with n vertices. Then $n < 2^d$.*

It is the objective of the present paper to replace this exponential bound by the following linear bound. Here $\lfloor d/x \rfloor$ denotes the largest integer less or equal to d/x.

THEOREM 2. *Let Σ be a $\lfloor d/x \rfloor$-neighborly unimodular d-dimensional fan with n vertices, where $x \ge 2$ is a real number and d is a sufficiently large integer. Then*

$$n \le \frac{4^x + 2.7}{10.8} d.$$

THEOREM 3. *Given any $\alpha > 0$, there exists an integer $D = D(\alpha)$, such that every neighborly unimodular fan of dimension $d \ge D$ has at most $n < (1+\alpha)d$ vertices.*

The main idea of the proofs is to associate a binary linear code C of length n with every unimodular fan Σ with n vertices. As a consequence of this construction, the degree of neighborliness of Σ is bounded above by the minimum distance of $C^\perp$, the dual code of C, which also has length n. Now any upper bound for the length n of a binary code in terms of its

minimum distance will imply an upper bound for the number of vertices of
a unimodular fan in terms of its neighborliness.

In §2 we explain our construction of binary codes from unimodular fans,
and we illustrate its use with an easy proof of Proposition 1. The proofs of
Theorems 2 and 3 are given in §3. We also demonstrate the limits of coding
theory for our approach by giving lower bounds for the length n of codes with
certain independency conditions. In §4 we comment on the identification of
unimodular fans and toric manifolds, and we state several open problems.

2. Constructing binary codes from unimodular fans

Suppose Σ is a d-dimensional unimodular fan with vertices $x_1, \ldots, x_n \in$
$\mathbf{Z}^d$. Let A be the $(d \times n)$-matrix over $\mathbf{Z}_2$ whose columns are the integer
vectors x_i modulo 2. Then A is a generator matrix of a binary linear
(n, d)-code C. For the basic concepts and facts of coding theory see e.g. [7].

From condition (v) for unimodular fans we see that $x_{i_1}, \ldots, x_{i_r}$ spans
a free $\mathbf{Z}$-module of rank r whenever $pos(x_{i_1}, \ldots, x_{i_r})$ is a face of Σ. In
this case the columns indexed $i_1, \ldots, i_r$ of the binary matrix A are linearly
independent. This implies that if Σ is r-neighborly then every r columns
of A are linearly independent over $\mathbf{Z}_2$. We call a matrix A r-neighborly if
every r columns of A are linearly independent but A has $r + 1$ columns
which are linearly dependent.

PROOF OF PROPOSITION 1. Suppose Σ is a 2-neighborly. Then the corresponding binary $(d \times n)$-matrix A is r-neighborly for some $r \geq 2$. This
implies that the columns of A are pairwise distinct, nonzero vectors in $\mathbf{Z}_2^d$.
Hence $n \leq \#(\mathbf{Z}_2^d \setminus \{0\}) = 2^d - 1$. □

Given A and C as above, let $A^\perp$ be a parity check matrix for C,
that is, $A^\perp$ is an $((n-d) \times n)$-matrix which is a generator matrix of the linear
$(n, n-d)$-code $C^\perp$ dual to C. Equivalently, $A^\perp$ is also the coordinate
matrix (modulo 2) of a *linear transform* (see e.g. [4, 6]) of the vector configuration $\{x_1, \ldots, x_n\}$. Note that $(C^\perp)^\perp = C$, and hence A is a parity check
matrix for the code $C^\perp$.

Fix a real number $x \geq 2$. Let $\mathscr{D}_x$ denote the set of all linear codes which
have a $\lfloor d/x \rfloor$-neighborly parity check matrix of rank d, for some d. Given
a linear code C, then its length will be denoted by n, its dimension by k,
and its minimum distance by δ. We also set $d := n - k$.

LEMMA 4. *If $C \in \mathscr{D}_x$ then $\lfloor d/x \rfloor + 1 = \delta \leq d$.*

PROOF. Pick a parity check matrix A for C which is $\lfloor d/x \rfloor$-neighborly.
Let $\omega \in C \subset \mathbf{Z}_2^n$ be any nonzero code word. Since $A \cdot \omega = 0$, it follows
that the columns of A indexed by $supp(\omega) = \{i : \omega_i \neq 0\}$ are linearly
dependent, and therefore $|supp(\omega)| \geq \lfloor d/x \rfloor + 1$. On the other hand, the
smallest linearly dependent column set of A has cardinality $\lfloor d/x \rfloor + 1$. This

implies the equation

$$\delta = \min\{|\operatorname{supp}(\omega)| : \omega \in C \setminus \{0\}\} = \lfloor d/x \rfloor + 1.$$

The inequality follows from the easy observation that $k + \delta \leq n$. □

In §3 we will combine Lemma 4 with known bounds from coding theory in order to get the desired estimates for the code length n of $C^{\perp}$.

3. Derivation of the bounds

In coding theory there are many different upper bounds for the dimension k of a linear code C in terms of its length n and its distance δ (cf. [7, Chapter 17]). We will use the following two bounds.

LEMMA 5 (Hamming Bound, [7, p. 19, Theorem 6]). *If* $\varepsilon = \lfloor \frac{\delta-1}{2} \rfloor$ *and* $V(n, \varepsilon) = \sum_{i=0}^{\varepsilon} \binom{n}{i}$, *then* $|C| = 2^k \leq 2^n/V(n, \varepsilon)$.

The *binary entropy function* $H_2 : (0, 1) \to \mathbf{R}$ is defined by

$$H_2(x) := -x \log_2(x) - (1 - x) \log_2(1 - x).$$

LEMMA 6 (McEliece-Rodemich-Rumsey-Welch Bound, [7, Theorem 35, p. 559]). *Let* $\beta \in (0, \frac{1}{2})$ *and suppose that* $\mathscr{C}_\beta$ *is a class of linear codes* C *such that* $\beta < \frac{\delta}{n} \leq \frac{1}{2}$. *Then there exists an integer* $\lambda > 0$ *such that*

$$\frac{k}{n} \leq H_2\left(\frac{1}{2} - \sqrt{\beta(1-\beta)}\right) \quad \text{for all } C \in \mathscr{C}_\beta \text{ with } n \geq \lambda.$$

PROOF OF THEOREM 2. Let $x \geq 2$ be a real number, and consider any code $C \in \mathscr{D}_x$ with n, d, k, ε as above. We first assume that $n < 2\delta$. In this case the Plotkin bound [7, Theorem 1, p. 41] shows that

$$k \leq \log_2\left(\frac{\delta}{\delta - n/2}\right). \tag{1}$$

Using Lemma 4 we find $\frac{\delta}{\delta-n/2} \leq n + 1 \leq 2\delta \leq 2d$. Since $k = n - d$ this implies

$$n \leq d + \log_2(d) + 1 \tag{2}$$

and hence the desired inequality.

We may now assume that $2\delta \leq n$. Using Lemma 4 we find

$$\frac{d}{\varepsilon} \leq \left(2 + \frac{1}{\varepsilon}\right)x. \tag{3}$$

Lemma 5 implies $V(n, \varepsilon) \leq 2^d$, and therefore

$$(n - \varepsilon + 1)^{\varepsilon} \leq \binom{n}{\varepsilon}\varepsilon! \leq V(n, \varepsilon)\varepsilon! \leq 2^d \varepsilon!. \tag{4}$$

Choosing $d > 2x$ we may assume that $\varepsilon \geq 1$. By Stirling's formula, we have $\varepsilon! \leq c(\varepsilon)\varepsilon^{1/2}(\varepsilon/e)^{\varepsilon}$ where $c(\varepsilon) := (2\pi e^{1/6\varepsilon})^{1/2}$. Combining this inequality with (3) and (4) we derive

$$n - \varepsilon + 1 \leq (\varepsilon!)^{\varepsilon} 2^{d/\varepsilon} \leq c(\varepsilon)^{1/\varepsilon} \varepsilon^{1/2\varepsilon} \frac{\varepsilon}{e} 2^{d/\varepsilon} \leq c(\varepsilon)^{1/\varepsilon} \varepsilon^{1/2\varepsilon} \frac{\varepsilon}{e} 2^{x/\varepsilon} 4^x. \tag{5}$$

If d tends to infinity then so does ε and the expression $c(\varepsilon)^{1/\varepsilon}\varepsilon^{1/2\varepsilon}2^{x/\varepsilon}$ tends to 1. Choose $d \gg 0$ sufficiently large (and hence $\varepsilon \gg 0$) so that this expression is bounded above by $e/2.7$. Now (5) implies

$$n - \varepsilon \leq \frac{\varepsilon 4^x}{2.7}. \tag{6}$$

We finally observe the inequality $\varepsilon \leq \frac{\delta-1}{2} \leq \frac{d}{2x}$, which follows from Lemma 4 and the definition of ε, and we get from (6)

$$n \leq \frac{4^x + 2.7}{5.4x}d. \tag{7}$$

The coefficient of d grows monotonically for $x \geq 1$, and hence inequality (7) holds for all $\lfloor d/y \rfloor$-neighborly matrices where $1 \leq y \leq x$. Therefore (7) holds for $\lfloor d/x \rfloor$-neighborly fans, which, in view of $x \geq 2$, implies the claimed inequality in Theorem 2. $\quad\square$

We note that Theorem 2 applied to the case $x = 2$ gives the estimate

$$n < \frac{18.7}{10.8}d < 1.75d \tag{8}$$

if d is large enough.

PROOF OF THEOREM 3. Let $x = 2$ and n, d, δ as before. For $n < 2\delta$ the desired inequality follows from (2), and so we continue to assume $n \geq 2\delta$. Using Lemma 4 and (8) we find

$$2\delta \leq n < \frac{z_0}{2}d < z_0\delta \quad \text{with } z_0 = 3.5. \tag{9}$$

Hence the codes under consideration form a class $\mathscr{C}_{1/z_0}$ as in Lemma 6. An application of Lemma 6 yields the inequality

$$\frac{n - d}{n} \leq H_2\left(\frac{1}{2} - \sqrt{\frac{1}{z_0}(1 - \frac{1}{z_0})}\right) \tag{10}$$

whenever d (or, equivalently, n) is large enough. Now consider the sequence $(z_i)_{i \geq 0}$ of real numbers defined by

$$z_{i+1} := 2\left(1 - H_2\left(\frac{1}{2} - \sqrt{\frac{1}{z_i}(1 - \frac{1}{z_i})}\right)\right)^{-1}. \tag{11}$$

Using (10), (11) and the inequality $d < 2\delta$ (cf. Lemma 4) we find

$$2\delta \leq n < dz_1/2 < z_1\delta. \tag{9'}$$

Iterating this process, we obtain the sequence of inequalities $2\delta < n < (z_i/2)d < z_i\delta$, each one for large enough d. It can be shown using elementary analysis that the sequence $(z_i)_{i \geq 0}$ is monotonically decreasing with limit 2. Given any $\alpha > 0$, there exists an i such that $z_i < 2 + \alpha$ and there exists an integer $D = D(\alpha)$ such that, for all codes with $d > D$,

$$n \leq \frac{z_i}{2}d < \frac{2 + \alpha}{2}d \leq (1 + \alpha)d. \tag{12}$$

This completes the proof of Theorem 2 . □

Theorems 1 and 2 are only concerned with the asymptotic behavior of n and d. If one is interested in an upper bound for n, given some specific value of d, then the inequalities derived above also provide an answer.

To determine the quality of our bounds, we would need to find $\lfloor d/x \rfloor$-neighborly unimodular fans with as many vertices as possible. Little is known about this question, which seems quite difficult. The following result for the case $x = 2$ attacks this problem on a different level.

PROPOSITION 7. *For any integer* $d \geq 2$ *there exists a* $\lfloor d/2 \rfloor$-*neighborly binary* $(n \times d)$-*matrix with* $n = d + \lfloor \log_2(d) - \log_2(\log_2(d)) \rfloor$.

The proof of Proposition 7 uses the techniques of [7, p. 550ff.]. Details are omitted here.

Considering arbitrary binary matrices, we disregard a lot of information available about matrices arising from unimodular fans. We may therefore expect better lower bounds for unimodular fans than for linear codes. In other words, Proposition 7 does not say anything about unimodular fans, but it does show the limitations of our coding theory approach. In addition, the matrices constructed in Proposition 7 may prove to be useful also for constructing neighborly unimodular fans with many vertices.

4. Remarks and open problems

In this section we relate our study of face numbers of unimodular fans to compact toric manifolds, and we close with several remarks and open problems.

A *toric d-variety* T is a G-invariant algebraic subvariety of a G-invariant completion of the d-dimensional complex torus $G = (\mathbf{C}^*)^d$. If T is an *affine* toric variety, then its coordinate ring is a finitely generated algebra of Laurent monomials. Such a monomial subalgebra of $\mathbf{C}[x_1, \ldots, x_d, x_1^{-1}, \ldots, x_d^{-1}]$ corresponds to a finitely generated submonoid σ of $\mathbf{Z}^d$, and we can visualize σ as a polyhedral cone in $\mathbf{Z}^d$. General toric varieties are obtained by gluing the spectra of the monomial algebras corresponding to a finite collection of polyhedral cones in $\mathbf{Z}^d$. This construction defines a certain correspondence between polyhedral fans in d-space and d-dimensional toric varieties. Here condition (v) in §1 guarantees that the toric variety corresponding to a fan Σ is nonsingular (i.e. it is a *toric manifold*), while condition (i) corresponds to compactness. For details we refer to [2] and [8].

The f-vector $(f_0, \ldots, f_d)$ of Σ corresponds via an invertible transformation to the so-called h-vector $(h_0, \ldots, h_d)$:

$$h_i = \sum_{j=0}^{i} (-1)^{i-j} \binom{d-j}{d-i} f_j \qquad (0 \leq i \leq d),$$

$$f_j = \sum_{i=0}^{j} \binom{d-i}{d-j} h_i \qquad (0 \leq i \leq d).$$

The invariants h_i can be interpreted in terms of the rational cohomology of a toric manifold X, namely, $h_i = \dim_{\mathbf{Q}} H^{2i}(X, \mathbf{Q})$. It follows from the Upper Bound Theorem for spheres [9] that for a unimodular d-fan Σ with n vertices:

$$h_i \leq \binom{n-d+i-1}{i} \qquad 0 \leq i \leq \lfloor \frac{d}{2} \rfloor.$$

In addition, the fan Σ is r-neighborly if and only if equality holds in the above inequality for $0 \leq i \leq r$. What we have shown in Theorems 2 and 3 is that r-neighborly unimodular fans cannot have "too many" vertices. Or equivalently, a unimodular fan with "many" vertices cannot be r-neighborly and hence its h_i must satisfy smaller upper bounds than the ones given by the Upper Bound Theorem. In summary, our results have the following algebraic-geometric interpretation: The rational Betti numbers of toric manifolds with "large" Picard number (which is equal to $n - d$) have smaller upper bounds than those of arbitrary toric varieties.

We expect that our asymptotic results can be drastically improved for specific Picard numbers. For instance, suppose that Σ is a d-dimensional neighborly unimodular fan with $n = d + 3$ vertices. Then it can be shown that $d \in \{0, 1, 2, 3, 5\}$, which means that the dimension of Σ can be at most 5. This global upper bound is found by inspection of the complete classification of all toric manifolds with Picard number 3. Such a classification follows from

- the complete classification due to Batyrev [1] of all unimodular d-fans with $d + 3$ vertices arising from projective varieties, and
- the fact that all d-dimensional unimodular fans with $d + 3$ vertices are projective [6].

We conjecture the existence of a global dimension upper bound independently of the Picard number. More precisely, it is conceivable that there are no neighborly d-dimensional unimodular fans with $n \geq d + 3 \geq 9$ vertices at all. (Such fans do exist for all $d \geq 2$ and $n = d + 2$.)

As a contrast to our conjecture, Kalai [personal communication] has shown that for given d and large n there exist unimodular fans which asymptotically have as many facets as neighborly fans. However, these fans are not even 2-neighborly.

For $d = 3$ all unimodular fans with n vertices have the same face numbers and it is easy to see that for each $n \geq 4$ there exists a unimodular fan. However, it is an open problem whether every combinatorial type can be realized by a unimodular fan.

REFERENCES

1. V. V. Batyrev, *On the classification of smooth projective toric varieties*, Tohôku Math. J. (to appear).
2. V. I. Danilov, *The geometry of toric varieties*, Math. USSR Izv. **17** (1981), 97–154.
3. M. W. Davis and T. Januszkiewicz, *Convex polytopes, Coxeter orbifolds and torus actions*, Mathematical Sciences Research Institute, preprint 08423-89, Berkeley, 1989.
4. B. Grünbaum, *Convex polytopes*, Interscience, New York, 1967.
5. J. Gretenkort, P. Kleinschmidt, and B. Sturmfels, *On the existence of certain smooth toric varieties*, Discrete Comput. Geom. **5** (1990), 255–262.
6. P. Kleinschmidt and B. Sturmfels, *Smooth toric varieties with small Picard numbers are projective*, Topology **30** (1991), 289–299.
7 F. J. McWilliams and N. J. A. Sloane, *The theory of error-correcting codes*, North-Holland, Amsterdam, 1978.
8. T. Oda, *Convex bodies and algebraic geometry*, Springer-Verlag, Berlin and Heidelberg, 1988.
9. R. P. Stanley, *The upper bound conjecture and Cohen-Macaulay rings*, Studies in Applied Math. **54** (1975), 135–142.

FAKULTÄT FÜR MATHEMATIK UND INFORMATIK, UNIVERSITÄT PASSAU, 8390 PASSAU, FEDERAL REPUBLIC OF GERMANY

DEPARTMENT OF MATHEMATICS, CORNELL UNIVERSITY, ITHACA, NEW YORK 14853

DIMACS Series in Discrete Mathematics
and Theoretical Computer Science
Volume 6, 1991

New Results for Simplicial Spherical Polytopes

PETER KLEINSCHMIDT AND ZEEV SMILANSKY

ABSTRACT. Let P be a convex, simplicial polytope with facets $F_1, \ldots, F_n$. A mapping $\Psi : F_i \to G_i$ from the set of facets into the set of faces of P, is called a *facet cover* if the sets $U_1, \ldots, U_n$, where $U_i = \{K$ a face of $F_i : K \supseteq G_i\}$ form a partition of the face lattice of P. For a convex simplicial polytope, the process of geometric shelling induces, geometrically, both a facet cover and a shelling order. We show that the geometric technique of constructing a facet cover generalizes to the case of spherical simplicial polytopes, while the shelling process may fail. One result is an elementary proof of the upper bound theorem for spherical polytopes. Another is the introduction to the concept of the σ-vector of a spherical simplex, which ties together angle-sums and h-vectors.

1. Introduction

The process of (geometric) shelling of a convex polytope and assignment of ranks to its facets is of basic importance in the combinatorial theory of polytopes. Briefly, the process is as follows: Given a simplicial convex polytope P with $0 \in \operatorname{int} P$, choose a *directing vector* $u \in S^{d-1}$ and use u to establish a shelling $F_1, \ldots, F_n$ of the facets of P. Also, use u to construct a *facet cover* of the lattice of faces of P; by this we mean a mapping $\Psi : F_i \to G_i$ from the set of facets into the set of faces of P, such that the sets $U_i = \{K$ a face of $F_i : K \supseteq G_i\}$ form a partition of the face lattice of P. The construction of the facet cover is achieved in the following way: Let v denote the intersection point of $\operatorname{aff} F_i$ with the line $\operatorname{lin} u$. Then, in $\operatorname{aff} F_i$, v lies beneath some j of the facets of F_i and beyond the other $d - j$, and G_i is the intersection of these j facets. We can now assign a rank to F_i, by $\operatorname{rank}(F_i) = d - j = \dim G_i + 1$. The connection between facet cover, rank assignment, and shelling order is that $\operatorname{rank}(F_i) = \#\{j : j < i, F_j \cap F_i$ is a facet of $F_i\}$, and $U_i = \{K$ a face of $F_i : K \nsubseteq F_j, j < i\}$. As observed by Kalai [**Kal**], there is a strong connection between an assignment of ranks and

---

1991 *Mathematics Subject Classification.* Primary 52B12.

Both authors were supported by a grant from German-Israeli Foundation for Scientific Research and Development.

a shelling order, even when the shelling is nongeometric.

Once the ranks are assigned, one may define $h_k(P)$ as the number of facets of P with rank k and

$$h(P) = (h_0(P), \dots, h_d(P)).$$

The h-vector so defined has the following properties:

1.1. It is independent of the shelling order.

1.2. It is related to the f-vector by the formula

$$f_j(P) = \sum_{k=0}^{j+1} \binom{d-k}{d-j-1} h_k(P), \qquad j = -1, \dots, d-1.$$

1.3. For a vertex v of P, put

$$h_k(v) = \#\{F \text{ a facet of } P, \, v \in F, \, \text{rank}(F) = k\}.$$

Then $h_k(v) \le h_k(P)$, $k = 0, \dots, d$.

In 1970, McMullen [McM] exploited properties 1.1–1.3 to obtain the following results for convex simplicial polytopes:

1.4. $h_i(P) = h_{d-i}(P)$, $i = 0, \dots, [d/2]$ (follows from (1.1)),

1.5. $h_i(P) \le \binom{n-d+i-1}{i}$, $i = 0, \dots, d$ (follows from (1.3)).

McMullen showed that properties 1.2, 1.4, and 1.5 imply the upper bound theorem for convex polytopes.

Our aim in this paper is to show that these results can be generalized to spherical polytopes (or, in other words, to starshaped polytopes). We shall show that the facet cover (and thus the rank assignment) can be constructed geometrically, using precisely the same procedure as in the convex case. In contrast, we shall show that the connection between facet cover and shelling order is broken and cite an example, due to Eikelberg [Eik], of a cover that cannot arise from any shelling. Since we do not use shellings, our proofs will be somewhat more general than other known proofs.

One motivation for our research is the fact that spherical polytopes have recently gained in interest, because they can serve as a model for toric varieties [Oda] and because they play a crucial role in the Gass-Saaty method for parametric linear programming [KlK].

Another motivation is an attempt to understand combinatorial results in polytope theory from a geometric point of view. Our contribution in this direction is the definition and study of the σ-vector of a spherical simplex (§4 below).

2. Definitions, notation, and the basic construction

Let Δ be a j-simplex in E^d, $0 \notin \text{aff}\Delta$. The set $S^{d-1} \cap \text{pos}\,\Delta$ is called a *spherical j-simplex*. A *simplicial spherical d-polytope* is a realization of a simplicial complex on S^{d-1}, where all cells are spherical simplices and their union covers S^{d-1}. Consider a spherical $(d-1)$-simplex Δ on S^{d-1}. Its facets determine d great circles $C_1, \ldots, C_d$ of S^{d-1}. Each C_i separates $S^{d-1} \backslash C_i$ into two (open) hemispheres, which we denote by C_i^+ and C_i^-, where by convention $\mathrm{int}\, \Delta \subseteq C_i^+$. These d great circles partition S^{d-1} into 2^d spherical simplices. A typical member is

$$\Delta' = \bigcap_{i=1}^{d} C_i^{\varepsilon_i},$$

with each ε_i being a "$+$" or a "$-$." We associate a rank with Δ' by $\mathrm{rank}(\Delta') = \#\{i : \varepsilon_i = \text{"} - \text{"}\}$.

For $k = 0, \ldots, d$, denote by $A_k = A_k(\Delta)$ the union of all (open) simplices, with rank k, in the partition induced by Δ. Consider a simplicial spherical d-polytope P. Write $\mathcal{N} = \mathcal{N}(P)$ for the union of all great circles determined by the subfacets of P. Clearly, $\mathcal{N}$ has Lebesque measure zero. Choose $u \in S^{d-1} \backslash \mathcal{N}$. We call u a *directing vector* and define, for every facet Δ of P,

$$\mathrm{rank}_u(\Delta) = k \Leftrightarrow u \in A_k(\Delta),$$
$$h_k(P, u) = \#\{\Delta \text{ a facet of } P : \mathrm{rank}_u(\Delta) = k\},$$
$$h(P, u) = (h_0(P, u), \ldots, h_d(P, u)).$$

Consider a facet Δ of P with rank k. This means that $u \in \bigcap_{i=1}^{d} C_i^{\varepsilon_i}$, with k of the ε_i's being a "$-$" sign. Put

$$F(\Delta) = \Delta \cap \bigcap \{C_i : \varepsilon_i = \text{"}+\text{"}\},$$
$$\mathcal{S}(\Delta) = \{F \text{ a face of } \Delta : F \supseteq F(\Delta)\}.$$

In what follows, we shall sometimes be interested in the rank of facet and sometimes in the specific set $\mathcal{S}(\Delta)$ associated with it. The reader should note these definitions and compare them with the discussion, in the previous section, of facet covers of convex polytopes. Specifically, $F(\Delta)$ corresponds to G_i, and $\mathcal{S}(\Delta)$ corresponds to U_i (see §3).

We shall now give two different lines of proof for properties 1.1–1.3. The two approaches use the same basic ingredients, but the recipes are different. Before we start, we need the following definitions.

Let F be a j-face of a $(d-1)$-spherical simplex Δ on S^{d-1}, $0 \le j < d-1$. Thus, F is the intersection of Δ with $(d-j-1)$ of the great circles determined by Δ, say $F = \Delta \cap C_{i_1} \cap \cdots \cap C_{i_{d-j-1}}$. Some stretching of notation makes this meaningful also for the case $j = d - 1$. Then

$$L(F, \Delta) = C_{i_1}^+ \cap \cdots \cap C_{i_{d-j-1}}^+$$

is called the *interior angle* of Δ at F. For a collection $\mathcal{W}$ of subsets of S^{d-1} we say that $\mathcal{W}$ is a *k-cover* of S^{d-1} if and only if every point $p \in S^{d-1} \backslash \mathcal{N}$ is contained in precisely k members of $\mathscr{W}$, where $\mathscr{N} \subseteq S^{d-1}$ has (Lebesgue) measure zero.

3. First approach

Fix a simplicial spherical d-polytope P with facets $\Delta_1, \ldots, \Delta_n$, and a directing vector $u \in S^{d-1} \backslash \mathscr{N}(P)$.

LEMMA 3.1. *The mapping* $\Delta \to F(\Delta)$ *is a facet cover of* P, *that is, every face of* P *belongs to* $\mathscr{S}(\Delta)$ *for precisely one* Δ.

PROOF. Suppose that F is a face of P and $\Delta_1, \ldots, \Delta_m$ are the facets of P that contain F. Obviously, $\mathscr{W} = \{L(F, \Delta_i): i = 1, \ldots, m\}$ is a 1-cover of S^{d-1}. Thus, there is a unique i_0 satisfying $u \in L(F, \Delta_{i_0})$. For Δ_{i_0}, there are $C_{i_1}, \ldots, C_{i_j}$ such that $F = \Delta_{i_0} \cap C_{i_1} \cap \cdots \cap C_{i_j}$ and $u \in C_{i_1}^+ \cap \cdots \cap C_{i_j}^+$. Since $F(\Delta)$ is the intersection of all C_i^+'s to which u belongs, $F(\Delta_{i_0}) \subseteq F$, so that $F \in \mathscr{S}(\Delta_{i_0})$.

On the other hand, suppose that Δ is another facet of P that contains F. Denote the great circles spanned by the facets of Δ by $C_1, \ldots, C_d$. Any F in $\mathscr{S}(\Delta)$ is an intersection of a partial set of those C_i^+'s to which u belongs. But, since $\Delta \neq \Delta_{i_0}$, $u \notin L(F, \Delta)$, so that some i_1 exists, $u \in C_{i_1}^-$, $F \subseteq C_{i_1}^+$. Thus, $F \not\supseteq F(\Delta)$ so that $F \notin \mathscr{S}(\Delta)$.

COROLLARY 3.2. *We have*

$$f_j(P) = \sum_{k=0}^{j+1} \binom{d-k}{d-j-1} h_k(P, u), \qquad j = -1, \ldots, d-1.$$

PROOF. If a facet Δ of P has rank k, then $F(\Delta)$ is the intersection of some $d-k$ facets of Δ while a j-face is the intersection of $d-j-1$. Thus, $\mathscr{S}(\Delta)$ contains $\binom{d-k}{d-j-1}$ j-faces, and the result follows.

COROLLARY 3.3. $h(P, u)$ *is invariant of the choice of the directing vector* u.

PROOF. The matrix relating the f-vector to the h-vector is triangular, hence invertible, so that the h-vector depends only on the numbers of faces of P.

LEMMA 3.4. *For every vertex* v *of* P, $h_i(v) \leq h_i(P)$.

PROOF. Let u be chosen on $S^{d-1} \backslash \mathscr{N}(P)$, sufficiently near v so that the $(d-2)$-sphere T on S^{d-1}, with center v, passing through u, intersects only those facets of P that contain v. Consider the spherical $(d-2)$-polytope Q on T, with facets $\{\Delta \cap T: \Delta$ a facet of $P\}$ and directing vector u. Clearly the rank of $\Delta \cap T$ in Q is equal to the rank of Δ in P. Thus, the contributions to $h(P)$ and to $h(Q)$ are equal. Since P may have facets that do not intersect T, the result follows.

REMARKS. 1. The proofs of the corollaries and of Lemma 3.4 are identical to the known proofs (cf. [McM]) and have been given to make the discussion complete.

2. The upper-bound theorem has been proved by Stanley [Sta] for spherical simplicial (not necessarily starshaped) complexes. In his proof, he used the fact that a certain ring associated with the complex is a Cohen-Macaulay-ring. In [KiK], a constructive approach to the Cohen-Macaulay property was given for shellable complexes. Even though our directing vector approach for spherical simplicial polytopes does not yield a shelling (see §5), we conjecture that the method of [KiK] carries over to spherical simplicial polytopes. In the sequel, we make this more precise:

Let $\mathscr{C}$ be a simplicial complex corresponding to a d-dimensional spherical simplicial polytope with vertices $x_1, \dots, x_m$ and facets $\Delta_1, \dots, \Delta_n$. Then the *Stanley-Reisner-ring* $A_{\mathscr{C}}$ is defined as follows:

Let K be an infinite field and $A := K[x_1, \dots, x_m]$ the polynomial ring in the independent variables $x_1, \dots, x_m$ over K (the vertices x_i are identified with the variables). Let I be the ideal of A which is generated by the square-free monomials $x_{i_1} \cdots x_{i_k}$ for which $\{x_{i_1}, \dots, x_{i_k}\}$ is not the vertex set of any face of $\mathscr{C}$. Then $A_{\mathscr{C}} := A/I$.

$A_{\mathscr{C}}$ is Cohen-Macaulay if and only if there are homogeneous polynomials $\Theta_1, \dots, \Theta_d$ of degree 1 (with respect to the natural grading inherited from A) s.t. $A_{\mathscr{C}}$ is a free $K[\Theta_1, \dots, \Theta_d]$-module.

Let I be the ideal in $A_{\mathscr{C}}$ generated by a such a sequence $\Theta_1, \dots, \Theta_d$.

Then the ring $B_{\mathscr{C}} := A_{\mathscr{C}}/I$ has a grading $B_{\mathscr{C}} := B_0 \oplus \cdots \oplus B_d$ where $\dim_k B_i = h_i(\mathscr{C})$. In the case that $\Delta_1, \dots, \Delta_n$ form a shelling order of $\mathscr{C}$, it has been shown in [KiK] that those monomials in $B_{\mathscr{C}}$, which correspond to the set of faces $\{F(\Delta_j) \mid \text{rank}(\Delta_j) = d - i\}$ as defined above, form a K-basis of B_i.

We conjecture that the same holds for the set of the $F(\Delta_j)$ obtained from a directing vector u of a (not necessarily shellable) simplicial spherical polytope.

This would be important in view of the fact that for a d-dimensional complete toric variety X, which has only quotient singularities, this explicit basis for $B_{\mathscr{C}}$ would yield an explicit basis for the rational cohomology of X. See [Oda] for details.

4. Second approach— σ-vectors and φ-vectors

Again, we fix a simplicial spherical d-polytope P with facets $\Delta_1, \dots, \Delta_n$ and a directing vector $u \in S^{d-1} \setminus \mathscr{N}(P)$. Let $\mu = \mu_{d-1}$ denote Lebesgue measure on S^{d-1}, normalized so that $\mu_{d-1}(S^{d-1}) = 1$. For a $(d-1)$-spherical simplex Δ, write $\sigma_k(\Delta) = \mu_{d-1}(A_k(\Delta))$ and define the σ-vector of

Δ by

$$\sigma(\Delta) = (\sigma_0(\Delta), \ldots, \sigma_d(\Delta)).$$

Also, write $\varphi_j(\Delta) = \sum_{\dim F = j} \mu_{d-1}(L(F, \Delta))$, the j th angle sum of Δ, and define the φ-vector of Δ by

$$\varphi(\Delta) = (\varphi_{-1}(\Delta), \ldots, \varphi_{d-1}(\Delta)).$$

LEMMA 4.1. *The h-vector $h(P, u)$ is invariant with respect to the choice of the directing vector u. Moreover, the set $\{A_k(\Delta_i): i = 1, \ldots, n\}$ is an $h_k(P)$-cover of P, for each $k = 0, \ldots, n$.*

REMARK. The claim holds trivially for $k = 0$.

PROOF. Suppose that $C_1, \ldots, C_m$ are the great circles of S^{d-1} spanned by all subfacets of P. The collection $\{C_1, \ldots, C_m\}$ partitions S^{d-1} into open cells (not necessarily simplices). In each cell, $h(P, u)$ is invariant. Consider a pair D_1 and D_2 of neighboring cells, that is, a sequence $\varepsilon_1, \ldots, \varepsilon_m$ of " $+$ " and " $-$ " signs that allows us to write, with no loss of generality,

$$D_1 = C_{i_0}^+ \cap \bigcap_{\substack{i=1 \\ i \neq i_0}}^m C_i^{\varepsilon_i}, \qquad D_2 = C_{i_0}^- \cap \bigcap_{\substack{i=1 \\ i \neq i_0}}^m C_i^{\varepsilon_i}.$$

Suppose that Δ_1 and Δ_2 are the facets of P whose intersection spans C_{i_0}. Pick $u_1 \in D_1$ and $u_2 \in D_2$. Clearly, for every other facet Δ', $\operatorname{rank}_{u_1}(\Delta') = \operatorname{rank}_{u_2}(\Delta')$, whereas for Δ_1 and Δ_2 we can write, again with no loss of generality,

$$\operatorname{rank}_{u_2}(\Delta_1) = \operatorname{rank}_{u_1}(\Delta_1) - 1,$$
$$\operatorname{rank}_{u_2}(\Delta_2) = \operatorname{rank}_{u_1}(\Delta_2) + 1,$$

Thus the total change cancels, and $h(P, u)$ remains unchanged. The first claim follows easily.

Recalling that $h_k(P, u) = \#\{\Delta_i: \operatorname{rank}_u(\Delta_i) = k\}$, we see that almost everywhere on S^{d-1}, u belongs to a fixed number of the sets $A_k(\Delta_i)$. That fixed number is $h_k(P, u)$ so the second claim is proved.

LEMMA 4.2. *We have*

$$h_k(P) = \sum_{i=1}^n \sigma_k(\Delta_i).$$

PROOF. For $u \in S^{d-1} \backslash \mathcal{N}(P)$, define

$$\chi_k(\Delta_i, u) = \chi(A_k(\Delta_i), u) = \begin{cases} 1 & u \in A_k(\Delta_i) \\ 0 & \text{otherwise.} \end{cases}$$

By definition,

$$h_k(P) = h_k(P, u) = \sum_{i=1}^n \chi_k(\Delta_i, u)$$

Thus,

$$
\begin{aligned}
h_k(P) &= \int_{S^{d-1}} h_k(P, u) d\mu_{d-1}(u) \\
&= \int_{S^{d-1}} \left(\sum_{i=1}^{n} \chi_k(\Delta_i, u) \right) d\mu_{d-1}(u) \\
&= \sum_{i=1}^{n} \int_{S^{d-1}} \chi_k(\Delta_i, u) d\mu_{d-1}(u) \\
&= \sum_{i=1}^{n} \mu_{d-1}(A_k(\Delta_i)) \\
&= \sum_{i=1}^{n} \sigma_k(\Delta_i).
\end{aligned}
$$

COROLLARY 4.3.

$$
f_j(P) = \sum_{k=0}^{j+1} \binom{d-k}{d-j-1} h_k(P), \qquad j = -1, \ldots, d-1.
$$

PROOF. It is easy to show, directly from the definitions, that the following holds:

$$
\varphi_j(\Delta) = \sum_{k=0}^{j+1} \binom{d-k}{d-j-1} \sigma_k(\Delta), \qquad j = -1, \ldots, d-1.
$$

As in the proof of Lemma 3.1, for every face F of P, $\bigcup_{\Delta \supseteq F} L(F, \Delta)$ is a 1-cover of S^{d-1}. Thus,

$$
\sum_{\Delta \supseteq F} \mu_{d-1}(L(F, \Delta)) = 1,
$$

and so

$$
\begin{aligned}
\sum_{i=1}^{n} \varphi_j(\Delta_i) &= \sum_{\dim F = j} \sum_{\Delta \supseteq F} \mu_{d-1}(L(F, \Delta)) \\
&= \sum_{\dim F = j} 1 = f_j(P).
\end{aligned}
$$

Combining the three equations

$$
f_j(P) = \sum_{i=1}^{n} \varphi_j(\Delta_i),
$$

$$
h_k(P) = \sum_{i=1}^{n} \sigma_k(\Delta_i),
$$

$$
\varphi_j(\Delta) = \sum_{k=0}^{j+1} \binom{d-k}{d-j-1} \sigma_k(\Delta),
$$

the result follows.

Note that in this approach, the proof of property 1.3 is identical to that given in the first approach.

REMARKS. 1. The φ-vector and the σ-vector can be considered as local versions of the f-vector and the h-vector. Thus, $\varphi(\Delta)$ and $\sigma(\Delta)$ can be interpreted as the contributions to $f(P)$ and $h(P)$ from the facet Δ, depending on the geometry of Δ and the choice of origin, but not of directing vector. These contributions are real numbers, which sum to integers. More precisely, $\sigma_k(\Delta)$ is the probability that Δ will have rank k when the directing vector is chosen at random. It is interesting to note that with this view Sommerville's formula for the volume of a spherical simplex

$$\sum_{j=0}^{d-1}(-1)^j\varphi_j(\Delta) = (1+(-1)^d)\sigma_0(\Delta)$$

turns out to be a local version of Euler's relation

$$\sum_{j=0}^{d-1}(-1)^j f_j(P) = 1+(-1)^d.$$

We remark that the similarities between the φ-vector and the f-vector were noticed long ago. One period of activity was the late 1960 s, when papers of Perles, Shepard, Grünbaum, and others appeared (see [Pe-Shep, Gru2, and Shep]). However, the connection with h-vectors (Lemma 4.2) seems new.

2. The local vectors φ and σ preserve, in a strong way, the properties of the global vectors f and h. We have seen examples above; another interesting example is the Dehn-Sommerville relations $h_k = h_{d-k}$, which follow, using Lemma 4.2, from the relations $\sigma_k(\Delta) = \sigma_{d-k}(\Delta)$, which, in turn, are a consequence of the congruences $A_k(\Delta) = -A_{d-k}(\Delta)$. It seems reasonable to look for further connections, the most interesting of which is the connection between the unimodality properties of the h-vector and the σ-vector.

3. In this spirit, we venture a conjecture. The connection between the shape of a spherical simplex and its σ-vector is, to say the least, obscure. Basic questions such as the extent to which the σ-vector determines the shape, or even the determination of the σ-vector in simple cases, are unanswered at present. In particular, the question of when a spherical simplex has a unimodal σ-vector is completely open. However, it seems that unimodality has something to do with regularity, and it is possible to show that a regular spherical simplex with edge length at most $1/2$ has a unimodal σ-vector.

It is also easy to construct spherical simplices with nonunimodal σ-vectors. Such simplices are almost degenerate in that some of their edges are very long and others very short. Drawing on the celebrated g-conjecture of McMullen [McM2], and its proof by Stanley [Sta2], we state the following:

CONJECTURE. *To every convex simplicial polytope* P *corresponds a spherical simplicial polytope* Q, *combinatorially equivalent to* P, *all of whose facets have a unimodal* σ*-vector.*

5. Rank Assignments and Shellings

We have shown how one can choose a directing vector u and with it construct a facet cover of a simplicial spherical polytope P. In the case of convex polytopes, a shelling order $F_1, \ldots, F_n$ is also obtained from u. This order is related to the facet cover, as discussed in the introduction.

A facet cover induces a partial order on the set of facets, in the following way:

$$F_i \cap F_j \in \mathscr{S}(F_i) \to F_i \prec F_j.$$

Thus, every subfacet $F_i \cap F_j$ carries a sense, either from F_i to F_j or vice versa. In the convex case, this partial order agrees with the complete shelling order determined by the directing vector u.

We come now to Eikelberg's example. Consider a regular triangular prism P in E^3, with the origin in its center, and choose u through the axis of P (Figure 1A). Rotate one of the bases slightly, and break the sides in the non-convex way (Figure 1B). The spherical image of the result is our example, indicated in Figure 1C. It is easy to check that the senses of the subfacets form a cycle and so cannot correspond to any shelling.

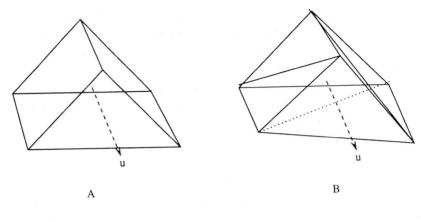

A   B

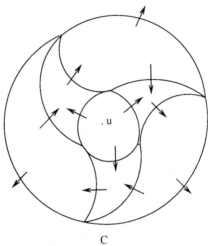

C

FIGURE 1

References

[Eik] M. Eikelberg, *Zur Homologie torischer Varietäten*, Ph.D. thesis, Fakultät für Mathematik der Universität Bochum, 1989.

[Gru] B. Grünbaum, *Convex Polytopes*, Wiley, New York, 1967.

[KiK] B. Kind and P. Kleinschmidt, *Schälbare Cohen-Macauley-Komplexe und ihre Parametrisierung*, Math. Z. **167** (1979), 173–179.

[Gru2] ____, *Grassman angles of convex polytopes*, Acta Math. **121** (1968), 293–302.

[Kal] G. Kalai, *A simple way to tell a simple polytope from its graph*, J. Combin. Theory Ser. A **49** (1988), 381–383.

[KlK] V. L. Klee and P. Kleinschmidt, *Geometry of the Gass-Saaty parametric cost LP algorithm*, Discrete Comput. Geom. **5** (1990), 13–26.

[Oda] T. Oda, *Convex bodies and algebraic geometry*, Springer-Verlag, Berlin and Heidelberg, 1988.

[McM] P. McMullen, *The maximum numbers of faces of a convex polytope*, Mathematika **17** (1970), 179–184.

[McM2] ____, *The numbers of faces of simplicial polytopes*, Israel J. Math. **9** (1971), 559–570.

[Pe-Shep] M. A. Perles and G. C. Shephard, *Angle sums of convex polytopes*, Math. Scand. **21** (1967), 199–218.

[Shep] G. C. Shephard, *Angle deficiencies of convex polytopes*, J. London Math. Soc. **43** (1968), 325–336.

[Sta] R. P. Stanley, *The upper bound conjecture and Cohen-Macaulay-rings*, Stud. Appl. Math. **54** (1975), 135–142.

[Sta2] ____, *The number of faces of a simplicial convex polytope*, Adv. in Math. **35** (1980), 236–238.

FAKULTÄT FÜR MATHEMATIK UND INFORMATIK, UNIVERSITÄT PASSAU, INNSTRASSE 33, 8390 PASSAU, GERMANY

ORBOT SYSTEMS LTD., 70651 YAVNE, ISRAEL

DIMACS Series in Discrete Mathematics
and Theoretical Computer Science
Volume **6**, 1991

Rational-function-valued Valuations on Polyhedra

JIM LAWRENCE

1. Introduction

Let $\mathscr{L}$ be a (distributive) lattice of subsets of a set X under intersection and union. Recall from [4] that a *valuation* on $\mathscr{L}$ is a function $\nu: \mathscr{L} \to \mathscr{A}$, where $\mathscr{A}$ is an abelian group (written additively), satisfying the *modular identity*

$$\nu(A \cup B) + \nu(A \cap B) = \nu(A) + \nu(B),$$

for A, $B \in \mathscr{L}$. As an example (with $\mathscr{A} = R$), one has the ordinary volume function on the lattice of finite unions of convex polytopes in R^d, and (with $\mathscr{A} = Z$), one has the Euler characteristic on this same lattice. See [1], where the Euler characteristic is described in a lattice-theoretic setting. For a survey of valuations on convex polytopes, see [3].

In this paper, we describe some valuations on the lattice of finite unions of convex polyhedra (or, in §5, rational convex polyhedra) for which the abelian group is $R(u_1, u_2 \ldots, u_d)$, the additive group of rational functions in d indeterminates over R. The domains of these valuations are not restricted to bounded sets. Furthermore, they share the property that when applied to a convex polyhedron containing a line, they yield 0.

The volume function is not defined on unbounded polyhedra. Because of this, one cannot use in the calculation of volume a variety of geometrical constructions—notably Gram's relation in its combinatorial form. See [6] or [2]. One of the valuations we describe agrees with volume on bounded sets. This valuation can be used in an algorithm based on Gram's relation and simplex-like pivoting to compute the volume of a convex polytope. See [2].

For all the valuations we consider, there will be a natural definition for the value on convex polyhedra not containing lines, that is, on convex polyhedra that have at least one vertex and on the empty set. From this definition, it will be clear that the modular identity is satisfied for convex polyhedra

---

1991 *Mathematics Subject Classification*. Primary 52A25.
The author was supported by NSF Grant #DMS-8711581.

containing no lines. Since every polyhedron can be written as a finite union of polyhedra, which do not contain lines, the principle of inclusion and exclusion, which holds for valuations, dictates the value on any polyhedron. It follows that there is at most one valuation defined on all polyhedra, which has the natural value given for convex polyhedra containing no lines. In the cases we consider, it will be proved that the value on convex polyhedra containing lines is 0.

Since the valuations are defined on all polyhedra, Gram's relation can be used. One has that, for any convex polyhedron P having at least one vertex, and for any valuation ν, which is defined on P and all the cones generated by P from its faces,

$$\nu(P) = \sum_F (-1)^{\dim(F)} \nu(\operatorname{cone}(F, P)),$$

where the sum is taken over all nonempty, bounded faces F of P, and $\operatorname{cone}(F, P)$ denotes the cone generated by P from F:

$$\operatorname{cone}(F, P) = \{\alpha x + \beta y : \alpha + \beta = 1, \beta \geq 0, x \in F, y \in P\}.$$

For the valuations ν considered here, Gram's relation is particularly simple: Since ν has value 0 on convex polyhedra which contain lines, one has

$$\nu(P) = \sum_v \nu(\operatorname{cone}(\{v\}, P)),$$

where the summation extends over all vertices v of P. For instance, when used in this formula the valuation of §3 yields a formula for the volume of the projective image of a convex polytope.

If it is desired to find an explicit *real*-valued valuation on polyhedra, which coincides with volume on polytopes, this can be done using either of the valuations in §§3 and 4: Simply compose the valuation with a homomorphism from $R(u_1, \ldots, u_d)$ to R, which has the property that it yields the value of the rational function at the origin when this is defined. (Such homomorphisms are easy to find.) Similarly, using the valuation in §5, one can find a valuation on rational polyhedra which, for polytopes P, gives the cardinality of the set $P \cap Z^d$.

2. Notes on valuations

In this paper, we identity R^d with the set of column vectors $x = [x_1, \ldots, x_d]^t$ of d real numbers.

We will be concerned with lattices of subsets of R^d. The elements of the lattice of finite unions of convex polytopes in R^d will be called, simply, *polytopes*, so that a polytope in this paper need not be convex. Similarly, the elements of the distributive lattice of finite unions of closed, convex polyhedra will be called *polyhedra*. In §5 we will be interested in sublattices of each of these: A *rational polytope* will be a finite union of convex polytopes whose vertices have rational coordinates; a *rational polyhedron* will be the finite union of convex polyhedra that are obtained as the intersection of finitely many halfspaces of the form $\{x \in R^d : a^t x \leq b\}$, where b is rational and a has rational entries.

A fact due to Volland [8] will be useful. Suppose $\mathscr{L}$ is a lattice generated by a family $\mathscr{S}$ of sets closed under (finite) intersection so that $\mathscr{L}$ is the collection of finite unions of elements of $\mathscr{S}$. Suppose $\nu_0 : \mathscr{S} \to \mathscr{A}$ is a function such that, whenever $S = S_1 \cup \cdots \cup S_k$, where $S_1, \ldots, S_k, S \in \mathscr{S}$, the principle of inclusion and exclusion holds for ν_0:

$$\nu_0(S) = \sum_J (-1)^{|J|} \nu_0 \left(\bigcap_{i \in J} S_i \right),$$

where the summation extends over nonempty subsets J of $\{1, \ldots, k\}$. Then ν_0 has a unique extension ν to a valuation on $\mathscr{L}$.

A theorem of Sallee [5] will also find a use in the proof of Theorem 1. Suppose ν_0 is a function on closed, convex polyhedra (or on closed, convex, rational polyhedra) having values in an abelian group $\mathscr{A}$. The function ν_0 is called a *weak valuation* if for each such polyhedron P and for each hyperplane H one has

$$\nu_0(P) + \nu_0(P \cap H) = \nu_0(P \cap H^+) + \nu_0(P \cap H^-),$$

where H^+ and H^- are the two closed halfspaces bounded by H. Sallee proved that a weak valuation extends uniquely to a valuation ν on polyhedra (or on rational polyhedra). (The values are dictated by the principle of inclusion and exclusion. Actually, Sallee proved his theorem in the case of the lattice of polytopes; however, similar arguments yield the results used here. Incidentally, an example of Groemer shows that the analogous statement for the lattice of finite unions of compact, convex sets is false: See [3, p. 173].)

Finally, we give a method of constructing a valuation from a collection of functions ν_p for $p \in R^d$ where ν_p is a valuation on the lattice of finite unions of closed, convex, polyhedral cones emanating from p. The following theorem describes the construction.

THEOREM 1. *Suppose that for each $p \in R^d$, ν_p is a valuation on polyhedral cones emanating from p, with values in an abelian group $\mathscr{A}$. Additionally, suppose that whenever the convex polyhedron P contains a line and $p \in R^d$, $\nu_p(U) = 0$. For a closed, convex polyhedron P in R^d define*

$$\nu(P) = \sum_p \nu_p(\mathrm{cone}(\{p\}, P)).$$

(Here, the summation extends over the vertices p of P.) Then ν extends uniquely to a valuation on polyhedra.

PROOF. By Sallee's theorem we need only show that ν is a weak valuation on convex polyhedra. Let P be a closed, convex polyhedron and H a hyperplane bounding closed halfspaces H^+ and H^-. We must show that

$\nu(P)+\nu(P\cap H)=\nu(P\cap H^+)+\nu(P\cap H^-)$. The vertices of $P\cap H^+$, $P\cap H^-$, and $P\cap H$ are of two types: Those that are vertices of P and those that are intersections with H of edges of P whose relative interiors intersect H. Denote the contributions to $\nu(P)$, $\nu(P\cap H)$, $\nu(P\cap H^+)$, and $\nu(P\cap H^-)$ at a vertex p of P by τ_p, τ_p^0, τ_p^+, and τ_p^-, respectively. For example, $\tau_p^+ = \nu_p(\text{cone}(\{p\}, P\cap H^+)$ if the vertex p of P is in H^+, and $\tau_p^+ = 0$ otherwise. Similarly, if E is an edge of P whose relative interior intersects H, denote the contributions at p, where $\{p\} = E\cap H$, by σ_E^0, σ_E^+, and σ_E^-. Then $\nu(P) = \Sigma\tau_p$, $\nu(P\cap H) = \Sigma\tau_p^0 + \Sigma\sigma_E^0$, $\nu(P\cap H^+) = \Sigma\tau_p^+ + \Sigma\sigma_E^+$, and $\nu(P\cap H) = \Sigma\tau_p^- + \Sigma\sigma_E^-$, so that it will suffice to show that

(a) $\tau_p + \tau_p^0 = \tau_p^+ + \tau_p^-$ for each vertex p of P, and
(b) $\sigma_E^0 = \sigma_E^+ + \sigma_E^-$ for each edge E of P crossing H.

If p is a vertex not in H^-, then $\tau_p^- = \tau_p^0 = 0$; also cone$(\{p\}, P) = $ cone$(\{p\}, P\cap H^+)$ in this case, so that $\tau_p = \tau_p^+$, and (a) is satisfied. Similarly (a) holds if p is not in H^+. If $p \in H$, then

(c) cone$(\{p\}, P) = $ cone$(\{p\}, P\cap H^+)\cup$ cone$(\{p\}, P\cap H^-)$, and
(d) cone$\{p\}, P\cap H) = $ cone$(\{p\}, P\cap H^+)\cap$ cone$(\{p\}, P\cap H^-)$.

So, by the modular identity applied to ν_p, we have (a). Finally, if E is an edge crossing H, and if $E\cap H = \{p\}$, then we again have (c) and (d). Since cone$(\{p\}, P)$ contains a line, ν_p applied to this set vanishes, so by the modular identity (b) holds. $\square$

3. Volume of projective image

In this section we describe a valuation φ that maps polyhedra in R^d to elements of the group $\mathscr{A} = R(u_1, \ldots, u_d)$ of rational functions under addition.

For a point $u \in R^d$, let $\tau_u(x) = x/(1 + u^t x)$. We consider this function to have as its domain the open halfspace

$$H_u = \{x \in R^d : u^t x > -1\}.$$

Let P be a convex polyhedron contained in this set such that the image $\tau_u(P)$ is bounded. Let $\varphi(P)(u)$ be the volume of $\tau_u(P)$. As we will see, for fixed P this is a rational function of u, and we will be able to view $\varphi(P)$ as an element of $\mathscr{A}$.

Let Q^d denote the nonnegative orthant in R^d:

$$Q^d = \{[x_1, \ldots, x_d]^t : x_i \geq 0 \text{ for } i = 1, \ldots, d\}.$$

We determine $\varphi(Q^d)$.

LEMMA 1. If $u = [u_1, \ldots, u_d]^t$ is in the interior of Q^d, then

$$\varphi(Q^d)(u) = 1/(d!u_1 \cdots u_d).$$

PROOF. In this case, $\tau_u(Q^d)$ is (up to interior) the convex hull of the origin and the points $(1/u_1)e_1, \ldots, (1/u_d)e_d$, where e_i is the ith coordinate vector. The volume of this simplex is $1/(d!u_1\ldots u_d)$. □

The following theorem describes how φ varies when a projective linear transformation is applied to P. Recall that an arbitrary projective linear transformation μ of a subset of R^d having the origin in its domain can be written in the form

$$\mu(x) = (Ax + b)/(c^t x + 1),$$

where A is a $d \times d$ matrix and b and c are column vectors in R^d. This can be rewritten as

$$\mu(x) = (A - bc^t)x/(c^t x + 1) + b,$$

a composition of the function τ_c followed by a linear transformation followed by a translation. Therefore, it suffices to handle each of these types of functions individually.

THEOREM 2. *Let A be a $d \times d$ matrix, b and c vectors in R^d, and P a polyhedron such that $P \subseteq H_c$ and $\tau_c(P)$ is bounded. Then:*

(a) $\varphi(AP)(u) = |\det A|\varphi(P)(A^t u)$;
(b) $\varphi(P + b)(u) = (1/(1 + u^t b)^{d+1})\varphi(P)(u/(1 + u^t b))$;
(c) $\varphi(\tau_c(P))(u) = \varphi(P)(u + c)$.

PROOF. We have that $\varphi(P)(u)$ is given by integration as

$$\varphi(P)(u) = \text{vol } \tau_u(P) = \int_{\tau_u(P)} dy = \int_P |J\tau_u(x)|\, dx$$

$$= \int_P dx/(1 + u^t x)^{d+1},$$

where $J\tau_u(x)$ denotes the Jacobian determinant of τ_u at x. For (a), we have

$$\varphi(AP)(u) = \int_{AP} dx/(1 + u^t x)^{d+1}$$

$$= \int_P |\det A|/(1 + u^t Ax)^{d+1}\, dx$$

$$= |\det A|\varphi(P)(A^t u).$$

For (b), we have

$$\varphi(P + b)(u) = \int_{P+b} dy/(1 + u^t y)^{d+1}$$

$$= 1/(1 + u^t b)^{d+1} \int_P dx/(1 + u^t x/(1 + u^t b)^{d+1})$$

$$= (1/(1 + u^t b)^{d+1})\varphi(P)(u/(1 + u^t b)).$$

Finally, for (c),

$$\varphi(\tau_c(P))(u) = \mathrm{vol}(\tau_u(\tau_c(P)))$$
$$= \mathrm{vol}\,\tau_{u+c}(P) = \varphi(P)(u+c). \quad \square$$

COROLLARY. *If T is the simplicial cone given by $T = AQ^d + b$, then*

$$\varphi(T)(u) = |\det A|/(d!(1 + u^t b)w_1 \ldots w_d),$$

where $[w_1, \ldots, w_d]^t = A^t u$.

PROOF. From parts (a) and (b) of the theorem we have

$$\varphi(AQ^d + b)(u) = 1/(1 + u^t b)^{d+1} \varphi(AQ^d)(u/(1 + u^t b))$$
$$= |\det A|/(1 + u^t b)^{d+1} \varphi(Q^d)(A^t u/(1 + u^t b))$$
$$= |\det A|/(d!(1 + u^t b)w_1 \ldots w_d). \quad \square$$

Any simplex, or projective image of a simplex, can be written as a projective image of Q^d (up to its boundary), so if T is such a polyhedron, then by Lemma 1 and Theorem 2 $\varphi(T)$ is a rational function. Any polyhedron P can be decomposed into a union of such projective simplexes with nonoverlaping interiors, and $\varphi(P)$ will be the sum of $\varphi(T)$ for T in the decomposition. It follows that $\varphi(P)$ is a rational function of u on its domain.

The domain of $\varphi(P)$ is easily seen to be an open set, which is nonempty if P contains no line. It follows that there is a unique rational function in $\mathscr{A}$ whose values agree with those of $\varphi(P)$ on the domain of $\varphi(P)$, and we henceforth identify $\varphi(P)$ with this element of $\mathscr{A}$. If P contains a line, then $\varphi(P)$ cannot be defined by vol $\tau_u(P)$; in this case, we define $\varphi(P)$ to be $0 \in \mathscr{A}$.

We show now that φ extends to a valuation on the lattice of polyhedra in R^d.

THEOREM 3. *The function φ extends uniquely to a valuation on the lattice of polyhedra.*

PROOF. The restriction of the function φ to the intersectional family of convex polyhedra, which contain no lines, clearly satisfies the principle of inclusion and exclusion in the form required for Volland's result. By that result, it extends uniquely to a valuation on the lattice of finite unions of such sets; but this is precisely the lattice of polyhedra.

It remains to show that on convex polyhedra containing lines, this extension (which we also denote by φ) has value 0. For a convex polyhedron P, let k denote the largest dimension of affine subspaces contained in P. Suppose $k \geq 1$. We must show that $\varphi(P) = 0$. Let L be an affine subspace in P having dimension k. Let H be a hyperplane not parallel to (or containing) L. Let H^+ and H^- be the two closed halfspaces bounded by H. Then $\varphi(P) = \varphi(P \cap H^+) + \varphi(P \cap H^-) - \varphi(P \cap H)$ and the largest dimension of affine spaces contained in each of $P \cap H^+$, $P \cap H^-$, and $P \cap H$ is $k - 1$; therefore, the question reduces to the case $k = 1$. In this case, L is a line, $P \cap H^+$, $P \cap H^-$, and $P \cap H$ contain no lines, and by using an affine linear transformation and Theorem 2, we may assume that

$$H = \{x \in R^d : x_d = 0\},$$

$$H^+ = \{x \in R^d : x_d \geq 0\}, \qquad H^- = \{x \in R^d : x_d \leq 0\},$$

and

$$L = \{x \in R^d : x_1 = \cdots = x_{d-1} = 0\}.$$

Also, since $P \cap H$ is not full-dimensional, $\varphi(P \cap H) = 0$, so we need only show that $\varphi(P \cap H^+) + \varphi(P \cap H^-) = 0$. Notice that since $L \subseteq P$, there is no u for which the appropriate volumes both exist, so we must prove this identity by considering rational expressions for $\varphi(P \cap H^+)$ and $\varphi(P \cap H^-)$. If $u = [u_1, \ldots, u_d]^t \in R^d$ and $\tau_u(P \cap H^+)$ is bounded, then $u_d > 0$ and the set is a pyramid having as its base $\tau_u(P \cap H) \subset H$ and apex $(1/u_d)e_d$, so its volume is $V/(du_d)$, where V is the $(d-1)$-dimensional volume of $\tau_u(P \cap H)$. Also, if $\tau_u(P \cap H^-)$ is bounded, then $u_d < 0$ and the set is a pyramid having the same base $\tau_u(P \cap H)$ and apex $(1/u_d)e_d$, so its volume is $-V/(du_d)$. It follows that $\varphi(P \cap H^+) + \varphi(P \cap H^-) = 0$. $\square$

Using Gram's relation with φ yields the next theorem.

THEOREM 4. *Let P be a convex polyhedron. Then*

$$\varphi(P) = \sum_p \varphi(\text{cone}(\{p\}, P)),$$

where the summation is over vertices of p of P.

PROOF. This is Gram's relation for φ. All of the terms involving higher-dimensional faces of P vanish since the cones generated by P from such faces contain lines. $\square$

In case P is a simple polytope, the sets $\text{cone}(\{p\}, P)$ are simplicial cones, so the corollary to Theorem 2 gives the value of $\varphi(\text{cone}(\{p\}, P))$. This yields a formula for the volume of the projective image of such a polytope. The example of the d-cube is worked out in detail in [2].

4. Extension of volume to a valuation on polyhedra

In this section we describe a valuation mapping polyhedra to $\mathcal{A}$ such that on bounded polyhedra the value is the volume of the polytope, a constant. We construct this valuation by utilizing Theorem 1.

In order to use Theorem 1, we need, for each $p \in R^d$, a valuation ν_p on cones emanating from p. The valuation φ of §3 will provide us with such valuations. For $p \in R^d$ and a cone U emanating from p, define

$$\nu_p(U)(u) = (-1)^d (u^t p)^d \varphi(U - p)(u).$$

If U is a cone emanating from the origin and $u \in R^d$ is such that $u^t x \geq 0$ for $x \in U$, then $\tau_u(U) = U \cap \{x \in R^d : u^t x < 1\}$. Hence, for u such that $u^t x > 0$ for $x \in U \setminus \{0\}$, so that $\tau_u(U)$ is bounded, we have that $\varphi(U)(u)$ is the volume of the set cut off by this halfspace. If $t > 0$ then $t^d \varphi(U)(u)$ is the volume of $U \cap \{x \in R^d : u^t x \leq t\}$. It follows that, for U emanating from p, $\nu_p(U)(u)$ is the constant term of the polynomial in t that gives the volume of $U \cap \{x \in R^d : u^t x \leq t\}$ for large t (when this set is bounded). This polynomial is $(t - u^t p)^d \varphi(U - p)(u)$.

Since φ has value 0 on convex polyhedra containing lines, the functions ν_p satisfy the hypotheses of Theorem 1. Let μ denote the valuation on polyhedra that Theorem 1 yields.

The proof of the next theorem will require a lemma.

LEMMA. *Let* $a_0, \ldots, a_d$ *be distinct complex numbers. Then*

$$\sum_{i=0}^{d} a_i^d / \prod_{j \neq i} (a_i - a_j) = 1.$$

PROOF. This follows from Vandermonde's identity, as follows. Let M be the $(d+1) \times (d+1)$ matrix having $i - j$th entry a_{j-1}^{i-1}. Then $\det(M) = \prod_{0 \leq j < i \leq d} (a_i - a_j)$, the Vandermonde determinant. Expanding the determinant of M by cofactors along the bottom row and then dividing both sides by $\prod(a_i - a_j)$ yields the equality. $\square$

THEOREM 5. *The valuation* μ *maps polyhedra in* R^d *to elements of* $\mathscr{A}$. *For polytopes* $P \subseteq R^d$, $\mu(P)$ *is a constant, namely, the volume of* P.

PROOF. We have only the second part left to prove.

The general case will follow if we can establish it when P is a d-simplex, for, since μ is a valuation which is 0 on sets of dimension less than d, its value on an arbitrary polytope is the sum of its values on the simplexes of a triangulation of the polytope. Suppose $P = \text{conv}\{b_0, b_1, \ldots, b_d\}$, where the points $b_0, \ldots,$ and b_d are in affine general position, so that P is a d-simplex and the b_i's are its vertices. Then $\text{cone}(\{b_i\}, P) = A_i Q^d + b_i$, where A_i is the $d \times d$ matrix having as its columns the vectors $b_0 - b_i, \ldots, b_{i-1} - b_i$, $b_{i+1} - b_i, \ldots, b_d - b_i$. By the definition of μ and Corollary 1, we have

$$\mu(\text{cone}(\{b_i\}, P))(u)$$

$$= (-1)^d | \det A_i | (u^t b_i)^d \left/ \left(d! \prod_{j \neq 1} (u^t b_j - u^t b_i) \right) \right. .$$

The number $| \det A_i | / d!$ is simply the volume of P, so it remains to show that $\sum (u^t b_i)^d / \prod_{j \neq i} (u^t b_j - u^t b_i) = (-1)^d$; but this follows from the lemma upon setting $a_i = u^t b_i$ for $i = 0, \ldots, d$. $\square$

5. Lattice point enumeration

Finally, we consider one more valuation, this one on the lattice of finite unions of polyhedra that are intersections of finitely many closed halfspaces of the form $\{x : a^t x \geq b\}$ where a is a vector of rational numbers and b is rational. For a vector $u = [u_1, \ldots, u_d]^t$ of indeterminates and $\alpha = [\alpha_1, \ldots, \alpha_d]^t$ in Z^d we write u^α for the monomial $\prod_i u_i^{\alpha_i}$. If P is a rational convex polyhedron containing no lines, then there is a nonempty open set on which $\sum_{\alpha \in P \cap Z^d} u^\alpha$ converges, and it is not difficult to show that the sum is a rational function of u in this set. Such functions have been studied extensively. See, e.g., [7].

Let $\varepsilon(P)$ denote this rational function (in $\mathscr{A}$) when P contains no lines, and let $\varepsilon(P) = 0$ if the rational convex polyhedron P contains a line.

THEOREM 6. *The function ε extends uniquely to a valuation on the lattice of finite unions of rational convex polyhedra.*

PROOF. The restriction of ε to rational convex polyhedra containing no lines clearly satisfies the principle of inclusion and exclusion as required for Volland's result. Therefore, that result applies to yield a unique extension to the lattice of finite unions of rational convex polyhedra. It remains to show that this extension has value 0 on rational convex polyhedra which contain lines.

Note that if P is a polyhedron not containing lines and $\beta \in Z^d$, then $\varepsilon(P + \beta) = u^\beta \varepsilon(P)$.

Now let P be a rational convex polyhedron containing a line L, which may be taken to be rational, as well. Then we may choose nonzero $\beta \in Z^d$ parallel to L. We may write $P = \bigcup_{i=1}^n P_i$, where none of the P_i's contain lines. Then

$$u^\beta \varepsilon(P) = \sum_J (-1)^{|J|} u^\beta \varepsilon \left(\bigcap_{i \in J} P_i \right)$$

$$= \sum_J (-1)^{|J|} \varepsilon \left(\bigcap_{i \in J} (P_i + \beta) \right) = \varepsilon(P + \beta) = \varepsilon(P).$$

From $(u^\beta - 1)\varepsilon(P) = 0$, it follows that $\varepsilon(P) = 0$. $\square$

REFERENCES

1. V. Klee, *The Euler characteristic in combinatorial geometry*, Amer. Math. Monthly **70** (1963), 119–127.
2. J. F. Lawrence, *Polytope volume computation*, Math. Comp. **57** (1991), 259–271.
3. P. McMullen and R. Schneider, *Valuations on convex bodies*, Convexity and Its Applications (P. M. Gruber and J. M. Wills, eds.), Birkhauser, Basel, 1983, pp. 170–247.
4. G. D. Rota, *On the combinatorics of the Euler characteristic*, Studies in Pure Mathematics (L. Mirsky, ed.), Academic Press, New York, 1971, pp. 221–233.

5. G. T. Sallee, *Polytopes, valuations, and the Euler relation*, Canad. J. Math. **20** (1968), 1412–1424.

6. G. C. Shephard, *An elementary proof of Gram's theorem for convex polytopes*, Canad. J. Math. **19** (1967), 1214–1217.

7. R. P. Stanley, *Combinatorial reciprocity theorems*, Adv. in Math. **14** (1974), 194–253.

8. W. Volland, *Ein Fortsetzungssatz für additive Eipolyederfunktionale im euklidischen Raum*, Arch. Math. **8** (1957), 144–149.

DEPARTMENT OF MATHEMATICAL SCIENCES, GEORGE MASON UNIVERSITY, FAIRFAX, VIRGINIA 22030

DIMACS Series in Discrete Mathematics
and Theoretical Computer Science
Volume **6**, 1991

Winding Numbers and the Generalized Lower-Bound Conjecture

CARL W. LEE

Dedicated to Günter Ewald on the occasion of his sixtieth birthday

ABSTRACT. For simplicial convex d-polytope P define

$$h_i(P) = \sum_{j=0}^{i}(-1)^{i-j}\binom{d-j}{d-i}f_{j-1}(P), \qquad i=0,\ldots,d,$$

and

$$g_k(P) = h_k(P) - h_{k-1}(P), \qquad k=0,\ldots,d+1,$$

where $f_j(P)$ is the number of j-dimensional faces of P and $h_{-1}(P) = h_{d+1}(P) = 0$. The polytope P is called k-stacked if P has a triangulation such that there is no interior face of dimension less than $d-k$. McMullen and Walkup made the following generalized lower-bound conjecture.

CONJECTURE. *Let P be a simplicial convex d-polytope. Then*

(i) $g_k(P) \geq 0$ *for* $k = 0, \ldots, \lfloor d/2 \rfloor$.
(ii) *Moreover, for* $1 \leq k \leq \lfloor d/2 \rfloor$, $g_k(P) = 0$ *if and only if P is $(k-1)$-stacked.*

Part (i) of the conjecture was proved by Stanley. McMullen and Walkup showed the "if" part of (ii). In this paper we demonstrate a connection between $g_k(P)$ and winding numbers of k-splitters. We use this to prove both (i) and (ii) when $f_0(P) \leq d+3$ and when $k < f_0(P)/(f_0(P) - d)$.

1. Introduction

The *f-vector* of a simplicial convex d-polytope P is $f(P) = (f_0(P), \ldots, f_{d-1}(P))$ where $f_j(P)$ denotes the number of j-faces (j-dimensional faces)

1991 *Mathematics Subject Classification.* Primary 52A25.
Key words and phrases. f-vector, *h*-vector, *k*-set, *k*-splitter, lower-bound conjecture, polytope, triangulation, stacked, winding number.
Research supported in part by NSF grant DMS-8802933, by NSA grant MDA904-89-H-2038, and by DIMACS (Center for Discrete Mathematics and Theoretical Computer Science), a National Science Foundation Science and Technology Center, NSF-STC88-09648.

of P. The *h-vector* of P is $h(P) = (h_0(P), \ldots, h_d(P))$ where

(1)
$$h_i(P) = \sum_{j=0}^{i} (-1)^{i-j} \binom{d-j}{d-i} f_{j-1}(P)$$

(taking $f_{-1}(P) = 1$). These relations are invertible:

(2)
$$f_{j-1}(P) = \sum_{i=0}^{j} \binom{d-i}{d-j} h_i(P), \qquad j = 0, \ldots, d.$$

Let

(3)
$$g_k(P) = h_k(P) - h_{k-1}(P), \qquad k = 0, \ldots, d+1,$$

using $h_{-1}(P) = h_{d+1}(P) = 0$. The polytope P is called *k-stacked* if P has a triangulation such that there is no interior face of dimension less than $d - k$. McMullen and Walkup [20] made the following generalized lower-bound conjecture (where their $g_k^{d+1}(P)$ is our $g_{k+1}(P)$):

CONJECTURE 1. *Let P be a simplicial convex d-polytope. Then*

    (i) $g_k(P) \geq 0$ *for* $k = 0, \ldots, \lfloor d/2 \rfloor$.
    (ii) *Moreover, for* $1 \leq k \leq \lfloor d/2 \rfloor$, $g_k(P) = 0$ *if and only if P is $(k-1)$-stacked.*

The conjecture is a generalization of Barnette's lower-bound theorem [1, 2] which proves (i) for $k = 2$. McMullen [17] verified (i) for polytopes with *few vertices*; i.e., $f_0(P) \leq d + 3$. McMullen and Walkup [20] remark that (i) can be proved when $f_0(P) \leq (kd - 1)/(k - 1)$, i.e, when $k < f_0(P)/(f_0(P) - d)$, for polytopes whose vertices are in sufficiently general position. The truth of (i) was settled in general when Stanley [23] established the necessity of McMullen's proposed characterization of f-vectors of simplicial polytopes.

Regarding (ii), the fact that $g_k(P) = 0$ if P is $(k-1)$-stacked is an easy consequence of equations (4) and (5) of §2 (see [20]). As for the converse, part (ii) for $k = 2$ is proved in [4] and generalized by Kalai [12]. For arbitrary k it is known that if $g_k(P) = 0$ then there exists a $(k-1)$-stacked polytope Q having the same f-vector as P [13].

In this paper we demonstrate a connection between $g_k(P)$ and the winding numbers of k-splitters in Gale diagrams of P. We use this to prove both (i) and (ii) when $f_0(P) \leq d + 3$. Then we show how the existence of centerpoints implies (i) and (ii) when $k < f_0(P)/(f_0(P) - d)$.

2. Shellings and winding numbers of k-splitters

Let $V = \{v_1, \ldots, v_n\}$ be a collection of at least $e + 1$ points in $\mathbf{R}^e$ where $e \geq 1$. Assume that these points are in affinely general position. Take $\mathscr{X}$ to be the collection of subsets X of V of cardinality e. A set $X \in \mathscr{X}$ is called a *k-splitter* if the hyperplane $H = \text{aff}(X)$ partitions the remaining

$n - e$ points of V into two sets, at least one of which has cardinality exactly k. (The set of cardinality k is known as a k-set.)

Choose any point p of $\mathbf{R}^e$ in affinely general position with respect to V and select any ray r from p in general direction such that for all $X \in \mathscr{X}$, if r meets conv(X) then r in fact meets relint(conv(X)). Almost all rays have this property. List the sets $X_1, \ldots, X_m$ of $\mathscr{X}$ whose convex hulls are pierced by r in the order that they are encountered as one moves from infinity to p along r. Put $W_i = V \setminus X_i$, $i = 1, \ldots, m$, and consider the simplicial d-complex $\Delta_r(p) = \bigcup_{i=1}^m \overline{W}_i$, where $d = n - e - 1$ and $\overline{W}_i$ denotes the power set of W_i.

THEOREM 1. *Assume $\Delta_r(p)$ is nonempty. Then*

(i) $\Delta_r(p)$ *is a shellable d-ball or d-sphere and $W_1, \ldots, W_m$ constitutes a shelling order.*

(ii) *For any subset U of V of cardinality d, U is a facet of $\partial \Delta_r(p)$ iff $p \in$ int(conv$(V \setminus U)$).*

PROOF. Let $X = X_j$ for some $j > 1$. The hyperplane H spanned by X partitions the facet $W = V \setminus X$ of $\Delta_r(p)$ into two sets F and G, where F is the set of points of W in the same open halfspace H^- as p, and G is the set of points in the opposite open halfspace H^+. Note that card(F)+card$(G) = d + 1$. The subfacets of W are of two types: (a) those of the form $W \setminus \{x\}$ for $x \in F$, and (b) those of the form $W \setminus \{x\}$ for $x \in G$.

Let $j' < j$ and consider any $X' = X_{j'}$. Put $W' = V \setminus X'$. Then $W \cap W'$ cannot contain G. For if it did then $X' \subset (X \cup F)$ and $X' \neq X$, so relint(conv(X')) lies in H^-, contradicting $j' < j$. As a consequence, the subfacets of W of type (a) do not lie in any preceding facet W'.

Now choose any $x \in G$ and consider the subfacet $W \setminus \{x\}$. As one moves through conv(X) along r one passes out of the simplex $S = $ conv$(X \cup \{x\})$. So at some earlier stage S was entered by passing through exactly one of the e other facets X' of S (each of which contains x). So $X' = X_{j'}$ for some $j' < j$. Let $W' = V \setminus X'$. Then $W \cap W' = W \setminus \{x\}$ and W' is the unique preceding facet with this property. So each subfacet of W of type (b) lies in exactly one preceding facet W'.

Therefore we can conclude that $\overline{W} \cap (\bigcup_{i=1}^{j-1} \overline{W}_i) = \bigcup_{x \in G} \overline{W \setminus \{x\}}$, so we have a shelling. Further, any subfacet of $\Delta_r(p)$ is contained in at most two facets, so $\Delta_r(p)$ is a shellable ball or sphere [6]. This verifies (i). The facets of $\partial \Delta_r(p)$ are the subfacets of $\Delta_r(p)$ that are contained in exactly one facet of $\Delta_r(p)$. But these are the type (a) subfacets of $\Delta_r(p)$ which are not later encountered as type (b) subfacets. Criterion (ii) now follows easily. □

COROLLARY 1. *Assume $\Delta_r(p)$ is nonempty.*

(i) $\Delta_r(p)$ *is a sphere iff $p \notin$ conv(V).*

(ii) $\partial \Delta_r(p)$ *is independent of the particular choice of r.*

Thus we can drop the subscript on $\partial\Delta_r(p)$ and refer simply to $\partial\Delta(p)$.

Equations (1) and (3) can be used to define h_i and g_k for any simplicial complex of dimension $d - 1$, or indeed for any collection Γ of sets for which $\max_{S\in\Gamma} \operatorname{card}(S) = d$. Given a triangulated d-ball Δ it is known [20] that

(4) $g_i(\partial\Delta) = h_i(\Delta) - h_{d+1-i}(\Delta),$ $i = 0, \dots, d+1,$

(5) $h_i(\Delta^o) = h_{d+1-i}(\Delta),$ $i = 0, \dots, d+1,$

where $\Delta^o = \Delta \setminus \partial\Delta$, the set of interior (nonboundary) faces of Δ.

Note that these equations can be used to provide a lower bound on the number of d-simplices in Δ in terms of the numbers of faces of $\partial\Delta$ (see [15, 21]):

$$f_d(\Delta) = \sum_{i=0}^{d+1} h_i(\Delta) \geq \sum_{i=0}^{\lfloor d/2\rfloor} h_i(\Delta) - h_{d+1-i}(\Delta)$$
$$= \sum_{i=0}^{\lfloor d/2\rfloor} g_i(\partial\Delta) = h_{\lfloor d/2\rfloor}(\partial\Delta).$$

Consider any $X \in \mathscr{X}$ and any ray r from p in general direction. Let u be the unit vector in the direction of r and let F and G be as in the beginning of the proof of Theorem 1. For $i = 0, \dots, d+1$ put

$$\delta(i, u, X) = \begin{cases} 1 & \text{if } i = \operatorname{card}(G) \text{ and } r \text{ intersects conv}(X), \\ 0 & \text{otherwise,} \end{cases}$$

and $\sigma(i, u, X) = \delta(i, u, X) - \delta(d+1-i, u, X)$. Then it is evident from equations (4) and (5) and the proof of Theorem 1 (see [19]) that

(6) $h_i(\Delta_r(p)) = \sum_{X\in\mathscr{X}} \delta(i, u, X)$

and

(7) $g_i(\partial\Delta(p)) = \sum_{X\in\mathscr{X}} \sigma(i, u, X)$

for $i = 0, \dots, d+1$.

Because $\partial\Delta(p)$ and hence $g_i(\partial\Delta(p))$ are independent of r we can define winding numbers. This was also known to Lawrence [14]. For X, F, and G as above set the *sign* of X (with respect to p) to be

$$\operatorname{sg}(X) = \begin{cases} +1 & \text{if } \operatorname{card}(G) < \operatorname{card}(F), \\ -1 & \text{if } \operatorname{card}(G) > \operatorname{card}(F), \\ 0 & \text{if } \operatorname{card}(G) = \operatorname{card}(F). \end{cases}$$

So sg(X) will agree with $\sigma(\mathrm{card}\,(G),u,X)$ when r meets conv(X). We say X is *positive* (respectively, *negative*) with respect to p if sg$(X) = +1$ (respectively, -1). Now let $\alpha(X)$ be the measure of the solid angle of the cone determined by conv(X) whose apex is p. The value of $\alpha(X)$ is normalized to equal the fraction of the surface area of a sphere centered at p that is intersected by the cone. Define the kth *winding number*

$$w_k(p) = \sum_{k\text{-splitters } X \in \mathcal{X}} \mathrm{sg}(X)\alpha(X).$$

THEOREM 2. $w_k(p) = g_k(\partial\Delta(p))$ *and hence is an integer.*

PROOF. Recalling that $\sigma(i,u,X)$ is defined for almost all unit vectors u we have

$$w_k(p) \;=\; \sum_X \mathrm{sg}(X)\alpha(X)$$

$$=\; \sum_X \int_{\mathbf{S}^{e-1}} \sigma(k,u,X)\,du$$

$$=\; \int_{\mathbf{S}^{e-1}} \sum_X \sigma(k,u,X)\,du$$

$$=\; \int_{\mathbf{S}^{e-1}} g_k(\partial\Delta(p))\,du$$

$$=\; g_k(\partial\Delta(p)). \quad \square$$

We remark that Theorem 2 can be proved more directly. It suffices to show that $\sum_X \sigma(k,u,X)$ is invariant under choice of u. As u is varied we need to check that this sum does not change when the ray determined by u is moved across a subset Y of V of cardinality $e-1$. The planar case follows from Corollary 2.6 of [8] in which it is proved that the planar k-splitters can be partitioned into a collection of directed cycles, and the higher-dimensional case is handled by projecting onto a plane orthogonal to Y.

COROLLARY 2. *Suppose V is translated to move the origin to p and the points of V are independently scaled by positive amounts to yield V' so that $V' \cup \{p\}$ is still in affinely general position. Then $w_k(p)$ remains unchanged even though the k-splitters may change.*

PROOF. Part (ii) of Theorem 1 implies that $\partial\Delta(p)$ will be unaffected by the scaling. $\square$

THEOREM 3. *When $\partial\Delta_r(p)$ is empty then the sphere $\Delta_r(p)$ is isomorphic to the boundary complex of a simplicial $(d+1)$-polytope. On the other hand,* *when* $\partial \Delta_r(p)$ *is nonempty* $\partial \Delta_r(p)$ *is isomorphic to the boundary complex of a simplicial d-polytope* P *and* $\Delta_r(p)$ *is isomorphic to a triangulation of* P.

PROOF. When $p \in \text{conv}(V)$ we can translate the origin of $\mathbf{R}^e$ to p and regard V as a positively scaled Gale transform [11, 18, 19] of a set of n points $\overline{V}$ in $\mathbf{R}^d$. Let $P = \text{conv}(\overline{V})$. Then (ii) of Theorem 1 is the familiar characterization of facets of P, which will be simplicial because p is in affinely general position. Using [3, 16] one can see that $\partial \Delta_r(p)$ is isomorphic to the boundary complex of P and that $\Delta_r(p)$ is a regular triangulation of $\overline{V}$.

On the other hand, if $p \notin \text{conv}(V)$ but $\Delta_r(p)$ is nonempty we can replace r by the line s spanned by r and translate V so that the origin lies on s but otherwise falls in affinely general position in the interior of $\text{conv}(V)$. Regarding the result as a positively scaled Gale transform of a set $\overline{V}$ of n points in $\mathbf{R}^d$, it is straightforward to verify that the facets of $\Delta_r(p)$ consist of the facets of two complementary regular triangulations of $\overline{V}$ and hence $\Delta_r(p)$ is itself isomorphic to the boundary complex of a simplicial $(d+1)$-polytope. □

Let $\Sigma_j = \bigcup_{i=1}^j \overline{W}_j$, $j = 1, \ldots, m$. It is easy to verify that $\partial \Sigma_j$ differs from $\partial \Sigma_{j-1}$ by a bistellar operation [9, 15, 21]. Further, by translating V so that the origin is moved along r, one also sees that Σ_1 is a d-simplex and $\partial \Sigma_j$ is always isomorphic to the boundary complex of a simplicial d-polytope. Suppose we start with a simplicial d-polytope P with n vertices and let V be a Gale diagram of the vertex set of P. By perturbing the points of V slightly we can assume that $V \cup \{\mathbf{O}\}$ is in affinely general position. Thus we have an alternate proof of the theorem of Ewald and Shephard [10] that the boundary complex of P is obtainable from that of a simplex by a sequence of bistellar operations such that the intermediate simplicial complexes are always polytopal.

Similarly, the case $p \notin \text{conv}(V)$ can be adapted to reprove that every simplicial polytope is shellable [5].

COROLLARY 3. $(w_0(p), \ldots, w_{\lfloor d/2 \rfloor}(p))$ *is an* M-*vector. In particular:*

(i) *If* $0 \le k \le \lfloor d/2 \rfloor$ *then* $w_k(p)$ *is nonnegative.*
(ii) *If* $w_k(p) = 0$ *for some* $0 \le k < \lfloor d/2 \rfloor$ *then* $w_\ell(p) = 0$, $\ell = k + 1, \ldots, \lfloor d/2 \rfloor$.

PROOF. An immediate consequence of [23]. □

THEOREM 4. *Suppose* $0 \le k \le \lfloor d/2 \rfloor$. *If there is a ray* r *from* p *that does not intersect the convex hull of any* k-*splitter then* $w_k(p) = 0$.

PROOF. Using equations (6) and (7) and Theorem 2,

$$h_k(\Delta_r(P)) = h_{d+1-k}(\Delta_r(p)) = 0 \quad \text{so } w_k(p) = g_k(\partial\Delta(p)) = 0. \quad \Box$$

CONJECTURE 2. *The converse of the above theorem holds. That is, suppose* $0 \le k \le \lfloor d/2 \rfloor$. *If* $w_k(p) = 0$ *then there is a ray in general direction from* p *that does not intersect the convex hull of any* k-*splitter.*

THEOREM 5. *Let* P *be a simplicial* d-*polytope with vertex set* $\overline{V}$ *such that* $\mathbf{O} = \sum_{v \in \overline{V}} v$ *and* $\overline{V} \cup \{\mathbf{O}\}$ *is in affinely general position. Then Conjecture 2 implies* (ii) *of Conjecture 1 for* P.

PROOF. Let $V \subset \mathbf{R}^e$ be a Gale transform of the vertex set of P, where $e = n - d - 1$. Choose p to be the origin. Then $V \cup \{p\}$ is in affinely general position. We are assuming $g_k(P) = 0$ so $w_k(p) = 0$ by Theorem 2. Conjecture 2 then implies there is a ray r from p that does not intersect the convex hull of any k-splitter. Therefore equation (6) implies $h_k(\Delta_r(p)) = 0$. But $(h_0(\Delta_r(p)), \dots, h_{d+1}(\Delta_r(p)))$ is an M-vector (O-sequence) [22] so $h_\ell(\Delta_r(p)) = 0$, $\ell = k, \dots, d+1$. So from equation (5) we have $h_{d+1-\ell}(\Delta_r^o(p)) = 0$, $\ell = k, \dots, d+1$. Hence $\Delta_r(p)$, which is isomorphic to a regular triangulation of P, has no interior face of dimension less than $d - k$. □

Note that Conjecture 1 does not imply Conjecture 2, since not all triangulations of polytopes are regular; i.e., induced by rays in a Gale transform.

3. Some special cases

We first give an elementary proof of the nonnegativity of $w_k(p)$ in $\mathbf{R}^1$ and $\mathbf{R}^2$ and use this to handle polytopes with few vertices.

PROOF OF (i) AND (ii) OF COROLLARY 3 WHEN $e \le 2$. First consider the case $e = 1$. Assume we have n_1 points to the right of p, and n_2 points to the left of p. Without loss of generality $n_1 \le n_2$. Let r be the ray from p that extends to the right. Then it is clear from equations (6) and (7) and Theorem 2 that

$$h_k(\Delta_r(p)) = \begin{cases} 1 & \text{if } k < n_1, \\ 0 & \text{if } k = n_1, \dots, d+1, \end{cases}$$

and

$$w_k(p) = g_k(\partial \Delta_r(p)) = \begin{cases} 1 & \text{if } k < n_1, \\ 0 & \text{if } k = n_1, \dots, \lfloor d/2 \rfloor. \end{cases}$$

Now suppose $e = 2$. Translate the points so that the origin is at p and scale them independently by positive amounts so that they lie on the circle of unit perimeter centered at p. By Corollary 2 this will not affect the value of $w_k(p)$. Assume that the points are numbered $v_0, \dots, v_{n-1}$ consecutively counterclockwise around the perimeter of the circle.

Fix an integer k such that $0 \le k \le \lfloor d/2 \rfloor$. The k-splitters in V consist of pairs of points $X_i = \{v_i, v_{i+k+1}\}$, $i = 0, \dots, n-1$, where the indices are computed mod n. Let $D_i = \text{conv}(X_i)$ and A_i be the arc of the circle containing $\{v_i, v_{i+1}, \ldots, v_{i+k+1}\}$, $i = 0, \ldots, n-1$. The D_i form the edges of a star n-gon. Define $\bar{\alpha}(X_i)$ to be the length of A_i, $i = 0, \ldots, n-1$. Then

$$\bar{\alpha}_i = \begin{cases} \alpha(X_i) & \text{if sg}(X_i) = 1, \\ 1 - \alpha(X_i) & \text{if sg}(X_i) = -1, \end{cases}$$

$i = 0, \ldots, n-1$.

Therefore

$$w_k(p) = \sum_{i=0}^{n-1} \text{sg}(X_i)\alpha(X_i)$$

$$= -z + \sum_{i=0}^{n-1} \bar{\alpha}(X_i),$$

where z is the number of X_i that are negative with respect to p.

Now $\sum_{i=0}^{n-1} \bar{\alpha}(X_i)$ is easily seen to equal $k+1$ since the A_i provide a $(k+1)$-fold covering of the circle. So we need to show that $z \le k+1$. If X_i and X_j are both negative then D_i and D_j must cross. Otherwise, $V \subset A_i \cup A_j$ and $A_i \cap A_j$ contains at least three points (D_i and D_j may share one endpoint but not two). Then

$$\text{card}(V) = \text{card}(V \cap A_i) + \text{card}(V \cap A_j) - \text{card}(V \cap A_i \cap A_j)$$
$$\le 2(k+2) - 3 \le d + 1 = n - 2 < n,$$

a contradiction. Hence either $j = i + \ell$ (mod n) for some $\ell \le k$ or vice versa. So to attain the largest number of mutually crossing k-splitters we must take a collection of the form $X_q, \ldots, X_{q+k}$ for some q (using again that $k \le \lfloor d/2 \rfloor$). Therefore $z \le k+1$.

Now suppose in fact that $w_k(p) = 0$. Take r to be the ray from p that intersects the arc $A = (v_{q+k}, v_{q+k+1})$ of the circle, where q is as above. Then it is easy to see that r does not intersect the convex hull of any k-splitter, nor does it intersect the convex hull of any positive j-splitter, $j > k$. Therefore by equation (6), $h_j(\Delta_r(p)) = 0$, $j = k, \ldots, d+1$, so $w_j(p) = g_j(\partial\Delta(p)) = h_j(\Delta_r(p)) - h_{d+1-j}(\Delta_r(p)) = 0$, $j = k, \ldots, \lfloor d/2 \rfloor$ by Theorem 2 and equation (4). $\square$

PROOF OF CONJECTURE 1 WHEN $f_0(P) \le d + 3$. Let P be a simplicial convex d-polytope such that $n \le d + 3$, where $n = f_0(P)$. If $n = d + 1$ then the result is trivial. Suppose $n = d + 2$. Construct a Gale transform of the vertex set $\bar{V}$ of P. It will be one-dimensional. Scale its points independently by positive amounts so that the resulting Gale diagram V consists of n distinct points. Choose p to be the origin. Now we are in the situation of the previous proof in the case that $e = 1$. Therefore $g_k(P) = w_k(p) \ge 0$ for $k = 0, \ldots, \lfloor d/2 \rfloor$. Using the same ray r as in the previous proof one can see that if $g_k(P) = 0$ for some $k \leq \lfloor d/2 \rfloor$ then $h_i(\Delta_r(p)) = 0$, $i = k, \ldots, d+1$, so $h_{d+1-i}(\Delta_r^o(p)) = 0$, $i = k, \ldots, d+1$ by equation (5). Thus $\Delta_r(p)$, a regular triangulation of P, has no interior face of dimension less than $d + 1 - k$.

Now assume $n = d + 3$. Construct a Gale transform of the vertex set $\overline{V}$ of P. It will be two-dimensional. Scale its points independently by positive amounts so that they all lie on the circle of unit perimeter centered at the origin. Call the resulting set V'. If multiple points occur *rotate diagonals* slightly to separate them so that a set V of n distinct points is obtained that is combinatorially equivalent to the original Gale transform and for which $V \cup \{\mathbf{O}\}$ is in affinely general position [**11, 18, 19**]. Choose p to be the origin. Now we are in the situation of the previous proof in the case that $e = 2$. Therefore $g_k(P) = w_k(p) \geq 0$ for $k = 0, \ldots, \lfloor d/2 \rfloor$.

Now suppose in fact that $g_k(P) = 0$, so $z = k + 1$. Consider the ray r as in the end of the previous proof. We then have by equation (5) that $h_{d+1-j}(\Delta_r^o(p)) = h_j(\Delta_r(p)) = 0$, $j = k, \ldots, d+1$ so $\Delta_r(p)$ has no interior face of dimension less than $d + 1 - k$. If we can show that $\Delta_r(p)$ is isomorphic to a regular triangulation of P then we will know that P is $(k - 1)$-stacked. It is sufficient to show we can rotate diagonals back again to restore V to V' without allowing any point to pass through r. The only obstacle to this is that r might separate two points u and v that originally coincided in V'. But if u and v originally coincided and were separated there can be no points in $-A$, the arc opposite to A that was used in the previous proof. On the other hand, the fact that D_q and D_{q+k} are negative crossing k-splitters implies that both x_q and x_{q+2k+1} are in $-A$, a contradiction. □

For the next two proofs we recall the notion of a centerpoint of $V \subset \mathbf{R}^e$ [**7**, Chapter 4]. A point c, not necessarily in V, is called a *centerpoint* if no closed halfspace that contains c contains less than $n/(e + 1)$ points of V. Every finite set of points V in $\mathbf{R}^e$, whether in general position or not, admits a centerpoint.

PROOF OF (i) AND (ii) OF COROLLARY 3 WHEN $k < n/(e + 1)$. Let c be a centerpoint of V. Let r be the ray from p in the direction opposite to c (if $p = c$ choose r to be arbitrary). Then r does not meet the convex hull of any negative k-splitter X. For suppose it did. Translate the hyperplane aff(X) parallel to itself a small amount toward p to get a hyperplane H. Then the closed halfspace determined by H containing p, and hence c, contains $k < n/(e + 1)$ points of V, a contradiction. Perturb r slightly, if necessary, so it is in general direction. Then by equations (4) and (6) and Theorem 2, $h_{d+1-k}(\Delta_r(p)) = 0$ so $w_k(p) = g_k(\partial\Delta(p)) = h_k(\Delta_r(p)) \geq 0$. □

PROOF OF CONJECTURE 1 WHEN $k < f_0(P)/(f_0(P) - d)$. Let P be a simplicial convex d-polytope and $0 \leq k < n/(n - d)$, where $n = f_0(P)$. Define

$e = n - d - 1$. Take $V \subset \mathbf{R}^e$ to be the set of points of a Gale transform of the vertices of P, scaled independently by positive amounts so that no hyperplane missing the origin contains more than e members of V. Let p be the origin. Then we are in the situation of the previous proof, except for the possibility that p is not in affinely general position with respect to V. Nevertheless we can still find a centerpoint c for V and consider the ray r above. By slightly perturbing r, if necessary, so that it lies in general direction, we will still have that r intersects no negative k-splitter. Our previous arguments will still go through to prove that $\Delta_r(p)$ is a shellable d-ball which is a regular triangulation of P such that $h_{d+1-k}(\Delta_r(p)) = 0$. Therefore $g_k(P) = h_k(\Delta_r(p)) \geq 0$.

Now suppose in fact that $g_k(P) = 0$ for some $0 \leq k < n/(n-d)$. Then $h_k(\Delta_r(p)) = 0$. We now proceed as in the proof of Theorem 5. □

4. Acknowledgments

We wish to thank Pankaj Agarwal, Boris Aronov, Marge Bayer, Fariba Bigdeli, Endre Boros, Zoltan Füredi, Jim Lawrence, William Steiger, and Rephael Wenger for helpful conversations, comments, and suggestions.

References

1. D. W. Barnette, *The minimum number of vertices of a simple polytope*, Israel J. Math. **10** (1971), 121–125.

2. ———, *A proof of the lower bound conjecture for convex polytopes*, Pacific J. Math. **46** (1973), 349–354.

3. L. J. Billera, P. Filliman, and B. Sturmfels, *Constructions and complexity of secondary polytopes*, Adv. in Math. **83** (1990), 155–179.

4. L. J. Billera and C. W. Lee, *A proof of the sufficiency of McMullen's conditions for f-vectors of simplicial convex polytopes*, J. Combin. Theory Ser. A **31** (1981), 237–255.

5. H. Bruggesser and P. Mani, *Shellable decompositions of cells and spheres*, Math. Scand. **29** (1971), 197–205.

6. G. Danaraj and V. Klee, *Shellings of spheres and polytopes*, Duke Math. J. **41** (1974), 443–451.

7. H. Edelsbrunner, *Algorithms in combinatorial geometry*, ETACS Monographs on Theoretical Computer Science, vol. 10, Springer-Verlag, Berlin and New York, 1987.

8. P. Erdös, L. Lovász, A. Simmons, and E. G. Straus, *Dissection graphs of planar point sets*, A Survey of Combinatorial Theory, North-Holland, Amsterdam, 1973, pp. 139–149.

9. G. Ewald, *Über stellare Äquivalenz konvexer polytope*, Resultate Math. **1** (1978), 54–60.

10. G. Ewald and G. C. Shephard, *Stellar subdivisions of boundary complexes of convex polytopes*, Math. Ann. **210** (1974), 7–16.

11. B. Grünbaum, *Convex polytopes*, Pure Appl. Math., vol. 16, Interscience, New York, 1967 (with the cooperation of Victor Klee, M. A. Perles, and G. C. Shephard).

12. G. Kalai, *Rigidity and the lower bound theorem. I*, Invent. Math. **88** (1987), 125–151.

13. P. Kleinschmidt and C. W. Lee, *On k-stacked polytopes*, Discrete Math. **48** (1984), 125–127.

14. J. Lawrence, 1988, personal communication.

15. C. W. Lee, *Two combinatorial properties of a class of simplicial polytopes*, Israel J. Math. **47** (1984), 261–269.

16. _____, *Regular triangulations of convex polytopes*, Applied Geometry and Discrete Mathematics: The Victor Klee Festschrift, DIMACS Ser. in Discrete Math. and Theoretical Comp. Sci., vol. 4, Amer. Math. Soc., Providence, RI, 1991.

17. P. McMullen, *The numbers of faces of simplicial polytopes*, Israel J. Math. **9** (1971), 559–570.

18. _____, *Transforms, diagrams and representations*, Contributions to Geometry (Proc. Geom. Sympos., Siegen, 1978), Basel, 1979, Birkhäuser, pp. 92–130.

19. P. McMullen and G. C. Shephard, *Convex polytopes and the upper bound conjecture*, London Math. Soc. Lecture Note Ser., vol. 3, Cambridge Univ. Press, London and New York, 1971 (prepared in collaboration with J. E. Reeve and A. A. Ball).

20. P. McMullen and D. W. Walkup, *A generalized lower-bound conjecture for simplicial polytopes*, Mathematika **18** (1971), 264–273.

21. U. Pachner, *Über die bistellare Äquivalenz simplizialer Sphären und Polytope*, Math. Z. **176** (1981), 565–576.

22. R. P. Stanley, *The upper bound conjecture and Cohen-Macaulay rings*, Stud. Appl. Math. **54** (1975), 135–142.

23. _____, *The number of faces of a simplicial convex polytope*, Adv. in Math. **35** (1980), 236–238.

DEPARTMENT OF MATHEMATICS, UNIVERSITY OF KENTUCKY, LEXINGTON, KENTUCKY 40506

DIMACS, RUTGERS UNIVERSITY, NEW BRUNSWICK, NEW JERSEY 08903

DIMACS Series in Discrete Mathematics
and Theoretical Computer Science
Volume **6**, 1991

Computing the Center of Planar Point Sets

JIŘÍ MATOUŠEK

ABSTRACT. Given a collection H of n lines in the plane, the *level* of a point x is the number of lines of H lying below x or passing through x. We show that for a given k, one can compute the convex hull of the set of points of level $\leq k$, in time $O(n \log^4 n)$. This implies that a description of the set of *centerpoints* of a given n-point set in the plane can be found within the same time bound, and a point of greatest *Tukey depth* (*Tukey median*) for an n-point set can be computed in time $O(n \log^5 n)$. We also mention the computation of an approximate centerpoint in higher dimension.

1. Introduction

Given an n-point set P in E^d and a point $x \in E^d$, one defines the *Tukey depth* (or depth for short) of x as the minimum number of points of P contained in a closed halfspace whose bounding hyperplane passes through x. The Helly theorem implies that for any P, there always exists a point x of depth at least $\lceil \frac{n}{d+1} \rceil$ (not necessarily belonging to P); such a point is called a *centerpoint* of P, and the set of all centerpoints is called the *center* of P. Since the condition that x is a centerpoint of P can be formulated as the containment of x in all halfspaces of a certain set (those containing at least $\lceil \frac{n}{d+1} \rceil$ points of P), the center is closed and convex.

A point of P of the largest depth is called a *Tukey median* of P. The Tukey median is one of many possible generalizations of the notion of median for a set of real numbers, and it has applications in statistics, see e.g. [GSW90].

An efficient algorithm computing a centerpoint in the plane was given by Cole et al. [CSY87]. This algorithm has complexity $O(n \log^5 n)$, and with the technique of Cole [Col87] this can be improved to $O(n \log^3 n)$ (this improved result, however, probably has not appeared anywhere in a written form). The

---

1991 *Mathematics Subject Classification*. Primary 68V05, 68P05.

This research was performed while the author was visiting at Georgia Institute of Technology, School of Mathematics, Atlanta.

algorithm has been extended to dimension 3 by Naor and Sharir [NS90], with complexity $O(n^2 \log^6 n)$. It is straightforward to determine the depth of all points of a given set in time $O(n^d)$ (by constructing the dual arrangement of hyperplanes), and thus the Tukey median can be determined within the same time bound.

In this paper we consider an efficient computation of the whole center and of the Tukey median in the plane, and we show the following:

THEOREM 1.1. *Given an n-point set P in the plane and a parameter k, the description of the set of points of depth at least k can be computed in time $O(n \log^4 n)$. The Tukey median of P can be found in time $O(n \log^5 n)$.*

We will reduce the problem to the finding of the convex hull of all points of level at most k in an arrangement of n lines, whose computation may be a result of independent interest.

2. A reduction: Level convex hull problem

Let H be a collection of n nonvertical lines in the plane. For a point x, we define its *level* as the number of lines of H lying strictly below x or passing through x. We will denote the closure of the set of all points of level at most k by $L_k(H)$ or simply L_k where H is understood. The boundary of L_k will be denoted by λ_k and called the *level k* (of H). It is not difficult to see that λ_k is a x-monotone piecewise linear curve, whose vertices are pairwise intersections of the lines of H and whose segments are parts of the lines of H. If the lines of H are in general position, then all points of λ_k have level $k + 1$ or $k + 2$.

We will also use the notation U_k for the closure of the set of points of level at least k.

These notions are related to the concepts in the introduction via a duality transform (this transform sends a point with coordinates (a, b) to a line with equation $y = 2ax - b$ and vice versa; see e.g., [Ede87]). Namely, a point x has depth at most k relative to a point set P, iff all points of the line dual to x have level at least k and at most $n - k$, relative to the collection of lines dual to P. We thus say that a line l has *depth at least k* if it intersects neither the interior of L_k nor the interior of U_{n-k}. A line of depth at least k is a dual notion to a point of depth at least k.

We observe that a line intersects the interior of L_k iff it intersects the interior of its convex hull. Thus the property "having depth $\geq k$" for a line can be decided with a knowledge of $\text{conv}(L_k)$ and $\text{conv}(U_{n-k})$. In the sequel, γ_k (or $\gamma_k(H)$) will denote the boundary of $\text{conv}(L_k)$.

An obvious way to compute γ_k is to compute all vertices of the boundary λ_k of L_k, and use some convex hull algorithm. As shown by Edelsbrunner and Welzl [EW86], a complete description of λ_k can be computed in time $O(n \log n + |\lambda_k| \log^2 n)$, where $|\lambda_k|$ denotes the number of segments forming

λ_k (this has been improved to $O(n \log n + |\lambda_k| \log^2 k)$ by Cole et al.[**CSY87**]). However, the worst-case complexity of λ_k is essentially unknown; the current best estimates are $\Omega(n \log k)$ from below and $O(n \sqrt{k} / \log^* k)$ from above (the upper bound is due to Pach et al. [**PSS90**]; see [**Ede87**] for older references) and tightening these bounds is one of the major open problems in combinatorial geometry.

On the other hand, we have the following easy observation:

LEMMA 2.1. *For any k, γ_k has at most n vertices.*

PROOF. Indeed, any vertex of γ_k is a vertex of λ_k, which in turn is an intersection of two lines of H. But every line may pass through at most two vertices of the convex chain γ_k, and two lines are needed at every vertex, thus the number of vertices is at most n. $\square$

The main result of this paper is an algorithm computing γ_k efficiently, without computing a complete description of λ_k:

THEOREM 2.2. *For a collection of n lines in the plane and a parameter k, the convex hull $\mathrm{conv}(L_k)$ can be computed in time $O(n \log^4 n)$.*

The computation of $\mathrm{conv}(U_{n-k})$ is completely symmetric. Let us indicate how this implies Theorem 1.1. If we know the convex hulls of L_k and of U_{n-k}, we can determine whether a given line intersects their interior, using an algorithm determining the position of a line relative to a given convex polygon (with running time $O(\log n)$ after a simple preprocessing of the polygon; see e.g., [**PS85**]). Hence once these convex hulls are computed, we can find all lines of H of depth at least k in additional $O(n \log n)$ time. The deepest line (dual to the Tukey median) is then found by a binary search on k, in $O(\log n)$ stages.

The set of all nonvertical lines of depth $\geq k$ can be described as the lines lying below all vertices of $\mathrm{conv}(U_{n-k})$ and above all vertices of $\mathrm{conv}(L_k)$. This gives the set of points of depth $\geq k$ in the primal plane as an intersection of $O(n)$ halfplanes.

3. Computing tangents to levels

In this section we describe several subroutines used in the main algorithm, which will be given in the next section. The material of this section (in a dual setting) is essentially contained in [**CSY87**]. For the reader's convenience, we briefly explain these parts of our algorithm. The results in this section are based on various versions of Megiddo's parametric search technique, and although we outline the necessary material, a familiarity with that technique might be helpful (see e.g., the original paper [**Meg83**] for a nice exposition).

For the sake of simplicity, we will assume that the given lines are in general position. This assumption could be removed either by so-called *simulation of simplicity* (see e.g., [**Ede87**]), or by a careful analysis of possible degenerate cases in our algorithm.

LEMMA 3.1. *Given a collection H of n lines and a line q, one can find all vertices of λ_k lying on q in time $O(n \log n)$.*

PROOF. We sort the intersections of the lines of H with q by increasing x-coordinate, which is also in the order of their appearance along q. The level of the leftmost intersection (or the bottommost one for a vertical q) is determined directly, and we then go along q and update the current level (by ± 1) appropriately at every intersection. □

LEMMA 3.2. *Given a collection H of n lines and a point x, one can find the tangent to γ_k passing through x and touching γ_k to the right of x (if it exists) in time $O(n \log^2 n)$.*

PROOF. We will actually compute a line τ^* such that its right semiline from x touches L_k (and does not cut it). In the end, we check whether the left semiline cuts L_k; if not, then τ^* is the answer, otherwise we infer that x lies below γ_k and no tangent exists. □

We begin by checking whether the level of x is at least k (otherwise no tangent exists). Then there comes our first application of the parametric search technique. We will search for the slope z^* of the desired tangent $\tau^* = \tau(z^*)$. To this end, we will run an algorithm sorting the intersections of τ^* with the lines of H, in left-to-right order along τ^* (to have a unique order, we formally consider the order of intersections along a line arising by lifting τ^* upwards by a sufficiently small amount). We do not know what τ^* actually is, but nevertheless for two given lines l, $l' \in H$ we can decide the order of their intersections along τ^* (we will refer to this as *answering the question (l, l')*): it suffices to know whether the intersection $y = l \cap l'$ lies above or below τ^*. This in turn can be decided by checking whether the line xy intersects L_k to the right of x, and thus a question (l, l') can be answered in time $O(n \log n)$ by the previous lemma.

With Megiddo's technique, one answers batches of questions, taking advantage of the batching; to this end, an efficient parallel algorithm is used for the sorting. In our case, suppose that we are given $m \le n/2$ questions $(l_1, l_1'), \dots, (l_m, l_m')$, $(l_i, l_i') \in H$. For every i, we compute the intersection $y_i = l_i \cap l_i'$ and the slope z_i of the line $\tau_i = xy_i$. We then find the median z of $z_1, \dots, z_m$ (i.e., half of the lines of $\tau_1, \dots, \tau_m$ have slope larger than z and half have slope smaller than z), and we find out whether the line $\tau(z)$ passing through x and with slope z intersects L_k to the right of x. Based on this, we can answer half of the questions $(l_1, l_1'), \dots, (l_m, l_m')$ in $O(n \log n)$ time. In order to apply an extension to Megiddo's technique due to Cole [Col87] (improving the efficiency by a logarithmic factor), we require the following technical modification: We are given m questions together with some nonnegative real weights $w_1, \dots, w_m$ (summing up to 1, say), and we want to answer questions of total weight at least $1/2$; let us call this a weighted questions batch problem. The above outlined method can be easily adapted to this, simply by taking the slope z to be a weighted median of the slopes $z_1, \ldots, z_m$ with weights $w_1, \ldots, w_m$ respectively. Cole's results show that the whole sorting of intersections along the unknown τ^* can be done by solving $O(\log n)$ weighted questions batch problems, giving a total time $O(n \log^2 n)$ in our case. It turns out that when the sorting is finished, the unknown τ^* must have been considered as one of the lines $\tau(z)$ used in solving the weighted questions batch problems, and thus we know τ^* explicitly. $\square$

Let us remark that a special case of the previous lemma (with x in the infinity) allows us to compute a tangent to γ_k with a prescribed slope.

LEMMA 3.3. *Given a collection H of n lines and a vertical line v, one can compute a tangent τ^* touching γ_k at its intersection with v, in time $O(n \log^3 n)$.*

PROOF. Among all tangents τ to γ_k, the desired τ^* minimizes the y-coordinate of the intersection with v. We will find such τ^* in two steps. $\square$

First we determine the order relation of the slope of τ^* to the slopes of all the lines of H. We will do a binary search locating the slope of τ^*, and all we need is an algorithm deciding the relation of the slope of τ^* to the slope of a given line q. Using Lemma 3.2, we find a tangent τ' to γ_k parallel to q. If τ' touches γ_k strictly to the left of v, we can push the intersection of τ' with v down by slightly lowering the right end of τ', and hence the slope of τ^* is smaller than the slope of q. We proceed similarly for τ' touching γ_k to the right of v. Thus the whole binary search can be performed in time $O(n \log^3 n)$.

The second step will already find τ^*, by a parametric search. We will again be sorting the intersections along τ^*. Since we know the relative slope of τ^* with respect to all lines of H, a question (l, l') (about the order of intersections of l and l' along τ^*) can be decided once we know whether the intersection $x = l \cap l'$ lies above or below τ^*. Suppose that x lies e.g., to the left of v. We compute the tangent τ to γ_k passing through x and touching γ_k to the right of x (if such a tangent does not exist then x is below γ_k and hence also below τ^*). Now it is easy to verify that if the touchpoint of τ is to the right of v, then x lies above τ^*, and otherwise it lies below τ^* (or on τ^* if there are touchpoints on both sides of v or at v).

We are now aiming at solving the weighted questions batch problem (see the proof of Lemma 3.2). Given a weighted collection $(l_1, l_1'), \ldots, (l_m, l_m')$ of questions, we compute the intersections $x_i = l_i \cap l_i'$, whose position relative to τ^* should be decided. We then pass to the dual plane, where we get the following situation: We have a weighted collection Q of lines (those dual to the points $x_1, \ldots, x_m$), and an unknown point t^* (the one dual to

τ^*), and for any given line r, we can decide the position of t^* relative to r (in time $O(n \log^2 n)$).

It is known that one can construct a constant number of lines $r_1, \ldots, r_c$ such that by deciding the position of t^* with respect to these lines, the position of t^* with respect to at least half of the lines of Q (in terms of weight) is determined. The lines $r_1, \ldots, r_c$ can be constructed in linear time. A construction of this type was first given by Megiddo (see the book [Ede87]). Other types of such constructions originated in the theory of random sampling, see e.g., [HW87, Mat90]. Using such a construction, we can solve the weighted questions batch problem in $O(n \log^2 n)$ time, and the sorting of intersections along τ^* can be thus performed in $O(n \log^3 n)$ time. Again one can show that when this algorithm finishes, the desired tangent τ^* must have been considered explicitly in answering the questions. $\quad \square$

4. Main algorithm

In this section we prove Theorem 2.2. We begin by some notation and simple observations.

For a point a in the plane, let $x(a)$ denote its x-coordinate. For a number x_0, let $v(x_0)$ stand for the vertical line $x = x_0$, and we write just $v(a)$ for $v(x(a))$.

We assume that H is a collection of n lines in general position. Let a, b be two points, $x(a) < x(b)$. We denote by $\gamma_k(H, a, b)$ the portion of $\gamma_k(H)$ lying between $v(a)$ and $v(b)$. The number of intersections of the lines of H with x-coordinate in an interval (x_0, x_1) will be denoted by $N(H, x_0, x_1)$. We will make use of the following elementary lemma:

LEMMA 4.1 [Mat90]. *Suppose that a, b have the same level in the arrangement of H. Then the open segment ab is intersected by at most*

$$2 \lfloor \sqrt{N(H, x(a), x(b))} \rfloor$$

lines of H. $\quad \square$

Let us now consider the situation with H, a, b as above. Put $N = N(H, x(a), x(b))$ and let us moreover suppose that a and b are vertices of $\gamma_k(H)$ (thus $a, b \in \lambda_k$, and a, b have the same level). Let τ_a, τ_b be tangents to $\gamma_k(H)$ touching at a, b respectively (for definiteness, we assume that τ_a has the smallest possible slope among all tangents at a, and τ_b the largest possible slope among all tangents at b). We introduce a classification of the lines of H based on their relation to a, b and related objects. First we distinguish the position of a line l with respect to the open segment ab (below, above, crossing). For the lines above ab, we introduce two finer classifications. We say that a line l above ab is *low above ab*, if at least one of the points $l \cap v(a)$, $l \cap v(b)$ has level at most $k + 1 + \lfloor \sqrt{N} \rfloor$; otherwise l is *high above ab*. We say that a line l above ab is *flat above ab* if the slope of l is between the slopes of τ_b and τ_a, otherwise l is *steep above* ab. Finally we call l *redundant* (with respect to a, b) if it is below ab or steep above ab or high above ab; otherwise l is *nonredundant*. The following lemma says just that redundant lines are really redundant:

LEMMA 4.1. *Let H' be the set of the nonredundant lines in H, and let k' be k minus the number of lines below ab. Then $\gamma_k(H, a, b) = \gamma_{k'}(H', a, b)$.*

PROOF. We proceed by induction, showing that by removing one redundant line and adjusting k appropriately, the portion of the convex hull in $x(a)$, $x(b)$ does not change. It is enough to show that each redundant line is completely above $\gamma_k(H, a, b)$ or completely below it. For the lines below ab it is clear, as well as for the lines steep above ab. Let l be a line high above ab. Let $a' \in v(a)$, $b' \in v(b)$ be points of level $k + \lfloor\sqrt{N}\rfloor + 2$. Since l is high above ab, the segment $a'b'$ lies below the line l. At the same time, no point of the segment $a'b'$ has level $k + 2$ or smaller (otherwise the segment $a'b'$ would be intersected by more than $2\lfloor\sqrt{N}\rfloor$ lines, contradicting Lemma 4.1), and thus all vertices of $\gamma_k(H, a, b)$ lie below $a'b'$. Therefore the whole convex arc $\gamma_k(H, a, b)$ lies below $a'b'$ and also below l. $\square$ $\square$

We need one more algorithmic lemma.

LEMMA 4.2. *Let c be a given constant, H be a collection of n lines and a, b given points, $x(a) < x(b)$. One can find x-coordinates $x_0 = x(a) < x_1 < \cdots < x_{c-1} < x_c = x(b)$ such that $N(H, x_{i-1}, x_i) \leq n^2/c$, $i = 1, 2, \ldots, c$, in time $O(n)$.*

This result is implied by [**Mat90**], and with the time bound $O(n \log n)$ instead of $O(n)$ it immediately follows from the results of Cole et al. [**CS$^+$89**]. $\square$

Let us now describe the algorithm which will be used for computing $\gamma_k(H)$. It will be a recursive divide-and-conquer type algorithm. Its input will be H, k and vertices a, b of γ_k, $x(a) < x(b)$. It will output (a suitable representation of) $\gamma_k(H, a, b)$. The main algorithm computing the whole γ_k is then obtained as follows: We compute the left and right unbounded rays of λ_k, we find tangents τ_l and τ_r to γ_k parallel to these rays and we use the vertices a, b where τ_l, τ_r touch γ_k in the initial call to the recursive algorithm.

In the recursive algorithm, we proceed as follows:

(i) If the size of H is smaller than a suitable constant, we compute the desired convex arc $\gamma_k(H, a, b)$ by some straightforward method. Otherwise we pass to the next step.

(ii) We use Lemma 4.2 with $c = 64$, and we find numbers $x_0 = x(a) < x_1 < \cdots < x_{c-1} < x_c = x(b)$ such that $N(H, x_{i-1}, x_i) \leq n^2/64$.

(iii) For every $i = 1, 2, \ldots, c - 1$, we apply Lemma 3.3 to compute a tangent τ_i touching γ_k at its intersection with $v(x_i)$. We let b_{i-1}

be the leftmost vertex in which τ_i touches γ_k, and a_i the rightmost such vertex. We also set $a_0 = a$ and $b_c = b$.

(iv) For every $i = 1, 2, \ldots, c$, we let H_i be the set of the lines of H which are nonredundant with respect to a_i, b_i (we, however, eliminate the pairs a_i, b_i with $x(a_i) \geq x(b_i)$). We let k_i be k minus the number of lines which are below $a_i b_i$.

(v) We call the algorithm recursively on H_i, k_i, a_i, b_i for every i, obtaining $\gamma_{k_i}(H_i, a_i, b_i)$, which by Lemma 4.1 is equal to $\gamma_k(H, a_i, b_i)$.

(vi) We merge the above obtained convex arcs, $\gamma_k(H, a_i, b_i)$ together with the segments $b_{i-1}a_i$, obtaining the desired convex arc $\gamma_k(H, a, b)$.

Let us analyze the running time of this algorithm. Excluding the time spent on the recursive calls, the running time of the algorithm with $|H| = n$ is $O(n \log^3 n)$, by our previous results. Let us estimate the size of the H_i's. Each nonredundant line is either crossing $a_i b_i$ or low above $a_i b_i$. The number of lines crossing $a_i b_i$ does not exceed $2\lfloor\sqrt{N(H, a_i, b_i)}\rfloor \leq 2\lfloor\sqrt{N(H, x_i, x_{i+1})}\rfloor \leq n/4$, by Lemma 4.1. The number of lines low above $a_i b_i$ is at most $2\lfloor\sqrt{N(H, a_i, b_i)}\rfloor \leq n/4$, so $|H_i| \leq n/2$. Hence the depth of recursive calls in our algorithm is $O(\log n)$.

Let us now view the computation as a rooted tree, with nodes corresponding to the recursive calls of the algorithm. If a node corresponds to a call with some a, b, then all lines entering that computation are nonredundant with respect to a, b, in particular they are all crossing ab or flat above ab. If we look at the points occurring as a, b at one level of the recursion tree, we see that they form a convex chain. Hence any given line occurs in at most two nodes at any given level of the recursion tree (either it is flat above in one node, or crossing in two nodes). Therefore the total computation time for every level of the recursion tree is $O(n \log^3 n)$, and the total running time is $O(n \log^4 n)$ as claimed. This finishes the proof of Theorem 2.2. □

Let us conclude this section by a remark about the implementability of our algorithm. As described, the algorithm includes quite heavy algorithmic tools (e.g., the AKS sorting network, hidden in the application of Megiddo's parametric search technique for sorting with Cole's improvement, and also in Lemma 4.2). However, if we permit randomization, this can be simplified considerably, essentially by replacing the parametric search subroutines by "randomized interval halving" (this idea is explained in [Mat91b]). The expected running time of the resulting randomized algorithm can be made no worse than the running time of the deterministic algorithm.

5. Approximate centerpoints

An obvious problem is the computation of centerpoints, of the center, and of the Tukey median for point sets in higher dimensions. Even if the methods used in the planar case could be extended to higher dimensions (in the spirit of [NS90]), one would obtain algorithm of complexity close to n^{d-1} at best. This does not look very satisfactory if all we want is a single point in the center, but achieving a more substantial improvement for this problem in higher dimensions probably requires quite a new insight.

Here we show that if we do not insist on exact determining of centerpoints or of Tukey median, but rather we admit some tolerance in the depth, we can get an efficient solution in any fixed dimension. Let us say that a point $x \in E^d$ is an ε-approximate centerpoint for P, if it has depth at least $(1 - \varepsilon)\frac{n}{d+1}$. Then for any fixed $\varepsilon > 0$, an ε-approximate centerpoint can be found in $O(n)$ time.

To this end, we use so-called ε-approximations. For our problem, we need the following definition: A point set $A \subseteq P$ is called an *ε-approximation for P (with respect to halfspaces)*, if for every halfspace h it is

$$\left| \frac{|A \cap h|}{|A|} - \frac{|P \cap h|}{|P|} \right| < \varepsilon.$$

Obviously if A is an ε-approximation for P and x is a centerpoint of A, then x is an ε-approximate centerpoint for P as well. Now for any $P \subseteq E^d$ (where the dimension d is fixed) and for every $\varepsilon > 0$, there exists an ε-approximation of size $O(\varepsilon^{-2} \log(1/\varepsilon))$ (see [VC71, HW87]) and for any *fixed* $\varepsilon > 0$, such an ε-approximation can be computed in time $O(n)$ (with the constant strongly depending on d and ε)—see [Mat91a]. For the constant-sized ε-approximation, we can then use a "brute-force" algorithm for finding a centerpoint; an $O(n^d)$ algorithm is straightforward. A problem with this approach is a large constant of proportionality hidden in the big-Oh notation.

REFERENCES

[Col87] R. Cole, *Slowing down sorting networks to obtain faster sorting algorithms*, J. Assoc. Comput. Mach. **34** (1987), 200–208.

[CS+89] R. Cole, J. Salowe, W. Steiger, and E. Szemerédi, *An optimal-time algorithm for slope selection*, SIAM J. Comput. **18** (1989), 792–810.

[CSY87] R. Cole, M. Sharir, and C. Yap, *On k-hulls and related problems*, SIAM J. Comput. **16** (1987), 61–67.

[Ede87] H. Edelsbrunner, *Algorithms in combinatorial geometry*, Springer-Verlag, Berlin and New York, 1987.

[EW86] H. Edelsbrunner and E. Welzl, *Constructing belts in two-dimensional arrangements with applications*, SIAM J. Comput. **15** (1986), 271–284.

[GSW90] J. Gil, W. Steiger, and A. Widgerson, *Geometric medians*, manuscript, 1990 (to appear in a volume dedicated to Z. Frolík (J. Nešetřil, ed.)).

[HW87]   D. Haussler and E. Welzl, *ε-nets and simplex range queries*, Discrete Comput. Geom. **2** (1987), 127–151.

[Mat90]   J. Matoušek, *Construction of ε-nets*, Discrete Comput. Geom. **5** (1990), 427–448.

[Mat91a]   _____, *Approximations and optimal geometric divide-and-conquer*, Proc. 23rd ACM Sympos. on Theory of Computing, 1991, pp. 506–511.

[Mat91b]   _____, *Randomized optimal algorithm for slope selection*, Inform. Process. Lett. (1991) (to appear).

[Meg83]   N. Megiddo, *Applying parallel computation algorithms in the design of serial algorithms*, J. Assoc. Comput. Mach. **30** (1983), 852–865.

[NS90]   N. Naor and M. Sharir, *Computing a point in the center of a point set in three dimensions*, Proc. 2nd Canad. Conf. on Computat. Geom., 1990.

[PS85]   F. Preparata and M. I. Shamos, *Computational Geometry—An Introduction*, Springer-Verlag, Berlin and New York, 1985.

[PSS90]   J. Pach, W. Steiger, and E. Szemerédi, *An upper bound on the number of planar k-sets*, Proc. 30th IEEE Sympos. on Theory of Comput. Sci., 1990, pp. 72–79.

[VC71]   V. N. Vapnik and A. Ya. Chervonenkis, *On the uniform convergence of relative frequencies of events to their probabilities*, Theory Probab. Appl. **16** (1971), 264–280.

DEPARTMENT OF APPLIED MATHEMATICS, CHARLES UNIVERSITY, MALOSTRANSKÉ NÁM. 25, 118 00 PRAHA 1, CZECHOSLOVAKIA

DIMACS Series in Discrete Mathematics
and Theoretical Computer Science
Volume **6**, 1991

Finite Quotients of Infinite Universal Polytopes

PETER McMULLEN AND EGON SCHULTE

ABSTRACT. It is well known that the euclidean and hyperbolic tessellations of type $\{p, q, r\}$ are the universal covering for infinitely many finite regular tessellations of the same type on closed compact 3-manifolds. This result is generalized to abstract regular polytopes. For example, if $\mathscr{P}_1$ and $\mathscr{P}_2$ are two toroidal regular maps for which the universal polytope $\{\mathscr{P}_1, \mathscr{P}_2\}$ is infinite, then there exist infinitely many finite regular polytopes with facet type $\mathscr{P}_1$ and vertex-figure type $\mathscr{P}_2$.

1. Introduction

In the classical theory of regular polytopes and tessellations it is a standard technique to construct new polytopes and tessellations by making suitable identifications. For example, the regular maps on surfaces can be obtained in this way from the regular tessellations $\{p, q\}$ of the euclidean 2-sphere or the euclidean or hyperbolic plane (cf. Coxeter [2,3], Coxeter-Moser [4]). Moreover, from each (infinite) euclidean or hyperbolic tessellation $\{p, q\}$ we can derive infinitely many regular maps of type $\{p, q\}$ which are finite maps on compact closed surfaces (cf. [4], Vince [21]). Analogous results are known for higher dimensions.

In this paper we extend these results to the theory of abstract regular polytopes; see §3. Abstract regular polytopes are combinatorial structures which resemble the classical regular polytopes and tessellations (cf. Danzer-Schulte [5]); for related notions see also McMullen [10], Grünbaum [7], Dress [6], Buekenhout [1], and Tits [20].

An important problem in the theory of abstract regular polytopes is the classification of the finite universal locally toroidal regular polytopes. Our results in §3 depend heavily on progress made on this problem in [12–15]. The proof of Theorem 1 generalizes a technique used in Vince [21, Theorem 6.3], to prove that for all Coxeter diagrams $\mathscr{D}$ belonging to an infinite irreducible Coxeter group there are infinitely many finite regular "combinatorial maps"

1991 *Mathematics Subject Classification.* Primary 51M20; Secondary 51E30, 52B15.

with diagram $\mathscr{D}$. The corollaries of our Theorem 1 solve problems posed in Grünbaum [7].

2. Basic notions

For a detailed introduction to the concept of abstract regular polytopes the reader is referred to [5,16,18]. An (*abstract*) *polytope* $\mathscr{P}$ *of rank* n is a partially ordered set with a strictly monotone rank function with range $\{-1, 0, \ldots, n\}$. The elements of rank i are called the *i-faces* of $\mathscr{P}$, or *vertices* or *facets* of $\mathscr{P}$ if $i = 0$ or $n - 1$, respectively. The *flags* (maximal totally ordered subsets) of $\mathscr{P}$ all contain exactly $n + 2$ faces, including the unique (least) (-1)-face F_{-1} and the unique (greatest) n-face F_n of $\mathscr{P}$. Further defining properties for $\mathscr{P}$ are the (global and local) *flag-connectedness* as well as the homogeneity property that for any $(i - 1)$-face F and any $(i + 1)$-face G with $F < G$ there are exactly two i-faces H of $\mathscr{P}$ with $F < H < G$.

If F and G are faces with $F < G$, we call $G/F := \{H | F \leq H \leq G\}$ a *section* of $\mathscr{P}$. There is little possibility of confusion if we identify a face F with the section F/F_{-1}. If F is a face, then F_n/F is said to be the *co-face* at F, or the *vertex-figure* at F if F is a vertex.

An n-polytope $\mathscr{P}$ is *regular* if its automorphism group $A(\mathscr{P})$ is transitive on the flags. If $\mathscr{P}$ is regular, then $A(\mathscr{P}) = \langle \rho_0, \ldots, \rho_{n-1} \rangle$ where ρ_i is the unique automorphism which keeps fixed all but the i-face F_i of some base flag of $\mathscr{P}$. These *distinguished generators* satisfy relations

(1) $(\rho_i \rho_j)^{p_{ij}} = 1$ $(i, j = 0, \ldots, n - 1)$;

here $p_{ii} = 1$, $p_{ji} = p_{ij} =: p_{i+1}$ if $j = i + 1$, and $p_{ij} = 2$ otherwise. The p_i's are the entries in the (*Schläfli-*) *type* $\{p_1, \ldots, p_{n-1}\}$ of $\mathscr{P}$. Also $A(\mathscr{P})$ has the intersection property

(2) $\langle \rho_i | i \in I \rangle \cap \langle \rho_i | i \in J \rangle = \langle \rho_i | i \in I \cap J \rangle$ for $I, J \subset \{0, \ldots, n - 1\}$.

A *C-group* is a group generated by involutions $\rho_0, \ldots, \rho_{n-1}$ such that (1) and (2) hold. The C-groups are precisely the groups of abstract regular polytopes.

Let $\mathscr{P}_1$ and $\mathscr{P}_2$ be regular n-polytopes such that the vertex-figures of $\mathscr{P}_1$ are isomorphic to the facets of $\mathscr{P}_2$. By $\langle \mathscr{P}_1, \mathscr{P}_2 \rangle$ we denote the set of all regular $(n + 1)$-polytopes $\mathscr{P}$ with facets isomorphic to $\mathscr{P}_1$ and vertex-figures isomorphic to $\mathscr{P}_2$. If $\langle \mathscr{P}_1, \mathscr{P}_2 \rangle \neq \emptyset$, then any such $\mathscr{P}$ is obtained from a *universal* member of $\langle \mathscr{P}_1, \mathscr{P}_2 \rangle$ by identifications; this universal polytope is denoted by $\{\mathscr{P}_1, \mathscr{P}_2\}$ (cf. [19]).

Call a regular 4-polytope $\mathscr{P}$ *locally of genus* (*at most*) g if its facets and vertex-figures are regular maps on orientable surfaces of genus at most g and if either the facets or vertex-figures are actually of genus g. We call $\mathscr{P}$ *locally spherical* or *locally toroidal* if $g = 0$ or 1, respectively. The Platonic

solids are the only spherical regular maps. The only toroidal regular maps
are $\{4, 4\}_{b,c}$, $\{3, 6\}_{b,c}$ and $\{6, 3\}_{b,c}$, with $b = c$ or $c = 0$ (cf. [4]).

Let U be any group. Then U is called *residually finite* if for any $\varphi \in U$, $\varphi \neq 1$, there exists a homomorphism f of U onto a finite group such that $f(\varphi) \neq 1$ (cf. [8]). It is easy to see that U is residually finite if and only if for each finite subset $\{\varphi_1, \ldots, \varphi_m\}$ of $U\backslash\{1\}$ there exists a homomorphism f of U onto a finite group such that $f(\varphi_j) \neq 1$ for $j = 1, \ldots, m$. By a central result in the theory of linear groups, due to Malcev [9], every finitely generated linear group is residually finite; see also [22]. Note that any (finitely generated) Coxeter group is a linear group and hence is residually finite.

3. Finite quotients

In this section we discuss finite quotients of infinite universal regular polytopes $\{\mathscr{P}_1, \mathscr{P}_2\}$. We begin with the following result.

THEOREM 1. *Let $\mathscr{P}_1$ and $\mathscr{P}_2$ be finite regular n-polytopes for which $\langle \mathscr{P}_1, \mathscr{P}_2 \rangle \neq \emptyset$. Let $\mathscr{P}$ be an infinite regular $(n + 1)$-polytope in $\langle \mathscr{P}_1, \mathscr{P}_2 \rangle$ whose group $A(\mathscr{P})$ is residually finite. Then $\langle \mathscr{P}_1, \mathscr{P}_2 \rangle$ contains infinitely many regular $(n + 1)$-polytopes which are finite and are covered by $\mathscr{P}$.*

PROOF. Let $A(\mathscr{P}) = \langle \rho_0, \ldots, \rho_n \rangle$, with $\rho_0, \ldots, \rho_n$ the distinguished generators of $A(\mathscr{P})$. Then $G_1 := \langle \rho_0, \ldots, \rho_{n-1} \rangle \simeq A(\mathscr{P}_1)$ and $G_2 := \langle \rho_1, \ldots, \rho_n \rangle \simeq A(\mathscr{P}_2)$. By our assumptions on $\mathscr{P}_1$ and $\mathscr{P}_2$ the groups G_1 and G_2 are finite, and so is the set $T_1 := (G_1 G_2)\backslash\{1\}$. Note that $G_1\backslash\{1\}, G_2\backslash\{1\} \subset T_1$. Since $A(\mathscr{P})$ is residually finite, there exists a surjective homomorphism $f_1 : A(\mathscr{P}) \mapsto A_1$ with A_1 a finite group such that $f_1(\varphi) \neq 1$ for all $\varphi \in T_1$. Let N_1 denote the kernel of f_1. Then $G_1 \cap N_1 = \{1\} = G_2 \cap N_1$, so that f_1 induces isomorphisms when restricted to the subgroups G_1 and G_2 of $A(\mathscr{P})$. To prove the intersection property

$$f_1(\langle \rho_i | i \in I \rangle) \cap f_1(\langle \rho_i | i \in J \rangle) = f_1(\langle \rho_i | i \in I \cap J \rangle) \quad \text{for } I, J \subset \{0, \ldots, n\}$$

it suffices to check $f_1(G_1) \cap f_1(G_2) = f_1(G_1 \cap G_2)$; all other cases follow from this and from (2) applied to $A(\mathscr{P}_1)$ and $A(\mathscr{P}_2)$. But now, if $\varphi_1 \in G_1$ and $\varphi_2 \in G_2$ such that $f_1(\varphi_1) = f_1(\varphi_2)$, then $\varphi_1^{-1}\varphi_2 \in G_1 G_2$ and $f_1(\varphi_1^{-1}\varphi_2) = 1$, so that by construction of T_1 we must have $\varphi_1 = \varphi_2 \in G_1 \cap G_2$, as required. This completes the proof that A_1 is a C-group with distinguished generators $f_1(\rho_0), \ldots, f_1(\rho_n)$.

It follows that A_1 is the group of a regular $(n + 1)$-polytope in $\langle \mathscr{P}_1, \mathscr{P}_2 \rangle$ which is finite and is covered by $\mathscr{P}$. To construct an infinite sequence of such $(n + 1)$-polytopes we proceed as follows.

Since $A(\mathscr{P})$ is infinite and A_1 is finite, N_1 is also infinite. Choose $\tau_1 \in N_1$ with $\tau_1 \notin T_1$ and define $T_2 := T_1 \cup \{\tau_1\}$. Then there is a surjective homomorphism $f_2 : A(\mathscr{P}) \mapsto A_2$ with A_2 a finite group such that $f_2(\varphi) \neq 1$ for all $\varphi \in T_2$. Since $T_1 \subset T_2$, the group $A_2 = \langle f_2(\rho_0), \ldots, f_2(\rho_n) \rangle$ is again a C-group and hence is the group of a finite regular $(n + 1)$-polytope in $\langle \mathscr{P}_1, \mathscr{P}_2 \rangle$ which is covered by $\mathscr{P}$. By construction the kernel N_2 of f_2 is distinct from N_1, so that the polytopes corresponding to A_1 and A_2 cannot be isomorphic. Note for this that any isomorphism between regular polytopes must necessarily correspond to an isomorphism of the group which maps distinguished generators to distinguished generators, and second that any two sets of distinguished generators of the same group are related by conjugation.

In a similar fashion we can construct an infinite sequence of homomorphisms $f_j : A(\mathscr{P}) \mapsto A_j$ with mutually distinct kernels N_j. The corresponding sets T_j are of the form $T_j = T_{j-1} \cup \{\tau_{j-1}\}$ with $\tau_{j-1} \in N_{j-1}$ but $\tau_{j-1} \notin T_{j-1}$. Note that for $j < k$ we have $f_j(\tau_j) = 1 \neq f_k(\tau_j)$. The polytopes corresponding to the groups A_j have the required properties. This completes the proof. □

As mentioned in §2, each finitely generated linear group is residually finite. Clearly the groups of regular polytopes are finitely generated. It follows that Theorem 1 applies if $\mathscr{P}$ is an infinite member in $\langle \mathscr{P}_1, \mathscr{P}_2 \rangle$ whose group is a linear group. This is the form in which Theorem 1 is used below. Note that Theorem 1 does not require that $\mathscr{P}$ equals the universal $\{\mathscr{P}_1, \mathscr{P}_2\}$.

4. Polytopes of rank 4

From now on we mainly restrict ourselves to the discussion of rank 4 polytopes though extensions of the results to higher ranks are possible. In the locally spherical case, $\mathscr{P}_1$ and $\mathscr{P}_2$ are Platonic solids and an infinite universal $\{\mathscr{P}_1, \mathscr{P}_2\}$ is necessarily a tessellation of euclidean or hyperbolic space. Here it is well known that there are infinitely many finite regular 4-polytopes in $\langle \mathscr{P}_1, \mathscr{P}_2 \rangle$. The result follows again from Theorem 1 applied to the universal $\{\mathscr{P}_1, \mathscr{P}_2\}$ whose group is a Coxeter group.

COROLLARY 1. *Let* $\{p, q, r\}$ = $\{4, 3, 4\}, \{3, 5, 3\}, \{5, 3, 5\},$ $\{4, 3, 5\},$ *or* $\{5, 3, 4\}$. *Then* $\langle \{p, q\}, \{q, r\} \rangle$ *contains infinitely many finite regular polytopes.*

In the locally toroidal case, $\mathscr{P}_1$ and $\mathscr{P}_2$ are spherical or toroidal but not both spherical. A complete classification of the finite universal locally toroidal regular 4-polytopes $\{\mathscr{P}_1, \mathscr{P}_2\}$ is known for the types $\{4, 4, 3\}$ and $\{6, 3, p\}$ with $3 \leq p \leq 6$, and for their dual types (cf. [12–15]). While the classification is almost complete for the type $\{4, 4, 4\}$, relatively little is known for the type $\{3, 6, 3\}$.

For the types $\{6, 3, p\}$ with $3 \leq p \leq 6$ and for the known cases of type $\{3, 6, 3\}$ it has been proved in [15, §7], that either the universal $\{\mathscr{P}_1, \mathscr{P}_2\}$ itself or at least one of its quotients $\mathscr{P}$ in $\langle \mathscr{P}_1, \mathscr{P}_2 \rangle$ has a group which can be realized as a linear group on a finite-dimensional complex space. It follows that Theorem 1 applies and proves the corresponding parts of Corollary 2 below.

The types $\{4, 4, 3\}$ and $\{4, 4, 4\}$ have been studied in [14]. Most of the corresponding polytopes were constructed from so-called twisting operations on Coxeter groups. The same technique as in [15, §7], can be used to prove that these polytopes have groups which can be realized as linear groups on finite-dimensional real spaces. This settles the case $\{4, 4, 3\}$ completely. For the type $\{4, 4, 4\}$ some of the polytopes are not directly constructed from twisting operations but are related to polytopes which are derived from twisting operations. Here it can be checked separately in all but two cases that Theorem 1 applies. The two exceptions are the classes $\langle\{4, 4\}_{s,s}, \{4, 4\}_{t,t}\rangle$ with s, t even and $\langle\{4, 4\}_{s,0}, \{4, 4\}_{t,0}\rangle$ with s, t odd and distinct. In the former case it is known that the universal $\{\{4, 4\}_{s,s}, \{4, 4\}_{t,t}\}$ is finite if and only if $s = t = 2$. We conjecture that Corollary 2 extends to this case. In the latter case the classification of the finite universal polytopes $\{\{4, 4\}_{s,0}, \{4, 4\}_{t,0}\}$ is not known yet.

COROLLARY 2. *Let $\mathscr{P}_1$ and $\mathscr{P}_2$ be regular toroidal maps for which the universal $\{\mathscr{P}_1, \mathscr{P}_2\}$ exists and is infinite. Assume that condition* (a) *or* (b) *holds if the type is* $\{3, 6, 3\}$ *or* $\{4, 4, 4\}$, *respectively*:

(a) *If $\{\mathscr{P}_1, \mathscr{P}_2\}$ is of type $\{3, 6, 3\}$, then $\{\mathscr{P}_1, \mathscr{P}_2\}$ equals one of the polytopes $\{\{3, 6\}_{m,m}, \{6, 3\}_{m,0}\}$ or $\{\{3, 6\}_{m,m}, \{6, 3\}_{3m,0}\}$ with $m \geq 2$, or the dual of one of these polytopes.*

(b) *If $\{\mathscr{P}_1, \mathscr{P}_2\}$ is of type $\{4, 4, 4\}$, then $\{\mathscr{P}_1, \mathscr{P}_2\}$ is not one of the polytopes $\{\{4, 4\}_{s,s}, \{4, 4\}_{t,t}\}$ with s, t even or $\{\{4, 4\}_{s,0}, \{4, 4\}_{t,0}\}$ with s, t odd and distinct.*

Under these assumptions $\langle\mathscr{P}_1, \mathscr{P}_2\rangle$ contains infinitely many regular polytopes which are finite.

We remark that the corollaries extend to many classes of 4-polytopes which are locally of higher genus. Also, the techniques generalize to higher ranks. This covers polytopes of types $\{3, 4, 3, 4\}$, $\{p, 3, 3, 5\}$ (with $p = 3, 4, 5$), $\{3, 3, 4, 3, 3\}$, $\{3, 4, 3, 3, p\}$ (with $p = 3, 4$), and their duals.

REFERENCES

1. F. Buekenhout, *Diagrams for geometries and groups*, J. Combin. Theory Ser. A **27** (1979), 121–151.

2. H. S. M. Coxeter, *Twisted honeycombs*, CBMS Regional Conf. Ser. in Math., no. 4, Amer. Math. Soc., 1970.

3. _____, *Regular polytopes*, 3rd ed., Dover, New York, 1973.

4. H. S. M. Coxeter and W. O. J. Moser, *Generators and relations for discrete groups*, 4th ed., Springer-Verlag, Berlin, 1980.

5. L. Danzer and E. Schulte, *Reguläre Inzidenzkomplexe*. I, Geom. Dedicata **13** (1982), 295–308.

6. A. W. M. Dress, *Regular polytopes and equivariant tessellations from a combinatorial point of view*, Conference on Algebraic Topology (Göttingen, 1984), Lecture Notes in Math., vol. 1172, Springer-Verlag, Berlin and New York, 1985, pp. 56–72.

7. B. Grünbaum, *Regularity of graphs, complexes and designs*, Problèmes Combinatoire et Théorie des Graphes, Coll. Int. CNRS No. 260, Orsay, 1977, pp. 191–197.

8. W. Magnus, A. Karrass, and D. Solitar, *Combinatorial group theory*, 2nd ed., Dover, New York, 1976.

9. A. I. Malcev, *On faithful representations of infinite groups of matrices*, Mat. Sb. **8** (1940), 405–422 (Russian); English transl. in Amer. Math. Soc. Transl. Ser. (2) **45** (1965), 1–18.

10. P. McMullen, *Combinatorially regular polytopes*, Mathematika **14** (1967), 142–150.

11. _____, *Realizations of regular polytopes*, Aequationes Math. **37** (1989), 38–56.

12. P. McMullen and E. Schulte, *Regular polytopes from twisted Coxeter groups*, Math. Z. **201** (1989), 209–226.

13. _____, *Regular polytopes from twisted Coxeter groups and unitary reflexion groups*, Adv. in Math., **82** (1990), 35–87.

14. _____, *Regular polytopes of type* $\{4, 4, 3\}$ *and* $\{4, 4, 4\}$, Combinatorica (to appear).

15. _____, *Hermitian forms and locally toroidal regular polytopes*, Adv. in Math., **82** (1990), 88–125.

16. _____, *Abstract regular polytopes*, monograph, in preparation.

17. B. Monson and A. I. Weiss, *Regular 4-polytopes related to general orthogonal groups*, Mathematika. **37** (1990), 106–118.

18. E. Schulte, *Reguläre Inzidenzkomplexe. II*, Geom. Dedicata **14** (1983), 33–56.

19. _____, *Amalgamations of regular incidence-polytopes*, Proc. London Math. Soc. (3) **56** (1988), 303–328.

20. J. Tits, *A local approach to buildings*, The Geometric Vein (The Coxeter-Festschrift) (Ch. Davis, B. Grünbaum, and F. A. Sherk, eds.), Springer-Verlag, Berlin, 1981, pp. 519–547.

21. A. Vince, *Regular combinatorial maps*, J. Combin. Theory **B35** (1983), 256–277.

22. B. A. F. Wehrfritz, *Infinite linear groups*, Springer-Verlag, New York, 1973.

23. A. I. Weiss, *Incidence-polytopes of type* $\{6, 3, 3\}$, Geom. Dedicata **20** (1986), 147–155.

DEPARTMENT OF MATHEMATICS, UNIVERSITY COLLEGE LONDON, LONDON WC1E 6BT, ENGLAND

DEPARTMENT OF MATHEMATICS, NORTHEASTERN UNIVERSITY, BOSTON, MASSACHUSETTS 02115

DIMACS Series in Discrete Mathematics
and Theoretical Computer Science
Volume **6**, 1991

The Universality Theorem on the Oriented Matroid Stratification of the Space of Real Matrices

NIKOLAI MNËV

The problem of describing the class of realization spaces of oriented matroids was posed by A. M. Vershik in the 1970s. This problem was solved in 1984 [**2, 3**]. The solution was unexpected. By definition, the set of all matrix realizations of rank three oriented matroids is a semialgebraic subset of the space of all $3 \times m$ real matrices. The following phenomenon was established: up to trivial stabilization, any elementary semialgebraic variety can be represented as the realization space of an oriented matroid of rank three.

Any open elementary variety (i.e., a variety defined by strict inequalities) can be represented as a realization space of a uniform, rank three oriented matroid. Moreover, it can be done in such a way that the number of points of the matroid is a linear function of the bit size for describing the polynomial system defining the variety. This means that given an arbitrary system of polynomial equations and inequalities, one can rewrite it in "oriented matroid" form, and in such a way as to preserve the algorithmic complexity class of various elementary problems relating to the geometry of the solutions.

Consider the space $\mathrm{Mx}^{3 \times m}$ of all $3 \times m$ real matrices. There is a natural stratification of this space—by signs of maximal minors. Each stratum is a realization space of some rank three oriented matroid on the set $\overline{1 : m}$. The aim of this paper is to announce a universality theorem for this class of stratified varieties. If this theorem is localized to one stratum, we get the universality theorem for the realization spaces of oriented matroids.

We begin with the definition of several notions that may be a little bit unfamiliar in the present context.

We are in the category of elementary semialgebraic varieties defined over $\mathbb{Z}$. The objects of this category are the sets of real solutions of polynomial systems of equations and strict inequalities. We suppose that the coefficients

---

1991 *Mathematics Subject Classification.* Primary 51A20, 14P10.

of the polynomials are the integers. The morphisms of our category are the maps of the sets which are generated by polynomial vector functions with integer coefficients.

1.

Let N be a variety. Denote by $\mathbb{Z}_+(N)$, the cone of all positive polynomial functions on N. Let ϕ be a finite subset of $\mathbb{Z}_+(N)$, $\phi = \{\varphi_1, \ldots, \varphi_k\}$ and consider the variety $\Gamma_+(\phi, N) = \{(x, y) \mid x \in N, 0 < y < \varphi_i(y) ; i \in \overline{1 : k}\}$. We are interested in the following two properties:

(i) The character of the natural projection $\Gamma_+(\phi, N) \xrightarrow{\pi} N$ (Figure 1a).

(ii) The character of the adjacency of $\Gamma_+(\phi, N)$ and N in the common space (Figure 1b).

(a)   (b)

FIGURE 1

1.1. Stabilization.

DEFINITION. The map $A \xrightarrow{\psi} B$ of varieties is called a 1-stabilization if it is equivalent to the projection $\Gamma_+(\phi, N) \xrightarrow{\pi} N$ for some N; $\phi \subset \mathbb{Z}_+(N)$. (By equivalence we mean here the existence of isomorphisms $A \xrightarrow{\alpha} \Gamma_+(\phi, N)$, $B \xrightarrow{\beta} N$ that make the diagram

$$
\begin{array}{ccc}
A & \xrightarrow{\psi} & B \\
\downarrow \alpha & & \downarrow \beta \\
\Gamma_+(\phi, N) & \xrightarrow{\pi} & N
\end{array}
$$

commutative.)

DEFINITION. The map of varieties is called a *stabilization* if it admits a decomposition into a chain of 1-stabilizations. Obviously a stabilization of varieties preserves the homotopical structure of the variety and the character of its singularities.

1.2. Stable adjacency. Let N be a subvariety of $\mathbb{R}^m$ and let $\mathbb{R}^m = H \oplus \ell$ be some representation of $\mathbb{R}^m$ as a direct sum of a linear hyperplane H and a one-dimensional linear subspace ℓ. Coordinatize ℓ by $t: \ell \to \mathbb{R}$. One can make a correspondence between the data $D = $ (decomposition, coordinatization) and the imbedding $\Gamma_+(N, \phi)_D$ of the variety $\Gamma_+(N, \phi)$

into the space

$$\mathbb{R}^m: \Gamma_+(N, \phi)_D = \{x + y \mid x \in N , \ 0 < t(y) < \min_{i \in 1 : k} \varphi_i(x)\} .$$

DEFINITION. A subvariety M of the space $\mathbb{R}^m$ is 1-*stably adjacent* to the subvariety N if there exists a family of functions $\phi \subset \mathbb{Z}_+[N]$ and $D =$ (decomposition of the space $\mathbb{R}^n$, coordinatization of the one-dimensional element of the decomposition) such that the subvariety M coincides with $\Gamma_+(\phi, N)_D$.

DEFINITION. A subvariety M of the space $\mathbb{R}^m$ is *stably adjacent* to the subvariety N if there exists a chain of subvarieties: $N = K_0 \subset \mathbb{R}^m$, $K_1 \subset \mathbb{R}^m, \dots, M = K_\ell \subset \mathbb{R}^m$ such that K_i is 1-stably adjacent to K_{i-1} for $i = 1, \dots, \ell$.

Obviously any decomposition from the last definition generates the stabilization $M \to N$. Denote by $\overset{\circ}{\mathbb{R}}{}^\ell_+$ the set

$$\{x = (x_1 \cdots x_\ell) \mid x_i \in \mathbb{R} , \ x_i > 0 ; \ i \in \overline{1 : \ell}\} .$$

PROPOSITION. *If a subvariety $M \subset \mathbb{R}^m$ is stably adjacent to the subvariety $N \subset \mathbb{R}^m$ then there exists* (a) *a linear subspace $H \subset \mathbb{R}^m$ such that $N \subset H$, and* (b) *a neighborhood T of M such that the triple $(N \subset H , M \cap T , T)$ is isomorphic to the triple $(N \subset H , N + \overset{\circ}{\mathbb{R}}{}^\ell_+ , H \oplus \mathbb{R}^\ell)$ for some $\ell \in \mathbb{N}$.*

2. Elementary semialgebraic partitions and stratifications

Consider the finite set of indices $\Phi = \{+1, 0, -1\}$ and some subset U of the set Φ^k, $k \in \mathbb{N}$.

DEFINITION. A *partition* of the variety N is the map $N \to U$ of type $(\text{sign } f_1, \dots, \text{sign } f_k)$, where $f_i \in \mathbb{Z}[N]$ for $i \in \overline{1 : k}$.

The element $(\epsilon_1 \cdots \epsilon_k) \in \Phi^k$ is called uniform if $\epsilon_i \neq 0$ for all $i \in \overline{1 : k}$.

DEFINITION. A *stratification* of the variety N is a finite, numbered covering of N by some disjoint set of subvarieties (i.e., a stratification is a map $I \overset{\sigma}{\longrightarrow} 2^N$, where I is a finite set, $\sigma(i)$ is a subvariety of N, $\sigma(j) \cap \sigma(i) = \varnothing$ when $i \neq j$, $i, j \in I$, $\bigcup_{i \in I} \sigma(i) = N$). An element $\sigma(i)$ of the stratification is called a *stratum*. Each partition $N \overset{p}{\longrightarrow} U$ corresponds to the stratification

$$U \overset{p^\sigma}{\longrightarrow} 2^N \quad \text{where } p^\sigma(j) = p^{-1}(j) \text{ for } j \in U .$$

The strata which correspond to the uniform elements of U are called *open*. These strata are really open subsets of N in the strong topology.

3. The main example

Let m be a natural number. Consider the set

$$\Lambda(m) = \{(i, j, k) \mid i < j < k , \ i, j, k \in \overline{1 : m}\} .$$

An oriented matroid is a function $\Lambda(m) \to \Phi$ which satisfies some requirements as given, for example, in [1]. So one can identify the set of all oriented matroids on $\{\overline{1 : m}\}$ with some subset $\mathscr{M}d\,(m)$ of the set $\Phi^{\Lambda(m)}$. Every $3 \times m$ matrix x with real coefficients defines the oriented matroid

$$\mathrm{Md}(x) \in \mathscr{M}d\,(m): \mathrm{Md}(x)_{i,j,k} = \text{ sign } \Delta_{i,j,k}(x) \;,$$

where $\Delta_{i,j,k}(x)$ is the determinant of the maximal minor of x with the columns $(i,\,j,\,k)$. So we have the partition

$$\mathrm{Mx}^{3 \times m} \xrightarrow{\mathrm{Md}} \mathscr{M}d\,(m) \subset \Phi^{\Lambda(m)} \;.$$

The stratification $\mathscr{M}d\,(m) \xrightarrow{\mathrm{Md}^{\sigma}} 2^{\mathrm{Mx}^{3 \times m}}$ associates to each oriented matroid the space of all its matrix (vector) representations. A uniform matroid χ is associated with an open stratum $\mathrm{Md}^{\sigma}(\chi)$.

4. The imbedding of partitions

By *imbedding of partition* $N \xrightarrow{\psi} U \subset \Phi^{k}$ into the partition $M \xrightarrow{\varphi} V \subset \Phi^{\ell}$ we shall understand the pair of imbeddings $(\alpha,\,\beta)$ where $N \xhookrightarrow{\alpha} M$, $U \xhookrightarrow{\beta} V$, and the square

$$\begin{array}{ccc} N & \xrightarrow{\psi} & U \\ \uparrow{\scriptstyle\alpha} & & \uparrow{\scriptstyle\beta} \\ M & \xrightarrow{\varphi} & V \end{array}$$

is cartesian. This means, for example, that the diagram

$$\begin{array}{ccc} U & \xrightarrow{\psi^{\sigma}} & 2^{N} \\ \uparrow{\scriptstyle\beta} & & \uparrow{\scriptstyle\bar{\alpha}} \\ V & \xrightarrow{\varphi^{\sigma}} & 2^{M} \end{array}$$

is commutative, where $\bar{\alpha}$ is the imbedding of 2^{N} into 2^{M} defined by the imbedding $N \xhookrightarrow{\alpha} M$. So, each stratum of the stratification ψ^{σ} becomes a stratum in the stratification φ^{σ} (Figure 2).

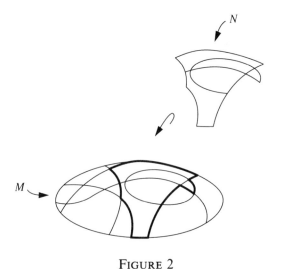

FIGURE 2

5. Partitions defined by partially oriented matroids

Let χ be a partially oriented matroid of rank three on the set $\overline{1:m}$. The space $[\chi]$ of all matrix representations of χ is naturally imbedded into $\mathrm{Mx}^{3\times m}$. Consider the set $\mathrm{Spec}(\chi) \subset \mathscr{Md}(m)$ of all oriented matroids on $\overline{1:m}$ whose orientations agree with χ.

The partition $p(\chi) = \mathrm{Md}\big|_{[\chi]} : [\chi] \to \mathrm{Spec}(\chi)$ is naturally imbedded into the partition Md. The stratification $\sigma(\chi) = p(\chi)^\sigma : \mathrm{Spec}(\chi) \to 2^{[\chi]}$ is a stratification of the set $[\chi]$ by complete oriented matroid type.

6. Stabilization of partitions

DEFINITION. A *stabilization of partitions* $M \xrightarrow{\psi} U$ and $N \xrightarrow{\varphi} V$ is a pair of maps $S = (\alpha, \beta)$ where $M \xrightarrow{\alpha} N$ is a stabilization of varieties, $U \xrightarrow{\beta} V$ is a bijection, and the diagram

$$
\begin{array}{ccc}
M & \xrightarrow{\psi} & U \\
\downarrow{\alpha} & & \downarrow{\beta} \\
N & \xrightarrow{\varphi} & V
\end{array}
$$

is commutative.

The stabilization S defines the map $V \xrightarrow{S^\sigma} 2^M$ with the property that for any $j \in V$, $S^\sigma(j) = \psi^\sigma(\beta^{-1}(j))$ and the map $\alpha\big|_{S^\sigma(j)} : S^\sigma(j) \to \varphi^\sigma(j)$ is a stabilization of strata.

7. Multiplication of partition by direct factor

Let $N \xrightarrow{\psi} U$ be a partition and let A be a semialgebraic set. Consider the variety $N \times A$ and the partition $\psi^{\times A} : N \times A \to U$ which is defined by the commutative diagram

$$
\begin{array}{ccc}
N \times A & \xrightarrow{\psi^{\times A}} & U \\
\pi \downarrow & \nearrow \psi & \\
N & &
\end{array}
$$

where π is the projection on the direct factor.

Obviously for any $j \in U$ the stratum $(\psi^{\times A})^\sigma(j)$ is isomorphic to the variety $\psi^\sigma(j) \times A$ (Figure 3).

$$N \times A$$

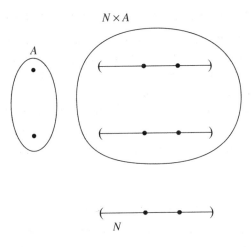

$$N$$

FIGURE 3

8.

An elementary semialgebraic partition of Euclidian space is a partition of type $\mathbb{R}^n \xrightarrow{\psi} \Phi^k$, where k, n are natural numbers, and

$$\psi = (\operatorname{sign} f_1, \ldots, \operatorname{sign} f_k), \qquad \{f_1, \ldots, f_k\} \subset \mathbb{Z}[\mathbb{R}^n].$$

Obviously, any semialgebraic partition can be imbedded into some partition of Euclidian space.

9.

Now we are able to formulate the universality theorem for oriented-matroid partitions (stratifications) of the spaces of real matrices. On the space $[\chi]$ (of all matrix representations of a rank three, partially oriented matroid χ), there is the natural action of the group $\mathrm{GL}_3(\mathbb{R})$. Modulo this action and some stabilization the partition $[\chi] \xrightarrow{p(x)} \operatorname{Spec}(\chi)$ can coincide with any semialgebraic partition of Euclidean space.

THEOREM. *Let $\mathbb{R}^n \xrightarrow{\psi} \Phi^k$ be an elementary semialgebraic partition of Euclidian space.*

(1) *There is a natural number m, a partially oriented matroid χ on the set $\overline{1 : m}$, and a pair of maps (α, β), $[\chi] \xrightarrow{\alpha} \mathbb{R}^n \times \mathrm{GL}_3(\mathbb{R})$, $\operatorname{Spec}(\chi) \xrightarrow{\beta} \Phi^k$, such that (α, β) is a stabilization of the partitions $\psi^{\times \mathrm{GL}_3}$ and $\sigma(\chi)$.*

Note that

    (a) *by §6, for any $j \in \Phi^k$, the realization space of the oriented matroid $\beta^{-1}(j)$ is a stratum $\sigma(\chi)(\beta^{-1}(j))$ of the stratification $\sigma(\chi)$ and this stratum is a stabilization of the stratum $(\psi^{\times \mathrm{GL}_3})^\sigma(j) = \psi^\sigma(j) \times \mathrm{GL}_3$.*

(b) *by* §5, *the partition* $p(\chi)$ *is naturally imbedded into the partition* $\mathrm{Mx}^{3 \times m} \xrightarrow{\mathrm{Md}(m)} \mathcal{M}d\,(m)$.

Moreover,

(2) *The partially oriented matroid* χ *can be chosen in such a way that for any stratum* $\Delta(j^*) = \sigma(\chi)(\beta^{-1}(j^*))$ *which corresponds to a uniform element* $j^* \in \Phi^m$ *there is an open stratum* $\beta(j^*)$ *of the stratification* $\mathrm{Md}^{\sigma}(m)$ *which is stably adjacent to* $\Delta(j^*)$. *(So,* $\beta(j^*)$ *is by definition the realization space of some uniform rank three oriented matroid.)*

The proof of this result is long and complicated. It will appear separately.

REFERENCES

1. J. Bokowski and B. Sturmfels, *Computational and synthetic geometry*, Lecture Notes in Math., vol. 1355, Springer-Verlag, Berlin and New York, 1989.
2. N. E. Mnëv, *Topology of the spaces of projective configurations*, Ph.D. thesis, Steklov Institute, Leningrad, 1986. (Russian)
3. _____, *The universality theorems on the classification problem of configuration varieties and convex polytope varieties*, Topology and Geometry—Rohlin Seminar (O. Y. Viro, ed.), Lecture Notes in Math., vol. 1346, Springer-Verlag, Berlin and New York, 1988.

KORABLESTROITELY STREET 401 187, ST. PETERSBURG, USSR

DIMACS Series in Discrete Mathematics
and Theoretical Computer Science
Volume **6**, 1991

The Densest Double-Lattice Packing
of a Convex Polygon

DAVID M. MOUNT

ABSTRACT. A lattice packing of a planar body is an infinite packing of the plane by translated copies of the body where the copies are translated to the points of a lattice. A double-lattice packing is a union of two lattice packings such that a $180°$ rotation about some point interchanges the two packings. We show that the densest double-lattice packing of an n-sided convex polygon can be computed in $O(n)$ time.

1. Introduction

Packing problems such as the knapsack problem and bin packing problem are well known in the fields of algorithm design and operations research because of their many applications to problems such as stock-cutting in computer-aided manufacturing. Because general formulations of these problems are known to be NP-complete [5], it is of interest to discover formulations of these problems which are solvable in polynomial time and yet are of general enough interest to be useful in applications. One such formulation is that of finding the densest (infinite) packing of congruent copies of a single polygon in the plane. More formally, given a simple polygon P, the problem is to determine an infinite collection of rigidly transformed copies of P in the plane having pairwise disjoint interiors, such that *density* of the system (intuitively the fraction of the plane covered by copies) is maximized. This problem is of interest in packing applications where a large number of identical 2-dimensional objects are to be packed into a large container. If the size of the objects is small relative to the size of the container, then a reasonable heuristic is to determine the densest infinite packing of the objects in the plane, and then truncate the packing to fit the container (see Figure 1(a)).

1991 *Mathematics Subject Classification.* Primary 68U05; Secondary 68Q25.

This research has been partially supported by National Science Foundation grant CCR–89–08901.

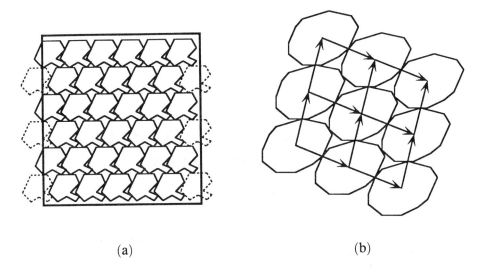

(a)   (b)

FIGURE 1. Lattice packings.

Mount and Silverman showed that for an n-sided convex polygon, the densest packing in the plane, allowing only translations not rotations, could be computed in $O(n)$ time. In proving this result they applied a classical theorem due to Rogers [14] which states that the densest packing by translates of a convex body is generated by a *lattice*, that is, a system of points defined by all integer linear combinations of two independent vectors (see Figure 1(b)). They showed how to compute the densest lattice packing in $O(n)$ time.

One major shortcoming in Mount and Silverman's result is that it does not allow objects to be rotated. Although this is reasonable for packing applications where the packing domain has a directional grain (e.g. when cutting fabric), better packing densities are achievable if rotation is allowed. For example, if the objects to be packed are triangles then the densest packing by translates has density 2/3 [2], while if rotation is allowed then a packing of density 1, a *tiling*, is possible by mating each triangle with a $180°$ rotation of itself to form a parallelogram, and then tiling the plane with these parallelograms.

When rotation is allowed we know of no general simple structure, such as the lattice, which is guaranteed to generate the densest packing. For example, Figure 2(a) shows an example due to Heesch of a pentagon which, if allowed to rotate, can tile the plane with a periodic structure [7] (see also [6, p. 31]). Perhaps the simplest packing structure which allows rotation is a *double-lattice* packing, which is the union of two lattice packings, such that a rotation by $180°$ about some point interchanges the two packings. The packing of Figure 2(b) is an example of a double-lattice packing of a regular pentagon, and the triangle tiling described earlier is also an example.

In this paper we describe an $O(n)$ algorithm for determining the densest

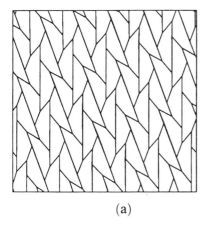

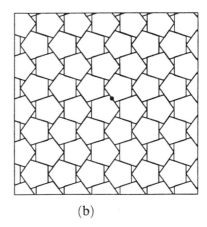

(a)   (b)

FIGURE 2. Packing with rotation.

double-lattice packing of a convex n-sided polygon P. Our result is based
on a reduction due to G. Kuperberg and W. Kuperberg [9]. They showed that
the densest double-lattice packing of a convex body can be derived by finding
a certain type of inscribed parallelogram, called an *extensive parallelogram*,
of minimum area.

The problem of finding the densest packings of a convex object in Euclidean space has a rich history. See Rogers [15] and Fejes-Toth [3] for surveys of this field. The related subject of tilings has been also been studied [6].
There have been relatively few computational results in this area. In addition
to Mount and Silverman's result, De Pano described a linear time algorithm
for packing congruent copies of a convex polygon in the plane (where rotations are allowed) such that the density of the resulting packing is at least
3/4 [1]. The packing generated by De Pano's algorithm (which is based on a
construction given by Kuperberg [10]) is also a double-lattice packing, but it
is not necessarily the densest double-lattice packing for the given input.

In §2 we present the geometrical underpinnings of the algorithm and show
how to reduce the packing problem to a problem of finding a certain minimum area inscribed parallelogram, called the *half-length* parallelogram. In
§3 we derive a rotating calipers algorithm for finding this parallelogram and
analyze the algorithm's running time.

2. Double-lattice packings and parallelograms

Throughout this paper P will denote an n-sided convex polygon in the
real plane, $\mathbf{R}^2$. For $v \in \mathbf{R}^2$, the *translate* of P by v, $P + v$, is the set of
points $\{p + v | p \in P\}$. Let $-P$ denote the set $\{-x | x \in P\}$, a rotation of
P through $180°$ about the origin. For a given pair of linearly independent
vectors u and v, the *lattice* generated by u and v is the set of vectors

$$L(u, v) = \{iu + jv \mid i \text{ and } j \text{ integers}\}.$$

The vectors u and v span a *basic parallelogram* of the lattice.

Consider an infinite system of bodies resulting by translating P by each vector in a lattice, $L(u, v)$. If the interiors of this system are pairwise disjoint, then we say that u and v define a *lattice packing* of P. The *density* of a lattice packing is the ratio of the area of P to the area of the basic parallelogram, the absolute value of det (u, v). The density of a packing is at most 1, where equality occurs if the packing is a tiling of the plane (implying that P is either a parallelogram or a centrally symmetric hexagon).

A *double-lattice packing* is the union of two lattice packings such that a $180°$ rotation about some point interchanges these two packings. (Figure 2(b) gives an example. A possible point of rotation is shown.) It is easy to see that the density of a double-lattice packing is the ratio of twice the area of P to the area of the basic parallelogram.

G. Kuperberg and W. Kuperberg [9] showed that there exists a double-lattice packing for any convex body P of density at least $\sqrt{3}/2$, matching existing lower bound for lattice packings of centrally symmetric convex bodies [11, 4]. For convex polygons with three or four sides, there exist double-lattice packings that tile the plane. The Kuperberg's also conjecture that the densest packings by congruent copies of regular pentagons and regular heptagons are double-lattice packings. It is not hard to show that any lattice packing of a convex body can be converted into a double-lattice packing of equal density, and hence double-lattice packings can achieve at least as good densities as lattice packings for any given convex polygon.

Given a convex polygon P, a *chord* of P is any line segment whose endpoints lie on the boundary of P. Define the *angle* of a line segment of nonzero length to be the arctangent of the slope of the line segment normalized to the interval $[0°, 180°)$. Given an angle θ, the *length* of P at angle θ is the length of the longest chord of P whose angle is θ, and the *width* of P at the angle θ is the perpendicular distance between the two parallel lines of support for P whose angle is θ. A chord is of maximal length for a given angle if and only if there exist two parallel lines of support for P that pass through the endpoints of the chord. We call such a chord a θ-*diameter*. The diameter chord at a given angle need not be unique, but it is easy to see that for any angle θ there exists a θ-diameter such that at least one endpoint of the chord coincides with a vertex of P.

An inscribed parallelogram is said to be *extensive* if the length of each of its sides is at least one-half the length of the diameter in the same direction as that side (see Figure 3(a)). Kuperberg and Kuperberg showed that there is a close relationship between dense double-lattice packings and extensive parallelograms [9]. They observed that if Q is an extensive parallelogram inscribed in P, then a double-lattice packing for P can be generated as follows. Translate Q and P simultaneously so that one of the vertices of Q coincides with the origin, and let u and v be the vectors that span Q. By

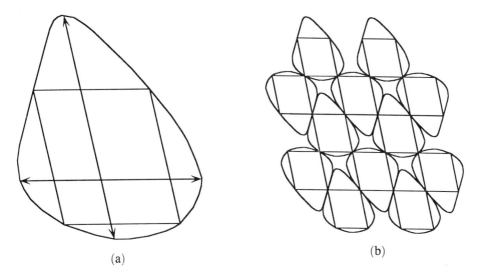

(a)   (b)

FIGURE 3. Extensive parallelogram and double-lattice packing.

translating $P \cup -P$ to each point of the lattice generated by $2u$ and $2v$, it follows from the convexity of P and the definition of extensive parallelogram that the resulting system is a double-lattice packing for P (see Figure 3(b)).

THEOREM (G. Kuperberg and W. Kuperberg). *If P is a convex body, there exists a densest double-lattice packing for P which is generated by a minimum area extensive parallelogram inscribed in P.*

Two distinct parallel chords of equal length in P define a parallelogram inscribed within P. If the angle of these chords is θ then we call this parallelogram a θ-parallelogram. The parallel chords are the *bases* of the θ-parallelogram. The *length* of a θ-parallelogram is the length of its bases, and the *width* of a θ-parallelogram is the perpendicular distance between the two lines containing its bases. The area of a θ-parallelogram is just the product of its length and width. A *half-length* θ-parallelogram for P is a θ-parallelogram whose bases are half the length of the θ-diameter of P (see Figure 4).

A convex planar body is said to be *strictly convex* if its boundary contains no line segments. Observe that for a strictly convex body (i.e., a convex body containing no line segment on its boundary) there are exactly two chords parallel to and of half the length of the θ-diameter, and thus there is a unique half-length θ-parallelogram. If a convex polygon has one or two edges which are parallel to and greater than half the length of the θ-diameter then there may be infinitely many half-length θ-parallelograms. These parallelograms arise by selecting the bases of the parallelogram to be any subsegment of the

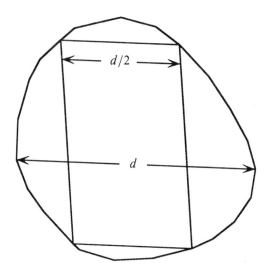

FIGURE 4. The half-length parallelogram.

appropriate length from one of these parallel edges (an example is shown in
Figure 8 on p. 256). Since all parallelograms generated in this manner have
equal area, for our purposes it suffices to select any one of them arbitrarily,
e.g. by sliding its bases as far counterclockwise along this edge as possible.
By assuming that all θ-parallelograms are slid into such a canonical configuration, we can talk about the *unique* half-length θ-parallelogram for a given
θ.

The Kuperberg's remark without proof [9] that in order to compute the
minimum area extensive parallelogram inscribed in P, it suffices to consider
the set of half-length θ-parallelograms for each θ between 0 and $180°$. (By
symmetry the θ-parallelograms repeat cyclically with period $180°$.) For the
sake of completeness we present a proof of this remark.

THEOREM 2.1. *A minimum area extensive parallelogram inscribed in a convex polygon P is achieved by a half-length θ-parallelogram for some angle θ
between 0 and $180°$.*

PROOF. It suffices to prove that for every θ, (1) every half-length θ-
parallelogram is extensive, and (2) an extensive θ-parallelogram is locally
minimal if and only if it is a half-length θ-parallelogram.

To show (1) let a and b be the bases of a half-length θ-parallelogram
Q, and let c be a θ-diameter. Consider a convex quadrilateral R which is
bounded by the two lines passing through the left endpoint of c and the left
endpoints of a and b, and the two lines passing through the right endpoint
of c and the right endpoints of a and b (see Figure 5(a)). Because a and
b are parallel to and of half the length of c, it follows that the endpoints of
a and b lie on the midpoints of the sides of R. Thus Q is extensive for
R since its width and length are exactly one-half the width and length of R,

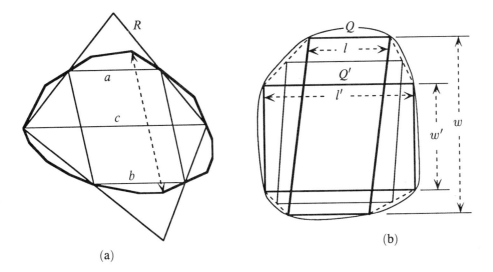

(a)

(b)

FIGURE 5. Minimality of the half-length parallelogram.

respectively. Consider the angle ϕ of the sides of Q other than a and b. It is a simple consequence of convexity that the longest chord for P at the angle ϕ lies entirely within R. It follows that the width of Q is at least half the width of P. Since P and R have the same θ-diameter, the length of Q is at least half the length of P. Therefore Q is extensive for P.

We will prove (2) for strictly convex bodies. (Although P is not strictly convex, it can be approximated by a convergent sequence of strictly convex bodies, and we can select a convergent subsequence of minimum extensive parallelograms.) If P is strictly convex, then there is a unique θ-parallelogram for each given length. As the length increases, the width of the θ-parallelograms decrease. We show that the area of the θ-parallelograms, as a function of length, is upward convex. Hence the minimum of this function over any interval is achieved at an endpoint of the interval. Thus the minimum extensive parallelogram must be a half-length θ-parallelogram.

Consider two lengths l and l', $0 < l < l' < |c|$. Let Q and Q' be the θ-parallelograms with these respective lengths. Let w and w' be the respective widths of Q and Q' (see Figure 5(b)). Consider the four lines segments which join each of the four vertices of Q to its corresponding vertex of Q'. For each p, $0 \le p \le 1$, let $q = 1 - p$. There is an interpolated parallelogram of length $pl + ql'$ and width $pw + qw'$ whose vertices lie on these four lines segments. (The endpoints of this parallelogram are weighted averages of corresponding pairs of endpoints of Q and Q'.)

By the convexity of P, each interpolated parallelogram is enclosed within P, and so the θ-parallelogram of the same length, $pl + ql'$, has area no smaller than this interpolated parallelogram. Hence it suffices to show that the area of the interpolated parallelogram is not less than the corresponding weighted average of the areas of Q and Q', that is

$$p(lw) + q(l'w') \leq (pl + ql')(pw + qw').$$

To prove this, first observe that because $l < l'$ and P is strictly convex we have $w > w'$. Clearly $0 \leq pq \leq 1$. Thus we have

$$0 \leq pq(l' - l)(w - w').$$

By simple manipulations and the facts that $1 - p = q$ and $1 - q = p$ we get

$$pq(l' - l)(w - w') = (pl + ql')(pw + qw') - (plw + ql'w')$$
$$0 \leq (pl + ql')(pw + qw') - (plw + ql'w')$$
$$p(lw) + q(l'w') \leq (pl + ql')(pw + qw'),$$

completing the proof. □

3. The algorithm

In the previous section we introduced the notion of a half-length parallelogram inscribed in the n-sided convex polygon P and showed that the problem of finding the densest double-packing can be reduced to computing the minimum area half-length parallelogram. In this section we show how to compute this minimum area half-length parallelogram. We employ the technique of *rotating calipers* [16]. For each angle θ, $0° \leq \theta < 180°$, we compute (explicitly or implicitly) a representative half-length θ-parallelogram (recalling our assumption that we can break ties among multiple half-length θ-parallelograms arbitrarily since they have equal area). It suffices to consider only 180 degrees of rotation because the half-length parallelogram in the directions θ and $180° + \theta$ are equal.

As in all rotating caliper algorithms, we define a finite set of critical angles θ at which we explicitly compute the half-length θ-parallelogram. An angle θ is said to be *critical* if either (1) both endpoints of the diameter chord in the direction θ coincide with vertices of the polygon (recall that we may always assume that at least one endpoint of the diameter chord coincides with a vertex of P), or (2) an endpoint of the θ-parallelogram coincides with a vertex of P. We will show that between any two consecutive critical angles, the diameter and the half-length parallelograms vary in a simple continuous way. This fact will allow us to compute the next critical angle in $O(1)$ time, and to determine the minimum area half-length parallelogram between a pair of consecutive critical angles in $O(1)$ time. In addition we will show that as θ increases, the points of contact between the half-length θ-parallelogram and the diameter in the direction θ will move monotonically counterclockwise around the boundary of P. From this property, which is called the *interspersing property*, it will follow that there are $O(n)$ critical angles.

The algorithm consists of three basic steps, where the second and third steps are repeated until $\theta \geq 180°$.

Initialization: Compute an initial half-length θ-parallelogram and a diameter for $\theta = 0°$.

Advancing to the next critical angle: Given an arbitrary θ, a half-length θ-parallelogram, and a θ-diameter, determine the next critical angle $\theta' > \theta$ and determine the half-length θ'-parallelogram and the θ'-diameter.

Minimizing between critical angles: Between a pair of consecutive critical angles θ and θ' determine the minimum area half-length parallelogram for all angles in the interval $[\theta, \theta']$.

Each of these steps will be treated separately in this section.

3.1. Initialization. Let us first consider the problem of finding the initial half-length θ-parallelogram and θ-diameter for $\theta = 0$. Consider a topmost vertex of P (with maximum y-coordinate) and a bottommost vertex of P (with minimum y-coordinate). These two vertices subdivide the boundary of P into a left side and a right side. By shooting a horizontal ray from each vertex on one side to the opposite side, we decompose the interior of P into a sequence of trapezoids with horizontal bases (where the topmost and bottommost trapezoids may degenerate to triangles) (see Figure 6). This trapezoidal decomposition can be computed in $O(n)$ time by merging the sorted lists of vertices on the left side of P with the vertices of the right side by y-coordinate.

By scanning through this list of trapezoids, it is an easy matter to find the maximum trapezoid base, which forms the horizontal diameter of P, and to find the two trapezoids, one lying above and one lying below the diameter, which contain the horizontal chords whose length is one half the length of the diameter. These chords are the bases for the initial half-length parallelogram.

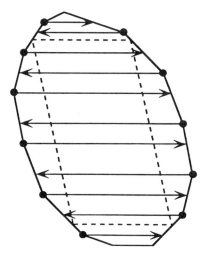

FIGURE 6. Computing the initial half-length parallelogram.

The total running time of this algorithm is clearly $O(n)$. (Observe that because the horizontal width function is convex, we could have computed the initial half-length parallelogram in $O(\log^2 n)$ time by performing a type of binary search on the y-coordinates of the polygon. However, this linear time algorithm is simpler and suffices for our purposes.)

3.2. Advancing to the next critical angle. Next we consider how vertices of the half-length parallelogram move incrementally as the directional angle θ rotates through a small angle starting with some initial value θ_0. We use the notation $\overrightarrow{ab}$ to denote the directed segment from point a to point b. By translating the tail of the segment to the origin, a directed segment $\overrightarrow{ab}$ can naturally be identified with the vector $b - a$. We will think of the edges of P as segments directed counterclockwise around the boundary of P, so that the *tail* and *head* of an edge are the clockwise and counterclockwise vertices of the edge, respectively. We consider each edge of P to be closed at its tail and open at its head, so that a vertex belongs to the edge following it in counterclockwise order about the boundary.

We begin by analyzing the movement of the diameter chord as θ increases. This analysis was given by Toussaint in describing his rotating calipers algorithm for finding the diameter of a convex polygon [16], but we repeat it here for completeness. Let $c_1(\theta)$ and $c_2(\theta)$ denote the endpoints of a θ-diameter (see Figure 7). Recall that there exist two parallel lines of support for P passing through the endpoints of this chord. If either endpoint lies in the interior of some edge, then these support lines are uniquely determined. If both of the endpoints lie on vertices of P, then there may be an infinite number of parallel support lines to choose from. We make the convention of selecting the extreme counterclockwise angle for these lines, and thus we are assured that at least one of the lines of support will be colinear with one of the edges following $c_1(\theta)$ or $c_2(\theta)$ in counterclockwise order. This selection can be made in constant time by analyzing the angles of the edges of P which lie clockwise of the current diameter chord's endpoints.

Suppose that the one of the support lines is colinear with the edge e of P which lies just counterclockwise of $c_1(\theta)$. As θ increases, the point $c_2(\theta)$ remains fixed and $c_1(\theta)$ travels counterclockwise along the edge e (see Figure 7(a)). The supporting lines for the diameter do not change until $c_1(\theta)$ reaches the vertex at the head of e. If this is not the case, then the other support line is colinear with the edge which lies just counterclockwise of c_2. In this case $c_1(\theta)$ remains fixed and $c_2(\theta)$ travels monotonically along this edge until reaching the next vertex (see Figure 7(b)). The angle at which this event occurs is denoted θ_c. Clearly $\theta_c > \theta_0$.

Next we consider how the endpoints of the bases of the θ-parallelogram vary with θ. (As we will see later, we do not deal with θ directly as an angle measured, say, in degrees. Rather the varying angle will be expressed parametrically and all quantities depending on angles will be simple linear

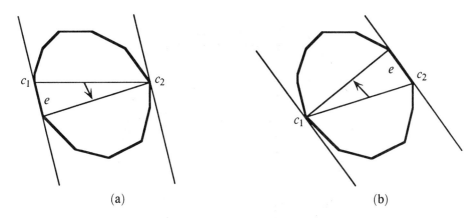

FIGURE 7. The movement of the diameter chord.

functions.) We first define two intermediate values that will be helpful. For $\theta_0 \leq \theta \leq \theta_c$, $C(\theta)$ denotes the vector which is one-half the length of the diameter chord at the angle θ, and let $D(\theta)$ be the net motion of head of this vector as θ increases from θ_0. That is,

$$C(\theta) = \frac{1}{2}(c_2(\theta) - c_1(\theta)), \qquad D(\theta) = C(\theta) - C(\theta_0).$$

For θ in the range $\theta_0 \leq \theta \leq \theta_c$, the quantities $c_1(\theta)$, $c_2(\theta)$, $C(\theta)$ and $D(\theta)$ can be computed in constant time. Clearly $D(\theta_0) = 0$. If c_1' and c_2' denote the heads of the edges on which c_1 and c_2 lie, then $D(\theta_c)$ is either equal to $(1/2)(c_1' - c_1)$ or $(1/2)(c_2 - c_2')$, depending on whether the c_1 ˙or c_2 is the moving endpoint of the diameter chord.

Let a_1 and a_2 denote the the endpoints of one of the base chords of the half-length θ_0-parallelogram. (An analogous construction can be applied to the other base chord from b_1 to b_2.) Let a_1' and a_2' denote the heads of the edges on which a_1 and a_2 lie. For the sake of illustration, assume that the chords $\overrightarrow{a_1 a_2}$ and $\overrightarrow{b_1 b_2}$ are horizontal and directed from left to right. Let a_1' and a_2' denote the heads of the edges on which a_1 and a_2 lie, respectively. For $i = 1, 2$, let h_i denote the directed segment $\overrightarrow{a_i a_i'}$. Notice that h_i is of nonzero length (by our convention that a vertex of P lies on the next edge in counterclockwise order) and h_i is directed counterclockwise about the boundary P.

For an angle θ, let $a_i(\theta)$, denote the position of the corresponding end-point for the half-length θ-parallelogram. We will show that as θ increases from θ_0, $a_i(\theta)$ travels monotonically along h_i, and furthermore the length of the motion vector $a_i(\theta) - a_i(\theta_0)$ is related to the length of $D(\theta)$ by a constant scale factor.

Suppose that the segments h_1 and h_2 are parallel to one another. Observe that because P is convex and $\overrightarrow{a_1 a_2}$ is parallel to and of strictly lesser length

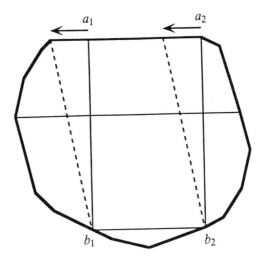

FIGURE 8. Sliding along an edge of P.

than the diameter, a_1 and a_2 must lie on the same edge of P (see Figure 8). This edge is parallel to the diameter chord. If this is the case, as mentioned earlier, we map this θ_0-parallelogram into canonical position by "sliding" the chord $\overrightarrow{a_1a_2}$ counterclockwise along this edge until the left endpoint a_1 is coincident with a vertex of P. (For the case of the chord $\overrightarrow{b_1b_2}$ it will be the right endpoint b_2 which becomes coincident with a vertex of P.) This transformation does not alter the area of the parallelogram. After sliding, we reevaluate h_1 and h_2. By our convention that a vertex belongs to the next counterclockwise edge, it follows that h_1 and h_2 are no longer parallel.

Let H_1 and H_2 denote the vectors corresponding to the directed segments h_1 and h_2. The parameters α_1 and α_2 which will define the incremental motion of $a_1(\theta)$ and $a_2(\theta)$ are introduced in the following lemma.

LEMMA 3.1. *There exist two unique nonnegative constants α_1 and α_2, at least one of which is nonzero, such that $D(\theta_c) = \alpha_2 H_2 - \alpha_1 H_1$.*

PROOF. Since h_1 and h_2 are not parallel to each other and are of nonzero length, H_1 and H_2 form a basis for R^2. Because $D(\theta_c)$ is nonzero there exist two unique α_1 and α_2 which cannot both be equal to zero such that $D(\theta_c) = \alpha_2 H_2 - \alpha_1 H_1$. To see that both α_1 and α_2 are nonnegative, imagine for the sake of concreteness that the diameter chord is horizontal and directed from left to right and that $\overrightarrow{a_1a_2}$ lies horizontally above this chord. Clearly a_1 lies on the left side of P's boundary and a_2 lies on the right side of P's boundary, and thus the vectors $-H_1$ and H_2 must each have nonnegative vertical components (see Figure 9). Because $D(\theta_c)$ is parallel to the supporting lines passing through the endpoints of the diameter chord, it follows from convexity that $D(\theta_c)$ lies within the minor angle subtended by $-H_1$ and H_2. Thus α_1 and α_2 are both nonnegative. $\square$

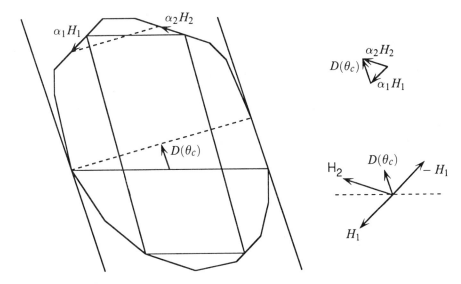

FIGURE 9. Rotating a base chord of the half-length parallelogram.

Rather than deal with angles and trigonometric functions directly, we can instead associate each angle θ in the range $\theta_0 \leq \theta \leq \theta_c$ uniquely with a parameter $0 \leq t \leq 1$. In particular, we can associate θ with the ratio $t(\theta)$ of the lengths of the (parallel) vectors $D(\theta_c)$ to $D(\theta)$. In other words, $t(\theta)$ is uniquely defined by the equation $t(\theta)D(\theta_c) = D(\theta)$. Observe that since $D(\theta_c)$ is nonzero, $t(\theta)$ is a monotonically strictly increasing function of θ.

Using this parametric representation of angles, for all sufficiently small angles (essentially up to the next critical angle), we can define the continuous motion of the endpoints of each base of the half-length θ-parallelogram as a simple linear function of this parameter.

LEMMA 3.2. *Let*

$$t_a = \frac{1}{\max(\alpha_1, \alpha_2)}.$$

For each angle θ for which $0 \leq t(\theta) \leq \min(t_a, 1)$, we have the following:

   (i) *The endpoints $a_1(\theta)$ and $a_2(\theta)$ lie on the respective segments h_1 and h_2.*

  (ii) *As θ increases these endpoints move monotonically counterclockwise along these segments, and in particular*

$$a_1(\theta) = a_1 + t(\theta)\alpha_1 H_1, \qquad a_2(\theta) = a_2 + t(\theta)\alpha_2 H_2.$$

 (iii) *For the value of θ for which $t(\theta) = t_a$, at least one of the endpoints of the base $\overrightarrow{a_1(\theta)a_2(\theta)}$ coincides with a vertex of P.*

PROOF. First observe that because α_1 and α_2 are both nonnegative and both cannot be zero, t_a is defined and positive. Define θ_a to be the angle for

which $t(\theta_a) = t_a$. Consider any θ, $\theta_0 \leq \theta \leq \min(\theta_a, \theta_c)$. For these values
of θ, we have $0 \leq t(\theta) \leq \min(1, t_a)$. For $i = 1, 2$, by simple substitution
we have $a_i(\theta) = a_i + t(\theta)\alpha_i H_i$. Clearly $t(\theta)\alpha_i \leq 1$ so $a_i(\theta)$ lies somewhere
along the segment h_i, establishing (i). Indeed, if $\theta = \theta_a$, then either $a_1(\theta)$
or $a_2(\theta)$ coincides with the head of its respective edge (depending on the
α-value which dominates in the definition of t_a), establishing (iii).

To show that $a_2(\theta)$ and $a_1(\theta)$ are the endpoints of the base chord of the
half-length θ-parallelogram, it suffices to show that $a_2(\theta) - a_1(\theta) = C(\theta)$.

$$
\begin{aligned}
a_2(\theta) - a_1(\theta) &= (a_2 + t(\theta)\alpha_2 H_2) - (a_1 + t(\theta)\alpha_1 H_1) \\
&= (a_2 - a_1) + t(\theta)(\alpha_2 H_2 - \alpha_1 H_1) \\
&= C(\theta_0) + t(\theta)D(\theta_c) \\
&= C(\theta_0) + D(\theta) \\
&= C(\theta).
\end{aligned}
$$

Together with the observation that this motion function is positive and linear
in t establishes (ii). $\square$

By applying a similar construction for the base chord $\overrightarrow{b_1 b_2}$ we derive an
angle $\theta_b > \theta_0$ at which the moving chord $\overrightarrow{b_1(\theta)b_2(\theta)}$ first encounters an
endpoint of P. Let

$$
\theta^* = \min(\theta_a, \theta_b, \theta_c).
$$

For all angles θ, $\theta_0 \leq \theta < \theta^*$, each endpoint of the θ-diameter and each
vertex of the half-length θ-parallelograms moves along a single edge of P.
Furthermore, the exact location of these endpoints and vertices can be determined by simple linear combinations as shown in Lemma 3.2. At the angle
θ^* at least one of these points coincides with a vertex of P, hence θ^* is
the next critical angle. The vertex which becomes critical depends on which
of θ_a, θ_b and θ_c is minimum. At this point the edges of contact with the
vertices must be reevaluated, and we are ready to repeat the process to find
the next critical angle.

We can now summarize the algorithm for advancing from the half-length
parallelogram at some angle θ_0 (not necessarily a critical angle) to the next
critical angle. Let $\overrightarrow{c_1 c_2}$ denote the current diameter chord and let $\overrightarrow{a_1 a_2}$ and
$\overrightarrow{b_1 b_2}$ denote the current base chords for the half-length parallelogram. We
observe that at no time is it necessary to compute the actual angles at which
the given events occur, but rather to compute the parameters t_a and t_b from
which these angles are derived. The parameter value $t = 1$ corresponds to
the motion which rotates the diameter chord to its next critical placement.

ALGORITHM. (Advancing to the next critical half-length parallelogram):

(1) Rotate the lines of support for P passing through c_1 and c_2 to their
most counterclockwise orientation. If one of the support lines is
colinear with the edge on which c_1 lies, then let $D := (1/2)(c_1 - c_1')$,

where c_1' is the head of the edge on which c_1 lies. Otherwise let $D := (1/2)(c_2' - c_2)$ where c_2' is the head of the edge on which c_2 lies. In the first case c_1 is the moving endpoint of the diameter, and in the other case c_2 is the moving endpoint. (D is $D(\theta_c)$ defined earlier.)

(2) If a_1 and a_2 both lie on the same edge of P, then slide these endpoints counterclockwise along this edge while maintaining a constant distance between the two of them until one reaches a vertex of P. Determine the new edge on which this vertex lies.

(3) Let a_1' and a_2' denote the heads of the edges on which a_1 and a_2 lie, respectively. Let $H_1 := a_1' - a_1$, and let $H_2 := a_2' - a_2$.

(4) Determine constants α_1 and α_2 such that $D = \alpha_2 H_2 - \alpha_1 H_1$. (These constants describe the relative movement of $a_1(\theta)$ and $a_2(\theta)$ along the segments h_1 and h_2 in terms of the movement of the diameter endpoint.)

(5) Let $t_a := 1/\max(\alpha_1, \alpha_2)$. (This is the maximum motion parameter before $a_1(\theta)$ or $a_2(\theta)$ reaches a vertex of P.)

(6) Repeat steps (2) through (5) replacing b_1 and b_2 for a_1 and a_2, respectively. Let β_1, β_2, and t_b denote the results.

(7) Let $t^* := \min(t_a, t_b, 1)$. This determines the motion to the next critical event.

(8) Let

$$a_1 := (1 - \alpha_1 t^*)a_1 + \alpha_1 t^* a_1' \qquad a_2 := (1 - \alpha_2 t^*)a_2 + \alpha_2 t^* a_2'$$
$$b_1 := (1 - \beta_1 t^*)b_1 + \beta_1 t^* b_1' \qquad b_2 := (1 - \beta_2 t^*)b_2 + \beta_2 t^* b_2'.$$

If c_1 is the moving endpoint of the diameter chord then let $c_1 := (1 - t^*)c_1 + t^* c_1'$ and otherwise let $c_2 := (1 - t^*)c_2 + t^* c_2'$.

A *convex combination* of two points (vectors) a and a' in the plane is a linear combination $(1 - t)a + ta'$, for some real t, $0 \le t \le 1$. Summarizing the above algorithm, we have

LEMMA 3.3. *Given a half-length θ-parallelogram Q and the corresponding θ-diameter, the next critical half-length parallelogram Q^* and diameter can be computed in constant time. Furthermore, the set of half-length parallelograms for all intermediate angles can be computed as convex combinations of the corresponding endpoints of Q and Q^*.*

Because each motion performed by the algorithm is locally counterclockwise we also have

(Interspersing Property). *As the angle θ rotates counterclockwise, the points of contact between the parallelogram move counterclockwise (or remain stationary) along the boundary of the P.*

As a consequence of the interspersing property we also have a bound on the number of critical events. Recall that by symmetry of half-length parallelograms, it suffices to consider rotating θ through $180°$.

LEMMA 3.4. *Given a convex polygon* P *with* n *vertices, the number of critical angles is* $3n$ *over a rotation of* $180°$.

PROOF. Let $\overrightarrow{a_1 a_2}$ and $\overrightarrow{b_1 b_2}$ denote the initial base chords. By Theorem 3.4, as θ rotates through $180°$, the vertex a_1 will rotate monotonically counterclockwise to b_2, and b_2 will rotate monotonically counterclockwise to a_1. Thus, between these two vertices, exactly n critical angles will be generated, corresponding to the moments at which these vectors become incident with vertices of P. Likewise, the two vertices a_2 and b_1 will generate a total of n critical events. Finally each of the endpoints of the diameter chord in the direction θ, will rotate counterclockwise about the boundary of P until they exchange places. Thus, the number of event points generated by the diameter chord will also be n. This yields $3n$ total event points. $\square$

From this last result and Lemma 3.3 we have the following corollary.

COROLLARY. *Given an* n-sided convex polygon P, in $O(n)$ time we can compute the ordered sequence of half-length parallelograms at each of the critical angles from $0°$ to $180°$.

3.3. Minimizing between two critical angles. To complete the algorithm it suffices to show how to compute the minimum area half-length parallelogram between two consecutive critical angles in constant time. Let θ and θ' be two angles such that there is no critical angle between them, and let $Q = \langle a_1, a_2, b_1, b_2 \rangle$ and $Q' = \langle a_1', a_2', b_1', b_2' \rangle$ be the vertices of the corresponding half-length parallelograms, respectively. Because there are no critical angles between θ and θ', corresponding pairs vertices (a_1 and a_1', for example) lie on the same edges of P. By Lemma 3.3 it follows that all half-length parallelograms between these two can be interpolated by considering convex combinations of corresponding vertices of Q and Q'. For $i = 1, 2$ and $0 \le t \le 1$, define $a_i(t) = (1 - t)a_i + ta_i'$ and $b_i(t) = (1 - t)b_i + tb_i'$. Let us assume that the vertices of Q are enumerated so that the chords $\overrightarrow{a_1 a_2}$ and $\overrightarrow{b_1 b_2}$ are parallel and similarly directed. Taking the point b_1 as the origin of the parallelogram, the vectors defining this parallelogram are $b_2(t) - b_1(t)$ and $a_1(t) - b_1(t)$. Thus the area of the interpolated parallelogram is given by the absolute value of the determinant

$$\det(b_2(t) - b_1(t), \; a_1(t) - b_1(t)).$$

This determinant is a polynomial of degree two in the invariant t, and this polynomial can be computed in $O(1)$ time given the endpoints of the two parallelograms Q and Q'. Therefore, by differentiating the polynomial symbolically, we can determine the minima in the interval $[0, 1]$ in constant time.

The entire algorithm for determining the densest double-lattice packing of a convex polygon is given below. The correctness and $O(n)$ running time of this algorithm follow from the previous discussion.

ALGORITHM. (Finding the maximum density double-lattice packing):

(1) Compute an initial half-length parallelogram Q for $\theta = 0$.
(2) While $\theta \le 180°$ do:
    (a) Advance to the next critical half-length parallelogram Q' using Algorithm 3.2.
    (b) Determine the minimum area half-length parallelogram between Q and Q' by the method described above.
    (c) Let $Q := Q'$.
(3) Let $Q = \langle a_1, a_2, b_1, b_2 \rangle$ be the minimum area half-length parallelogram found in step (2b). Translate Q so that b_1 coincides with the origin. Let $u = b_2 - b_1$ and let $v = a_1 - b_1$. Pack P in the plane using the lattice generated by $2u$ and $2v$ and pack $-P$ using this same lattice.

4. Concluding remarks

We have shown that the problem of computing the densest double-lattice packing of a convex polygon is solvable in linear time. There are a number of interesting open problems suggested by this work. The first is to generalize the problem to the densest double-lattice packings of other types of shapes in the plane, for example, the connected union of two convex polygons, or star-shaped polygons. Unfortunately, the characterization of the densest packing in terms of extensive parallelograms relies heavily on convexity. A second problem is to consider other periodic structures analogous to the double-lattice for packings which allow rotation and which may be more economical for other types of convex polygons.

REFERENCES

1. N. Adlai and A. DePano, *Polygon approximation with optimized polygonal enclosures: Applications and algorithms*, Ph.D. Thesis, Dept. of Computer Science, University of New Orleans, 1987.

2. I. Fáry, *Sur la densité des réseaux de domaines convexes*, Bull. Soc. Math. France **78** (1950), 152–161.

3. G. Fejes Tóth, *New results in the theory of packing and covering*, Convexity and Its Applications (P. M. Gruber and J. M. Wills, eds.), Birkhäuser Verlag, Basel, 1983, pp. 318–359.

4. L. Fejes Tóth, *One the densest packing of domain*, Proc. Kon. Ned. Akad. Wet. **51** (1948), 189–192.

5. M. R. Garey and D. S. Johnson, *Computers and intractibility: A guide to the theory of NP-completeness*, W. H. Freeman and Company, New York, 1979.

6. B. Grünbaum and G. C. Shephard, *Tilings and patterns*, W. H. Freeman and Company, New York, 1987.

7. H. Heesch, *Reguläres Parkettierungsproblem*, Westdeutscher Verlag, Cologne and Opladen, 1968.

8. V. Klee and M. L. Laskowski, *Finding the smallest triangles containing a given convex polygon*, J. Algorithms **6** (1985), 359–375.

9. G. Kuperberg and W. Kuperberg, *Double-lattice packings of convex bodies in the plane*, Discrete and Comput. Geom. **5** (1990), 389–397.

10. W. Kuperberg, *Covering the plane with congruent copies of a convex body*, Bull. London Math. Soc. **21** (1989), 82–86.

11. K. Mahler, *The theorem of Minkowski-Hlawka*, Duke Math. J. **13** (1946), 611–621.

12. D. M. Mount and R. Silverman, *Packing and covering the plane with translates of a convex polygon*, J. Algorithms **11** (1990), 564–580.

13. J. O'Rourke, A. Aggarwal, S. Maddila, and M. Baldwin, *An optimal algorithm for finding minimal enclosing triangles*, J. Algorithms **7** (1986), 258–269.

14. C. A. Rogers, *The closest packing of convex two-dimensional convex sets*, Acta Math. **86** (1951), 309–321.

15. C. A. Rogers, *Packing and covering*, Cambridge Univ. Press, London, 1964.

16. G. T. Toussaint, *Solving geometric problems with the 'rotating calipers,'* Proc. IEEE Melecon 83, 1983.

DEPARTMENT OF COMPUTER SCIENCE AND INSTITUTE FOR ADVANCED COMPUTER STUDIES, UNIVERSITY OF MARYLAND, COLLEGE PARK, MARYLAND 20742

DIMACS Series in Discrete Mathematics
and Theoretical Computer Science
Volume 6, 1991

Arrangements in Topology

PETER ORLIK

1. Introduction

The aim of this paper is to give a brief survey of how and why topologists got interested in arrangements, and to describe a problem which motivates their current research.

We begin with a familiar story. Recall that braids and the braid group were defined by Artin [4]. Braids with ℓ strands may be composed by juxtaposition. There is a suitable notion of isotopy of braids which makes this an associative multiplication. The inverse of a braid is the braid which untangles it. Isotopy classes of braids on ℓ strands form a group called the *braid group*, $B(\ell)$. For details, see Birman's book [6]. Let $\mathrm{Sym}(\ell)$ denote the symmetric group on ℓ letters. There is a natural surjection $B(\ell) \to \mathrm{Sym}(\ell)$ which sends each braid to the permutation of its ends. The kernel of this map is called the pure braid group $PB(\ell)$. The corresponding pure braids have the property that each strand returns to its point of origin. The fact that the pure braid group is the fundamental group of the *complement of an arrangement* first appeared in a paper by Fox and Neuwirth [16] in 1962. To make this statement precise we need a few definitions. For undefined terms see [1] or [22].

DEFINITION 1.1. Let $\mathbf{K}$ be a field and let V be a vector space of dimension ℓ. A *hyperplane* H in V is an affine subspace of dimension $(\ell - 1)$. An *arrangement* $\mathscr{A} = (\mathscr{A}, V)$ is a finite set of hyperplanes in V.

In this paper $\mathbf{K} = \mathbf{R}$ or $\mathbf{K} = \mathbf{C}$. We call $\mathscr{A}$ an ℓ-*arrangement* when we want to emphasize the dimension of V. Let V^* be the space of linear forms on V. Let $S = S(V^*)$ be the symmetric algebra of V^*. Choose a basis $\{x_1, \dots, x_\ell\}$ in V^*. We may identify $S(V^*)$ with the polynomial algebra $S = \mathbf{K}[x_1, \dots, x_\ell]$. Each hyperplane $H \in \mathscr{A}$ is the kernel of a polynomial α_H of degree 1, defined up to a constant.

1991 *Mathematics Subject Classification.* Primary 05B35, 14B05, 32C40, 57N65; Secondary 14F40, 14J99, 51A05.

DEFINITION 1.2. The product

$$Q = Q(\mathscr{A}) = \prod_{H \in \mathscr{A}} \alpha_H$$

is called a *defining polynomial* for $\mathscr{A}$.

We call $\mathscr{A}$ *central* if $\bigcap_{H \in \mathscr{A}} H \neq \emptyset$, and noncentral otherwise. If $\mathscr{A}$ is central, then we choose coordinates so that each hyperplane is a vector subspace. Then α_H is a linear form and Q is a homogeneous polynomial.

DEFINITION 1.3. Let $\mathscr{A}$ be an arrangement. Define the *variety* of $\mathscr{A}$ by

$$N(\mathscr{A}) = \bigcup_{H \in \mathscr{A}} H = \{ v \in V \mid Q(v) = 0 \}.$$

In general this variety has complicated singularities. There is an active area of research within arrangements devoted to its study, using tools of algebraic geometry.

DEFINITION 1.4. Let $\mathscr{A}$ be an arrangement. Define its *complement* by

$$M(\mathscr{A}) = V \setminus N(\mathscr{A}).$$

If $\mathbf{K} = \mathbf{R}$ then M is a disjoint union of open convex submanifolds of $\mathbf{R}^\ell$. If the arrangement is central, then they are cones, otherwise some may be bounded. They present many interesting counting problems. See Zaslavsky's work [30] for solutions. If $\mathbf{K} = \mathbf{C}$ then M is a connected, open, smooth manifold. This brings us back to the braid groups.

DEFINITION 1.5. For $1 \leq i < j \leq \ell$ let $H_{i,j} = \ker(x_i - x_j)$. The *braid arrangement*, $\mathscr{A}_\ell$, is defined by

$$Q_\ell = \prod_{1 \leq i < j \leq \ell} (x_i - x_j).$$

Let $\mathbf{K} = \mathbf{C}$. The variety $N_\ell = N(\mathscr{A}_\ell)$ is called the *superdiagonal* and its complement $M_\ell = M(\mathscr{A}_\ell)$ is the *pure braid space*. A braid on ℓ strands may be viewed as the graph of the motion, between times $t = 0$ and $t = 1$, of ℓ distinct points in the complex line, subject to the condition that the points remain distinct throughout the motion. Thus we have a map $f : [0, 1] \to \mathbf{C}^\ell$ such that for each t the image point $(f_1(t), \ldots, f_\ell(t))$ satisfies the condition $f_i(t) \neq f_j(t)$. The braid is pure if $f(0) = f(1)$. Thus a pure braid is the image of a circle in M_ℓ. It represents an element of the fundamental group $\pi_1(M_\ell)$. For the converse choose a base point $x \in M_\ell$. An element of $\pi_1(M_\ell, x)$ is represented by a map $f : (I, \{0, 1\}) \to (M_\ell, x)$ which we may assume to be a smooth embedding. The coordinate functions of f are the strands of the pure braid. Thus $\pi_1(M_\ell) = PB(\ell)$. Note that M_ℓ admits a free action of $\mathrm{Sym}(\ell)$ by permuting the coordinates. Let $B_\ell = M_\ell / \mathrm{Sym}(\ell)$ be the orbit space and let $p : M_\ell \to B_\ell$ be the projection of this covering. A similar argument shows that $\pi_1(B_\ell) = B(\ell)$.

Fox and Neuwirth were trying to compute the cohomology of the braid group. Recall that one way to compute the cohomology groups $H^q(\pi; G)$ of π with coefficients in G is to construct a CW-complex X with fundamental group π and no higher homotopy groups. Then X is called a $K(\pi, 1)$ space, and $H^q(\pi; G) = H^q(X; G)$. In general X is an infinite-dimensional complex. It turns out that the braid space is a $K(B(\ell), 1)$ space. Since $p : M_\ell \to B_\ell$ is a covering, it suffices to show that M_ℓ is a $K(PB(\ell), 1)$ space; see [12].

THEOREM 1.6. *The complement of the braid arrangement* $\mathscr{A}_\ell$ *is a* $K(\pi, 1)$ *space.*

PROOF. The projection map $\mathbf{C}^\ell \to \mathbf{C}^{\ell-1}$ defined by $(x_1, \dots, x_\ell) \to (x_1, \dots, x_{\ell-1})$ induces a locally trivial fibration $M_\ell \to M_{\ell-1}$. The fiber over $(\xi_1, \dots, \xi_{\ell-1})$ is $\mathbf{C} \setminus \{\xi_1, \dots, \xi_{\ell-1}\}$. Thus the fiber retracts onto a wedge of $(\ell - 1)$ circles, so it is a $K(\pi, 1)$-space. Since $M_2 = \{(x_1, x_2) \mid x_1 \neq x_2\} = \mathbf{C} \times \mathbf{C}^*$ we are done by induction. □

Given a topological space X, let $b_p(X) = \operatorname{rank} H^p(X)$ be the Betti numbers of X. The Poincaré polynomial of X is $P(X, t) = \sum_{p \geq 0} b_p(X) t^p$. It follows from (1.6) that $H^q(B(\ell); \mathbf{Z}) = H^q(B_\ell; \mathbf{Z})$. The torsion in these groups turned out to be very hard to compute. It also follows that $H^q(M(\ell); \mathbf{Z}) = H^q(M_\ell; \mathbf{Z})$. In 1969 Arnold [2] calculated the Poincaré polynomial of the pure braid space M_ℓ and the cohomology ring structure of $H^*(M_\ell)$. Arnold showed that

$$P(M_\ell, t) = (1 + t)(1 + 2t) \cdots (1 + (\ell - 1)t).$$

Arnold also showed that $H^*(M_\ell)$ is generated by the 1-dimensional elements

$$\omega_{p,q} = \frac{1}{2\pi i} \frac{dz_p - dz_q}{z_p - z_q}$$

and that all relations among these generators are consequences of the relations

$$\omega_{p,q}\omega_{q,r} + \omega_{q,r}\omega_{r,p} + \omega_{r,p}\omega_{p,q} = 0.$$

Arnold stated two conjectures for an arbitrary arrangement $\mathscr{A}$. The first said that $H^*(M, \mathbf{Z})$ is torsion free. The second may be stated as follows. Define holomorphic differential forms $\omega_H = (1/2\pi i)(d\alpha_H/\alpha_H)$ for $H \in \mathscr{A}$ and let $[\omega_H]$ denote the corresponding cohomology class. Let $R = \bigoplus_{p=0}^{\ell} R_p$ be the graded $\mathbf{C}$-algebra of holomorphic differential forms on M generated by the ω_H and 1. Arnold conjectured that the natural map $\eta \to [\eta]$ of $R \to H^*(M, \mathbf{C})$ is an isomorphism of graded algebras. These results and conjectures are considered the genesis of the topological study of arrangements.

2. The $K(\pi, 1)$ problem

We noted above that the complement of the braid arrangement is a $K(\pi, 1)$ space. It is not known in general which arrangements have $K(\pi, 1)$ complements. An arrangement $\mathscr{A}$ is called a *general position* arrangement if for every subset $\{H_1, \dots, H_p\} \subseteq \mathscr{A}$ with $p \le \ell$, $\dim H_1 \cap \dots \cap H_p = \ell - p$, and when $p > \ell$, $H_1 \cap \dots \cap H_p = \emptyset$. Note that if $\mathscr{A}$ is a central general position ℓ-arrangement then $|\mathscr{A}| \le \ell$. Thus the only interesting general position arrangements are noncentral. Hattori [18] proved that the complement of a general position arrangement is not $K(\pi, 1)$.

DEFINITION 2.1. Let $\mathbf{n} = \{1, \dots, n\}$. If $I \subseteq \mathbf{n}$ let $|I|$ be its cardinality. Define the subtorus T_I of T^n by

$$T_I = \{(z_1, \dots, z_n) \in T^n \mid z_j = 1 \text{ for } j \notin I\}.$$

THEOREM 2.2. *Let $\mathscr{A}$ be a noncentral ℓ-arrangement in general position and assume that $n = |\mathscr{A}| \ge \ell + 1$. Then $M = M(\mathscr{A})$ has the homotopy type of $M_0 = \bigcup_{|I|=\ell} T_I$.*

Hattori [18] also proved that $\pi_1(M)$ is free abelian of rank n, and that the universal covering space $\tilde{M}$ of M has trivial homology in dimensions $\neq 0, \ell$. He also gave a free $\mathbf{Z}(\pi_1 M)$ resolution of $H_\ell(\tilde{M}, \mathbf{Z})$. In particular, if $n = \ell + 1$ then $H_\ell(\tilde{M}, \mathbf{Z})$ is a free $\mathbf{Z}(\pi_1 M)$-module of rank 1.

DEFINITION 2.3. Let $\mathscr{A}$ be a central ℓ-arrangement with $\ell \ge 2$. Call $\mathscr{A}$ a *generic* arrangement if the hyperplanes of every subarrangement $\mathscr{B} \subseteq \mathscr{A}$ with $|\mathscr{B}| = \ell$ are linearly independent.

COROLLARY 2.4. *The complement of a generic arrangement is not a $K(\pi, 1)$ space.*

Since general position arrangements form a dense open subset of all arrangements, and generic arrangements form a dense open subset of all central arrangements, it follows that the set of $K(\pi, 1)$ arrangements is "thin." It is remarkable that in spite of this, so many "interesting" arrangements happen to be $K(\pi, 1)$. The representation of the pure braid space as the total space of a sequence of fibrations has been generalized by Falk and Randell [14].

DEFINITION 2.5. Let $\mathscr{A}$ be a central ℓ-arrangement. Call $\mathscr{A}$ *strictly linearly fibered* if after a suitable linear change of coordinates the restriction of the projection of $M(\mathscr{A})$ to the first $(\ell - 1)$ coordinates is a fiber bundle projection whose base space B is the complement of an arrangement in $\mathbf{C}^{\ell-1}$, and whose fiber is the complex line $\mathbf{C}$ with finitely many points removed.

DEFINITION 2.6. (i) The 1–arrangement $\{0\}$ is *fiber type*.

(ii) For $\ell \ge 2$ the ℓ-arrangement $\mathscr{A}$ is *fiber type* if $\mathscr{A}$ is strictly linearly fibered with base $B = M(\mathscr{B})$ and $\mathscr{B}$ is an $(\ell - 1)$-arrangement of fiber type.

PROPOSITION 2.7. *If $\mathscr{A}$ is fiber type then $M(\mathscr{A})$ is a $K(\pi, 1)$ space.*

In [29] Terao gave a combinatorial characterization of strictly linearly fibered arrangements. In order to state his result we need a definition.

DEFINITION 2.8. Let $\mathscr{A}$ be an arrangement and let $L = L(\mathscr{A})$ be the set of all intersections of elements of $\mathscr{A}$. Define a *partial order* on L by:

$$X \leq Y \Longleftrightarrow Y \subseteq X.$$

Define a rank function on L by $r(X) = \operatorname{codim} X$. Thus $r(V) = 0$ and $r(H) = 1$ for $H \in \mathscr{A}$. Call H the *atoms* of L. Define the *join* by $X \vee Y = X \cap Y$ and the *meet* by $X \wedge Y = \cap \{Z \in L \mid X \cup Y \subseteq Z\}$.

Note that this is *reverse* inclusion. Thus V is the unique minimal element. The advantage of this order is that for central arrangements L is a geometric lattice.

THEOREM 2.9. *Let $\mathscr{A}$ be a complex central arrangement. Then $\mathscr{A}$ is strictly linearly fibered if and only if $L(\mathscr{A})$ is supersolvable.*

In a 1971 Bourbaki Seminar talk, Brieskorn [8] generalized Arnold's results. He replaced the symmetric group and the braid arrangement by a Coxeter group W acting in an ℓ-dimensional real vector space $V_{\mathbf{R}}$. Let V be the complexification of $V_{\mathbf{R}}$. Then W acts as a reflection group in V. Let $\mathscr{A} = \mathscr{A}(W)$ be its reflection arrangement. Brieskorn conjectured that $\mathscr{A}(W)$ is a $K(\pi, 1)$-arrangement for all Coxeter groups W. He proved this for some of the groups by representing M as the total space of a sequence of fibrations. Some of these fibrations are not strictly linear. Deligne [9] settled the question by proving the much stronger result stated below.

DEFINITION 2.10. Let $(\mathscr{A}_{\mathbf{R}}, V_{\mathbf{R}})$ be a real arrangement. Call $\mathscr{A}_{\mathbf{R}}$ a *simplicial* arrangement if every component of $M(\mathscr{A}_{\mathbf{R}})$ is an open simplicial cone.

THEOREM 2.11. *Let $(\mathscr{A}_{\mathbf{R}}, V_{\mathbf{R}})$ be a simplicial arrangement. Then the complement of its complexification $(\mathscr{A}, V)$ is a $K(\pi, 1)$ space.*

This result proves Brieskorn's conjecture because the arrangement of a Coxeter group is simplicial [7]. There are arrangements with $K(\pi, 1)$ complements, which are neither strictly linearly fibered, nor the complexifications of simplicial real arrangements. The definitive result has yet to be found.

3. The cohomology algebra

In the same 1971 paper Brieskorn [8] showed that $H^*(M; \mathbf{Z})$ is torsion free, and that the $\mathbf{Z}$–subalgebra of R generated by the forms ω_H and 1 is isomorphic to the singular cohomology ring $H^*(M; \mathbf{Z})$. Brieskorn's description of this ring is combinatorial. Systematic use of $L(\mathscr{A})$ originated with [24], where we obtained a formula for the Poincaré polynomial of M, and a presentation of $H^*(M; \mathbf{C})$. Recall the *Möbius function* of L, $\mu_{\mathscr{A}} = \mu : L \to \mathbf{Z}$.

It is defined recursively: $\mu(V) = 1$, and $\sum_{Z \leq X} \mu(Z) = 0$ for $X > V$. Let t be an indeterminate. Define the *Poincaré polynomial* of $\mathscr{A}$ by

$$\pi(\mathscr{A}, t) = \sum_{X \in L} \mu(X)(-t)^{r(X)}.$$

This polynomial is closely related to the characteristic polynomial of a lattice. The virtue of this definition is that $\pi(\mathscr{A}, t)$ has nonnegative coefficients.

THEOREM 3.1. *Let $\mathscr{A}$ be a complex arrangement with complement $M(\mathscr{A})$. Then $P(M(\mathscr{A}), t) = \pi(\mathscr{A}, t)$.*

The proof of (3.1) uses a technique called *deletion and restriction*. Let $\mathscr{A}$ be a nonempty arrangement and let $H_0 \in \mathscr{A}$. Call $\mathscr{A}' = \mathscr{A} \setminus \{H_0\}$ the *deleted* arrangement and call $\mathscr{A}'' = \{H \cap H_0 \mid H \in \mathscr{A}'\}$ the *restricted* arrangement with respect to H_0. Roughly speaking, properties of arrangements which are inherited by deletion and restriction may be proved by induction on the number of hyperplanes. Here are two equivalent formulations of the fact that the cohomology of the complement may be computed this way.

THEOREM 3.2. *Let $i : R(\mathscr{A}') \to R(\mathscr{A})$ be the inclusion map and define $j : R(\mathscr{A}) \to R(\mathscr{A}'')$ by $j(\phi) = \text{res}(\phi)$ for $\phi \in R(\mathscr{A})$, where $\text{res}(\phi)$ is the residue of ϕ along H_0. Then there is an exact sequence:*

$$0 \to R(\mathscr{A}') \overset{i}{\to} R(\mathscr{A}) \overset{j}{\to} R(\mathscr{A}'') \to 0.$$

Let $M = M(\mathscr{A})$, $M' = M(\mathscr{A}')$, $M'' = M(\mathscr{A}'')$. The spaces M, M' are open complex manifolds of complex dimension ℓ. The map $i : M \to M'$ is the inclusion of a submanifold, and M'' is a submanifold of M' of complex codimension one. It has a tubular neighborhood, so by the Thom isomorphism theorem $H^k(M', M) \cong H^{k-2}(M'')$. This may be inserted in the cohomology long exact sequence of the pair (M', M). The special nature of the topology of arrangements gives:

THEOREM 3.3. *For $k \geq 0$ there exist split short exact sequences:*

$$0 \to H^k(M') \overset{i^*}{\to} H^k(M) \overset{\phi}{\to} H^{k-1}(M'') \to 0.$$

Next we describe a presentation of the cohomology algebra following [24]. Here we require that $\mathscr{A}$ be central. With some additional effort the noncentral case may be treated along these lines.

DEFINITION 3.4. Let $\mathscr{A}$ be a central arrangement. Let $E_1 = \bigoplus_{H \in \mathscr{A}} \mathbf{K} e_H$ and let $E = E(\mathscr{A}) = \Lambda(E_1)$ be the exterior algebra of E_1.

Note that E_1 has a basis consisting of elements e_H in one-to-one correspondence with the hyperplanes of $\mathscr{A}$. Write $uv = u \wedge v$ and note that

$e_H^2 = 0$, $e_H e_K = -e_K e_H$ for $H, K \in \mathcal{A}$. The algebra E is graded. If $|\mathcal{A}| = n$ then

$$E = \bigoplus_{p=0}^{n} E_p$$

where $E_0 = \mathbf{K}$, E_1 agrees with its earlier definition, and E_p is spanned over $\mathbf{K}$ by all $e_{H_1} \cdots e_{H_p}$ with $H_k \in \mathcal{A}$. Define a linear map $\partial_E = \partial : E \to E$ by $\partial 1 = 0$, $\partial e_H = 1$, and for $p \geq 2$

$$\partial(e_{H_1} \cdots e_{H_p}) = \sum_{k=1}^{p} (-1)^{k-1} e_{H_1} \cdots \widehat{e_{H_k}} \cdots e_{H_p}$$

for all $H_1, \ldots, H_p \in \mathcal{A}$. Given a p-tuple of hyperplanes $S = (H_1, \ldots, H_p)$ write $|S| = p$, $e_S = e_{H_1} \cdots e_{H_p} \in E$, and $\cap S = H_1 \cap \cdots \cap H_p \in L$. If $p = 0$ we agree that $S = (\ )$ is the empty tuple, $e_S = 1$ and $\cap S = V$. Since the rank function on L is codimension, it is clear that if $|S| = p$ then $r(\cap S) \leq p$. Let $S = (H_1, \ldots, H_p)$. Call S *independent* if $r(S) = p$ and *dependent* if $r(S) < p$. Thus S is dependent if and only if it is a *circuit*. The terminology has geometric significance. The tuple S is independent if the corresponding linear forms $\alpha_1, \ldots, \alpha_p$ are linearly independent. Equivalently, the hyperplanes of S are in general position. Let $\mathbf{S}_p$ denote the set of all p-tuples $(H_1, \ldots, H_p)$ and let $\mathbf{S} = \bigcup_{p \geq 0} \mathbf{S}_p$. Let $I = I(\mathcal{A})$ be the ideal of E generated by ∂e_S for all dependent $S \in \mathbf{S}$. Since I is generated by homogeneous elements, it is a graded ideal. Let $I_p = I \cap E_p$. Then $I = \bigoplus_{p=0}^{n} I_p$.

DEFINITION 3.5. Let $\mathcal{A}$ be an arrangement. Let $A = A(\mathcal{A}) = E/I$. Let $\varphi : E \to A$ be the natural homomorphism and let $A_p = \varphi(E_p)$. If $H \in \mathcal{A}$ let $a_H = \varphi(e_H)$ and if $S \in \mathbf{S}$ let $a_S = \varphi(e_S)$.

THEOREM 3.6. *Let $\mathcal{A}$ be an arrangement over $\mathbf{C}$, let $M = M(\mathcal{A})$, and let $A = A(\mathcal{A})$. The map $a_H \to [\omega_H]$ induces an isomorphism $A \to H^*(M, \mathbf{C})$ of graded $\mathbf{C}$-algebras.*

Call an invariant of $\mathcal{A}$ *combinatorial* if it is determined by $L(\mathcal{A})$. Since $A(\mathcal{A})$ is defined in terms of $L(\mathcal{A})$ it follows that the cohomology ring is a combinatorial invariant of $\mathcal{A}$. Using broken circuits, Jambu and Terao [19] constructed a subspace $C \subset E$ isomorphic to A, and hence a section for the map $\varphi : E \to A$.

4. Beyond combinatorial invariants

It is widely believed that not all topological invariants of $\mathcal{A}$ are combinatorial, but we have yet to find one that is not. The closest result along these lines is about a space related to $\mathcal{A}$, called the associated Milnor fiber.

Let $f = f(x_1, \ldots, x_\ell)$ be a homogeneous polynomial of degree $n \geq 2$ in ℓ complex variables. The Milnor fibration [21] of f is usually defined in a neighborhood of the origin. Since f is homogeneous there is a global fibration

$$f : \mathbf{C}^\ell \setminus f^{-1}(0) \to \mathbf{C} \setminus \{0\}$$

and $F = f^{-1}(1)$ is the *Milnor fiber* of the map f. Consider all homology and cohomology with complex coefficients. Since F is a Stein manifold of dimension $(\ell - 1)$, it has the homotopy type of a finite CW-complex of dimension $(\ell - 1)$. If f has an isolated singularity, it is known from Milnor's work that F has the homotopy type of a wedge of $(n-1)^\ell$ spheres of dimension $(\ell - 1)$. If the singularity of f is not isolated, very little is known about the cohomology of F. Special cases have been studied by Dimca [10], Esnault [11], Randell [25], Siersma [28], and others.

If $\mathscr{A}$ is a central arrangement then $Q(\mathscr{A})$ is homogeneous of degree $n = |\mathscr{A}|$ and for $\ell \geq 3$ the singularity is not isolated. The hyperplane complement $M = V \setminus Q^{-1}(0)$ is the total space of the Milnor fibration with fiber $F = Q^{-1}(1)$. Randell and I showed [23] that for generic arrangements the cohomology of F depends only on the number of hyperplanes. In a recent paper, Artal–Bartolo [3] constructed examples of arrangements whose lattices are almost isomorphic, but the associated Milnor fibers have different Betti numbers. Since a space may be fibered over the circle with different fibers, it is still possible that their complements are homeomorphic.

The fact that M is a Stein manifold implies that it has the homotopy type of a finite CW-complex. If we want to find an invariant which is not combinatorial, it must be at least on the level of homotopy groups. In addition, it must be computable by some algorithm which uses the special nature of arrangements. The stratified Morse theory approach of Goresky and MacPherson [17] is promising to yield homotopy information. So is the minimal model work of Kohno [20] and Falk [13]. We refer to [15] for a survey. The most obvious candidate is the fundamental group of the complement, $\pi_1(M)$. It is surely complicated enough, just think of the example of the braid arrangement. But is it computable?

Due to a general position argument of Zariski it suffices to consider 2-arrangements. Randell [26] and Salvetti [27] solved the problem independently in case $\mathscr{A}$ is the complexification of a real arrangement $\mathscr{A}_\mathbf{R}$. A presentation of $\pi_1(M)$ is obtained from the graph $\Gamma_\mathbf{R}$ of the real 2-arrangement. Thus the edges of $\Gamma_\mathbf{R}$ represent the complex lines of $\mathscr{A}$, and the vertices of $\Gamma_\mathbf{R}$ are the intersections of the lines of $\mathscr{A}$. I know of several serious attempts to construct real arrangements $\mathscr{A}$, $\mathscr{B}$ with $L(\mathscr{A}) \approx L(\mathscr{B})$ such that for the complexified arrangements, $\pi_1 M(\mathscr{A}) \not\approx \pi_1 M(\mathscr{B})$. The failure of these attempts was attributed to "lack of space" in $\mathbf{R}^2$. Further progress was hampered by the fact that no algorithm existed for the presentation of $\pi_1(M)$ for arbitrary 2-arrangements.

In his forthcoming thesis Avola [5] has solved this problem. He uses the defining polynomial of $\mathscr{A}$ to construct a *marked* graph $\Gamma \subset \mathbf{R}^2$. The edges of

Γ represent the lines of $\mathscr{A}$. There are two kinds of vertices in Γ. The first are intersection points of the lines of $\mathscr{A}$. These are unmarked. The second are "virtual" intersection points. These are marked with $+$ or $-$ depending on an orientation convention. The marked graph Γ contains sufficient information for a presentation of $\pi_1(M)$. In case $\mathscr{A}$ is the complexification of a real arrangement, we may choose $\Gamma = \Gamma_{\mathbf{R}}$. We know how difficult it is to prove that two groups given by generators and relators are not isomorphic. Yet I am optimistic that we will succeed in finding topological invariants of arrangements which are not combinatorial.

References

1. M. Aigner, *Combinatorial theory*, Grundlehren Math. Wiss. **234** (1979).

2. V. I. Arnold, *The cohomology ring of the colored braid group*, Mat. Zametki **5**(1969), 227–231; Math. Notes **5** (1969), 138–140.

3. E. Artal–Bartolo, *Sur le premier nombre de Betti de la fibre de Milnor du cône sur une courbe projective plane et son rapport avec la position des points singuliers*, preprint, 1989.

4. E. Artin, *Theorie der Zöpfe*, Hamb. Abh.2 **4** (1925), 47–72.

5. W. Arvola, *The fundamental group of the complement of an arrangement*, Ph. D. thesis, University of Wisconsin–Madison, 1990.

6. J. Birman, *Braids, links, and mapping classes*, Ann. of Math. Stud. **87** (1974).

7. N. Bourbaki, *Groupes et algèbres de Lie*, Chapters 4, 5, and 6, Hermann, Paris, 1968.

8. E. Brieskorn, *Sur les groupes de tresses*, Séminaire Bourbaki 1971/72, Lecture Notes in Math., vol. 317, Springer-Verlag, Berlin, Heidelberg, and New York, 1973, pp.21–44.

9. P. Deligne, *Les immeubles des groupes de tresses généralisés*, Invent. Math. **17** (1972), 273–302.

10. A. Dimca, *On the Milnor fibrations of weighted homogeneous polynomials*, preprint, 1989.

11. H. Esnault, *Fibre de Milnor d'un cône sur une courbe plane singulière*, Invent. Math. **68** (1982), 477–496.

12. E. Fadell and L. Neuwirth, *Configuration spaces*, Math. Scand. **10** (1962), 111–118.

13. M. Falk, *The minimal model of the complement of an arrangement of hyperplanes*, Trans. Amer. Math. Soc. **309** (1988), 543–556.

14. M. Falk and R. Randell, *The lower central series of a fiber-type arrangement*, Invent. Math. **82** (1985), 77–88.

15. ———, *On the homotopy theory of arrangements*, Complex Analytic Singularities, Adv. Stud. Pure Math., vol. 8, North-Holland, Amsterdam 1987, pp.101–124.

16. R. H. Fox and L. Neuwirth, *The braid groups*, Math. Scand. **10** (1962), 119–126.

17. M. Goresky and R. MacPherson, *Stratified Morse theory*, Springer-Verlag, Berlin, Heidelberg, and New York, 1988.

18. A. Hattori, *Topology of C^n minus a finite number of affine hyperplanes in general position*, J. Fac. Sci. Univ. Tokyo **22** (1975), 205–219.

19. M. Jambu and H. Terao, *Arrangements of hyperplanes and broken circuits*, Singularities (Richard Randell, ed.), Contemp. Math., vol. 90, Amer. Math. Soc., Providence, RI, 1989, pp. 147–162.

20. T. Kohno, *On the minimal algebra and $K(\pi, 1)$-property of affine algebraic varieties*, preprint.

21. J. Milnor, *Singular points of complex hypersurfaces*, Ann. Math. Stud. **61**, (1968).

22. P. Orlik, *Introduction to arrangements*, CBMS Regional Conf. Ser. in Math., no. 72, Amer. Math. Soc., Providence, RI, 1989.

23. P. Orlik and R. Randell, *The Milnor fiber of a generic arrangement*, preprint, 1990.

24. P. Orlik and L. Solomon, *Combinatorics and topology of complements of hyperplanes*, Invent. Math. **56** (1980), 167–189.

25. R. Randell, *On the topology of non–isolated singularities*, Geometric Topology, Proceedings of the Georgia Topology Conference, 1977, Academic Press, NY, 1979, pp. 445–473.

26. _____, *The fundamental group of the complement of a union of complex hyperplanes*, Invent. Math. **69**(1982), 103–108. Correction in Invent. Math. **80** (1985), 467–468.

27. M. Salvetti, *Topology of the complement of real hyperplanes in C^N*, Invent. Math. **88** (1987), 603–618.

28. D. Siersma, *Singularities with critical locus a 1-dimensional complete intersection and transversal type A_1*, Topology Appl. **27** (1987), 51–73.

29. H. Terao, *Modular elements of lattices and topological fibration*, Adv. in Math. **62** (1986), 135–154.

30. T. Zaslavsky, *Facing up to arrangements: Face-count formulas for partitions of space by hyperplanes*, Mem. Amer. Math. Soc., no. 154, Amer. Math. Soc., Providence, RI, 1975.

DEPARTMENT OF MATHEMATICS, UNIVERSITY OF WISCONSIN, MADISON, WISCONSIN 53706
E-mail address: orlik@math.wisc.edu

DIMACS Series in Discrete Mathematics
and Theoretical Computer Science
Volume **6**, 1991

Notes on Geometric Graph Theory

JÁNOS PACH

ABSTRACT. A geometric graph (hypergraph) is a pair (V, E), where V is a set of points in the plane (in space) in general position, and E is a set of closed segments (simplices) induced by the elements of V. We survey some recent results in geometric graph and hypergraph theory, with special emphasis on the following problem. What is the maximum number of edges that a geometric graph (hypergraph) of n vertices can have without containing some given forbidden configuration? One of the new results we prove is that any complete d-dimensional hypergraph with n vertices contains $n^{1/d}$ disjoint edges (d-dimensional simplices) such that no hyperplane supporting a facet of an edge intersects any other edge.

1. Introduction

Most textbooks on graph theory start with the discussion of the Königsberg Bridge Problem (Eulerian and Hamiltonian graphs) or problems on planar graphs (the Four-Color Conjecture, Euler's Polyhedral Formula, Kuratowski's Criterion, etc.). However, these questions did not only lead to the birth of graph theory, but they also inspired the development of another discipline that greatly influenced modern mathematics: topology. These two fields stem from the same root. Therefore, it was perfectly natural that the first genuine monograph on graph theory had the subtitle "Kombinatorische Topologie der Streckenkomplexe" [**Kö**]. In the last few decades the connection between graph theory and topology has somewhat faded away. In the most prolific new areas of graph theory (Ramsey theory, extremal graph theory, random graphs, etc.), graphs are regarded as abstract binary relations rather than one-dimensional simplicial complexes. More than 20 years ago Turán [**T1**] remarked that the part of graph theory "mainly relevant for applications belongs essentially to logic." This statement is still valid, even for applications to problems in plane topology, because very few general results are known about graphs embedded in the plane. Geometric graph theory is a badly

1991 *Mathematics Subject Classification.* Primary 52C10, 05C35; Secondary 05C10, 05C65.

The author was supported by NSF grant CCR–8901484 and by the Hungarian National Foundation for Scientific Research grant OTKA–1812.

underdeveloped field of research. To illustrate our ignorance, we recall two famous unsolved problems from this area.

1.1. THE BRICK FACTORY PROBLEM [T2]. Let $K_{m,n}$ denote the complete bipartite graph with m and n vertices in its classes. Determine the *crossing number* $\text{cross}(K_{m,n})$, i.e., the minimum number of crossings in an embedding of $K_{m,n}$ in the plane such that the vertices are represented by points and the edges are represented by Jordan arcs joining the appropriate pairs of points. (Two distinct arcs define a crossing if they have an interior point in common.) Zarankiewicz (see [G]) conjectured that

$$\text{cross}(K_{m,n}) = \left\lfloor \frac{m}{2} \right\rfloor \left\lfloor \frac{m-1}{2} \right\rfloor \left\lfloor \frac{n}{2} \right\rfloor \left\lfloor \frac{n-1}{2} \right\rfloor ,$$

which is known to be true if $\min\{m, n\} \leq 6$ [K]. The problem of deciding whether the crossing number of an arbitrary graph exceeds an integer k is NP-complete [GJ]. This suggests that it might be difficult to settle the above problem.

1.2. THE THRACKLE CONJECTURE [C]. A *thrackle* is an embedding of a graph in the plane in which the vertices are represented by points and the edges are represented by Jordan arcs with the property that two distinct arcs either meet at exactly one common vertex or they have an interior point in common such that each arc crosses from one side to the other of the other arc. J. H. Conway conjectured that the number of arcs in a thrackle is at most as large as the number of vertices. Assuming the truth of this conjecture, Woodall [W] classified all graphs that have a representation as a thrackle. According to his result, a graph G is thrackleable if and only if it has at most one odd cycle, no cycle of length 4, and each of its connected components is a tree or has exactly one cycle.

In the last 15 years many questions similar to the above ones have been raised in connection with a large variety of problems in computational geometry, motion planning, and VLSI design. They require new techniques concerning arrangements of curves and surfaces, combining combinatorial and topological ideas (see, e.g., [E, GS, PA]). Many of the difficulties are present already in the two-dimensional case for arrangements of straight-line segments.

1.3. DEFINITION. A *geometric graph* G is a graph drawn in the plane by straight-line segments, i.e., it is defined as a pair $(V(G), E(G))$, where $V(G)$ is a set of points in the plane in general position and $E(G)$ is a set of closed segments whose endpoints belong to $V(G)$. $V(G)$ and $E(G)$ are called the vertex set and the edge set of G, respectively. If $V(G)$ is the set of vertices of a convex polygon, then G is called a *convex geometric graph*.

The aim of this paper is to survey some recent results about geometric graphs, to pose a few new problems, and to present a couple of new proofs.

2. Forbidden configurations

One of the fundamental problems in extremal graph theory is to characterize those graphs which must necessarily occur in every graph with n vertices and m edges. The first instance of such a problem was solved by Turán [T3], who determined the maximum number of edges that a graph with n vertices can have without containing a complete subgraph of $k + 1$ vertices. The excellent monograph [B] and review articles [Si, F] give almost encyclopedic accounts of Turán type results for graphs and hypergraphs.

The investigation of similar problems for geometric graphs was initiated by Erdős, Kupitz [Ku1], and Perles. Given two geometric graphs G and H, we say that H is a (geometric) subgraph of G (in notation, $H \subseteq G$) if $V(H) \subseteq V(G)$ and $E(H) \subseteq E(G)$. For any class $\mathscr{H}$ of so called *forbidden* geometric subgraphs, let $t(\mathscr{H}, n)$ $(t_c(\mathscr{H}, n))$ denote the maximum number of edges that a geometric graph (a convex geometric graph) with n vertices can have without containing a subgraph which belongs to $\mathscr{H}$.

The first result of this type is due to Erdős [Er1], but it can also be deduced using the arguments in [HP]. Two edges of a geometric graph are said to be *independent* if the corresponding closed segments are disjoint. Let $\mathscr{I}_{k+1}$ be the family of all geometric graphs consisting of $k + 1$ pairwise independent edges.

2.1. THEOREM [Er1]. *Any geometric graph without two independent edges has at most as many edges as vertices, and equality can hold. In other words,*

$$t(\mathscr{I}_2, n) = n \quad \text{for every} \quad n \geq 3.$$

Note that this result can be regarded as an affirmative answer to the Thrackle Conjecture (1.2) for straight-line arcs.

The cases of equality in 2.1 have been characterized by Woodall.

2.2. THEOREM [W]. *Let G be a geometric graph without two independent edges. If $|E(G)| = |V(G)|$, then G consists of an odd cycle C with some extra vertices, all of which are joined to exactly one point of C.*

The problem of determining or estimating $t(\mathscr{I}_3, n)$ was explicitly stated in [AH]. Improving a somewhat weaker result of Alon and Erdős [AE], O'Donnel and Perles [OP] have recently obtained the following bounds.

2.3. THEOREM [OP]. *For every $n \geq 5$,*

$$\left\lfloor \tfrac{5}{2}(n - 1) \right\rfloor \leq t(\mathscr{I}_3, n) \leq 3.6n + c,$$

where c is a suitable constant.

For $k \geq 3$ it is not known whether $t(\mathscr{I}_{k+1}, n)$ grows linearly with n. For convex geometric graphs the situation is much simpler.

2.4. PROPOSITION [Ku2]. *For every k and $n > 2k$,*

$$t_c(\mathscr{I}_{k+1}, n) = kn.$$

An abstract graph consisting of a simple path P with some extra vertices, all of which are connected to exactly one point of P, is called a *caterpillar*. The following result, which has been proved recently by Perles (personal communication), is a far-reaching generalization of 2.4.

2.5. THEOREM (Perles). *Let $\mathcal{H}$ be the class of all geometric graphs whose underlying graphs are isomorphic to a given caterpillar with k vertices, and which do not contain two crossing edges. Then, for every $n \geq k$,*

$$t_c(\mathcal{H}, n) = \lfloor n(k/2 - 1) \rfloor .$$

This theorem also settles an interesting special case of a well-known conjecture of Erdős and Sós (see [Er2]) stating that any (abstract) graph with n vertices and more than $\lfloor n(k/2 - 1) \rfloor$ edges contains every tree of k vertices as a subgraph. Another special case of this conjecture was confirmed by Sidorenko [S].

A *geometric graph* is called *planar* if no two of its edges are crossing, i.e., they have no interior points in common. The connected components of the complement of the union of all edges of a planar geometric graph G are called the *faces* of G. If all vertices of G lie on the boundary of the unbounded face, then G is said to be *outerplanar*. Two planar geometric graphs are *isomorphic* if there is a one-to-one correspondence between their vertices, edges, and faces, respectively, preserving the incidence relations. A geometric graph is *complete* if its edge set contains every segment connecting a pair of vertices.

Gritzmann, Mohar, Pach, and Pollack [GMPP] (see also [FPP]) and independently Perles (personal communication) have shown that any complete geometric graph has a subgraph isomorphic to any outerplanar geometric graph with the same number of vertices. Roughly speaking, this means that any outerplanar graph with n vertices can be redrawn in the plane by non-crossing straight-line segments using any set of n points in general position as its vertices. This immediately implies the following result.

2.6. THEOREM. *Let $\mathcal{H}$ be the class of all geometric graphs isomorphic to a given outerplanar geometric graph H with k vertices. Then, for every $n \geq k$,*

$$t(\mathcal{H}, n) \leq T(k, n) ,$$

where $T(k, n)$ is the maximum number of edges that an abstract graph with n vertices can have without containing a complete subgraph with k vertices. Moreover, $t_c(\mathcal{H}, n) = t(\mathcal{H}, n) = T(k, n)$, provided that H is two-connected.

3. Crossing segments

Let $\mathcal{C}_{k+1}$ denote the class of all geometric graphs consisting of $k + 1$ pairwise crossing edges. It follows from Euler's Polyhedral Formula that the maximum number of edges of a (convex) geometric graph with n vertices is $3n - 6$ (resp. $2n - 3$). That is, for every $n \geq 3$,

$$t(\mathscr{C}_2, n) = 3n - 6, \qquad t_c(\mathscr{C}_2, n) = 2n - 3.$$

3.1. THEOREM [ACNS]. *Every geometric graph with n vertices and $m \geq 4n$ edges contains at least $\frac{1}{100} \frac{m^3}{n^2}$ pairs of crossing edges.*

The following construction shows that the order of magnitude of this bound is best possible. Distribute n points as evenly as possible on $n^2/2m$ disjoint circles in the plane, and join two points by a segment if they lie on the same circle.

This result can be used to obtain subquadratic upper bounds for $t(\mathscr{C}_{k+1}, n)$ for every fixed $k \geq 2$.

3.2. PROPOSITION. *For every n, $t(\mathscr{C}_3, n) < 13n^{3/2}$.*

PROOF. Given a geometric graph G with n vertices and $m \geq 4n$ edges, by 3.1 we can find an edge crossing at least $\frac{1}{50} \frac{m^2}{n^2}$ other edges of G. If G has no three pairwise crossing edges, then

$$\frac{1}{50} \frac{m^2}{n^2} \leq t(\mathscr{C}_2, n - 2) = 3n - 12.$$

Hence $m < 13n^{3/2}$, as required. $\square$

Iterating this argument we see that $t(\mathscr{C}_{k+1}, n) = O(n^{2-(1/2^{k-1})})$ for any fixed k. In what follows, for large values of k we shall slightly improve this naive upper bound, but we remain short of answering the following simple question.

3.3. PROBLEM (B. Gärtner). Let $k \geq 2$ be fixed. Is $t(\mathscr{C}_{k+1}, n) = O(n)$?

A first step toward the clarification of this question may be to characterize the "unavoidable" subgraphs occurring in every complete geometric graph of n vertices. A result of this kind has been recently established by Aronov, Erdős, Goddard, Kleitman, Klugerman, Pach, and Schulman [AEGKKPS].

3.4. THEOREM [AEGKKPS]. *Let $K_{p,p}$ denote a complete bipartite geometric graph with p vertices in its classes, i.e., $V(K_{p,p}) = V_1 \cup V_2$, where $|V_1| = |V_2| = p$ and $E(K_{p,p})$ consists of all segments connecting a point of V_1 with a point of V_2. Then $K_{p,p}$ contains at least $\sqrt{p}/5$ pairwise crossing edges.*

3.5. COROLLARY. *For any fixed k and $n \to \infty$,*

$$t(\mathscr{C}_{k+1}, n) = O\left(n^{2-(1/25(k+1)^2)}\right).$$

PROOF. Let G be a geometric graph with n vertices and m edges, containing no $k+1$ pairwise crossing edges. By 3.4, G has no complete bipartite geometric subgraph $K_{p,p}$ with $p = 25(k + 1)^2$. An old theorem of Kővári,

Sós, and Turán [KST] states that $|E(G)| = m = O(n^{2-(1/p)})$, provided that $G \not\supseteq K_{p,p}$. □

Another interesting consequence of 3.4 can be deduced by a standard duality argument. Two closed segments are *strongly independent* if they are disjoint and the four vertices defining them are in convex position.

3.6. COROLLARY [AEGKKPS]. *Any complete (bipartite) geometric graph with r vertices (in its classes) contains at least $\sqrt{r}/5$ pairwise strongly independent edges.*

This yields, as above, that for any fixed k

$$t(\mathscr{I}^*_{k+1}, n) = O\left(n^{2-(1/25(k+1)^2)}\right),$$

where $\mathscr{I}^*_{k+1}$ stands for the class of all geometric graphs consisting of $k+1$ pairwise strongly independent edges.

An affirmative answer to the following question would obviously lead to some improvement in both of the above corollaries.

3.7. PROBLEM. Does there exist a constant $c > 0$ such that any complete (bipartite) geometric graph with r vertices (in its classes) contains at least cr pairwise crossing edges?

In the next section we shall present a simple alternative argument to prove a generalization of 3.6.

The analogue of Problem 3.3 for convex geometric graphs has been recently settled by Capoyleas and Pach [CP].

3.8. THEOREM [CP]. *For any k and n,*

$$t_c(\mathscr{C}_{k+1}, n) = \begin{cases} \binom{n}{2} & \text{if } n \le 2k+1, \\ 2kn - \binom{2k+1}{2} & \text{if } n \ge 2k+1. \end{cases}$$

One can easily extend the proof of [ACNS] to establish a generalization of 3.1 for convex geometric graphs.

3.9. THEOREM. *Every convex geometric graph with n vertices and $m \ge (2k-1)n$ edges contains at least*

$$\frac{1}{10^k k^{2k}} \frac{m^{2k-1}}{n^{2k-2}}$$

k-tuples of pairwise crossing edges.

Furthermore, the order of magnitude of this bound cannot be improved if k is fixed and n tends to infinity.

PROOF. Let $f_k(n, m)$ denote the minimum number of k-tuples of pairwise crossing edges in a geometric graph with n vertices and m edges.

It follows immediately from 3.8 that

$$f_k(n, m) \ge m - t_c(\mathscr{C}_k, n) \ge m - 2(k-1)n.$$

We are going to show by induction on n that

$$f_k(n, m) \geq \frac{1}{30^k}\binom{n}{2k}\left(\frac{m}{\binom{n}{2}}\right)^{2k-1} \quad \text{whenever } m \geq (2k-1)n.$$

For $m \leq 2kn$, this is weaker than the previous inequality.

Let G be a convex geometric graph with n vertices and $m > 2kn$ edges, and let $f_k(G)$ stand for the number of k-tuples T of pairwise crossing edges in G. Counting the pairs (x, T), where $x \in V(G)$ is not an endpoint of any segment of T, we obtain

$$(n - 2k)f_k(G) \geq \sum_{x \in V(G)} f_k(G - x) \geq \sum_{x \in V(G)} f_k(n - 1, m - d(x)).$$

(Here, as usual, $d(x)$ denotes the degree of x in G, and $G - x$ stands for the graph obtained from G by the deletion of x.) By the induction hypothesis, we have

$$f_k(G) \geq \frac{1}{n - 2k}\frac{1}{30^k}\binom{n-1}{2k}\frac{\sum_x(m - d(x))^{2k-1}}{\binom{n-1}{2}^{2k-1}}.$$

Using the fact that $\sum_x(m-d(x)) = (n-2)m$, it follows by Jensen's Inequality that

$$\sum_x(m - d(x))^{2k-1} \geq n\left(\frac{(n-2)m}{n}\right)^{2k-1}.$$

Hence,

$$f_k(G) \geq \frac{1}{30^k}\binom{n}{2k}\left(\frac{m}{\binom{n}{2}}\right)^{2k-1},$$

as required, and the bound in the theorem follows by simple calculation. □

Note that an affirmative answer to Problem 3.3 would imply a similar lower bound on the number of k-tuples of pairwise crossing edges in any geometric graph $(k \geq 3)$.

4. Geometric hypergraphs

A *d-dimensional geometric r-hypergraph* H is defined as a pair $(V(H), E(H))$, where $V(H)$ is a set of points in $\mathbb{R}^d$ in general position, and $E(H)$ is a set of closed $(r - 1)$-dimensional simplices whose vertices belong to $V(H)$. (Obviously, $r \leq d + 1$.) Thus, a geometric graph can be regarded as a two-dimensional geometric two-hypergraph. The elements of $V(H)$ and $E(H)$ are called the *vertices* and *(hyper)edges* of H, respectively. We shall use the same notation as for geometric graphs, i.e., given a family $\mathscr{F}$ of *forbidden configurations* (d-dimensional geometric r-hypergraphs), $t(\mathscr{F}, n)$ is the maximum number of hyperedges that a d-dimensional geometric r-hypergraph can have without containing a subhypergraph which belongs to $\mathscr{F}$.

Two edges of H are called *independent* if the corresponding simplices are disjoint.

The following result of Akiyama and Alon [**AA**] is a surprising consequence of the Borsuk-Ulam (or ham-sandwich) theorem [**Bo**].

4.1. THEOREM [**AA**]. *Let $K_d(k)$ be a complete d-partite d-dimensional geometric d-hypergraph with k elements in its classes, i.e., $V(K_d(k)) = V_1 \cup V_2 \cup \cdots \cup V_d$, where $|V_i| = k$ for every i, and $E(K_d(k))$ consists of all $(d-1)$-dimensional simplices having precisely one vertex from each V_i.*

Then $K_d(k)$ contains k pairwise independent edges.

A well-known result of Erdős [**Er3**] states that any abstract d-hypergraph with n vertices and $m \geq n^{d-(1/k^{d-1})}$ hyperedges (d-tuples) contains a complete d-partite subhypergraph with k elements in each of its classes. Combining this with the previous result, we obtain at once the following theorem.

4.2. THEOREM [**AA**]. *Let $\mathscr{S}_k^d$ denote the family of all d-dimensional geometric d-hypergraphs consisting of k pairwise independent edges. Then*

$$t(\mathscr{S}_k^d, n) \leq n^{d-(1/k^{d-1})}.$$

However, it is conjectured that the true order of magnitude of $t(\mathscr{S}_k^d, n)$ is n^{d-1}. Theorems 2.1 and 2.3 show that this is the case for $d = 2$ and $k = 2, 3$.

Two d-dimensional simplices in $\mathbb{R}^d$ are said to be *strongly independent* if no hyperplane supporting a facet of one of them intersects the other. In particular, two strongly independent triangles in the plane are not only disjoint, but picking one side of each triangle, their four vertices are always in convex position. Therefore, the following result can be regarded as a generalization of (the non-bipartite case of) 3.6 in two different directions.

4.3. THEOREM. *Let K be a complete d-dimensional geometric $(d+1)$-hypergraph consisting of all full-dimensional simplices induced by an r-element point set $V = V(K)$. Then K has at least $c_d r^{1/d}$ pairwise strongly independent edges, where $c_d > 0$ is a suitable constant.*

PROOF. According to a theorem of Chazelle and Welzl [**CW**], the points of V can be connected by a simple polygonal path $P = p_1 p_2 \cdots p_r$ such that any hyperplane intersects at most $c_d' r^{1-(1/d)}$ edges of P.

Let us cut P into disjoint pieces of length $d + 1$,

$$P_i = p_{(i-1)(d+1)+1} p_{(i-1)(d+1)+2} \cdots p_{i(d+1)}, \quad i = 1, 2, \ldots, \lfloor r/(d+1) \rfloor.$$

Define an abstract graph G on the vertex set $\{P_1, P_2, \ldots\}$ by joining P_i and P_j with an edge if and only if their convex hulls are not strongly independent.

Any hyperplane h supporting a facet of conv P_i can intersect only at most $c_d' r^{1-(1/d)}$ polygonal chains P_j, and h intersects conv P_j if and only if it intersects P_j. Since each simplex $\mathrm{conv}\, P_i$ has $d+1$ facets, it follows that the number of edges of our graph is

$$|E(G)| \le \left\lfloor \frac{r}{d+1} \right\rfloor (d+1)\, c'_d r^{1-(1/d)} \le c'_d r^{2-(1/d)}.$$

It is a well-known consequence of Turán's theorem [**T3**] that any graph G contains an empty subgraph with at least

$$\frac{|V(G)|^2}{2|E(G)| + |V(G)|}$$

vertices, which completes the proof. $\square$

Similarly, one can try to estimate the maximum number of edges in a given geometric hypergraph that have an interior point in common. Generalizing a result of Boros and Füredi [**BF**], Bárány [**Bá**] established the following theorem.

4.4. THEOREM [Bá]. *There exists a constant $c_d > 0$ such that any complete d-dimensional geometric $(d+1)$-hypergraph with n vertices contains at least $c_d n^{d+1}$ hyperedges which have a common interior point.*

Živaljević and Vrećica [**ZV**] have recently succeeded in proving a beautiful related result using elegant topological arguments. It can be viewed as a multipartite version of a famous theorem of Tverberg [**Tv**].

4.5. THEOREM [ZV]. *Let K be a complete $(d+1)$-partite d-dimensional geometric $(d+1)$-hypergraph with $4k$ points in each of its classes. Then K has at least k edges with disjoint vertex sets, which have an interior point in common.*

Aronov et al. [**ACEGSW**] have extended some ideas of [**PSSz**] to strengthen 4.4, at least for $d = 2$.

4.6. THEOREM [ACEGSW]. *Let H be a 2-dimensional geometric 3-hypergraph with n vertices*

    (i) *For every $c > 0$ there exists $c' > 0$ such that, if $|E(H)| \ge cn^3$, then H has at least $c'n^3$ edges with a common interior point.*

    (ii) *If $|E(H)| \ge n^{3-\varepsilon}$ for some $\varepsilon > 0$, then H has at least $n^{3-3\varepsilon}/ (512 \log^5 n)$ edges with a common interior point.*

Of course, 4.6(ii) provides a reasonable bound only as long as ε is small. For $\varepsilon > \frac{1}{2}$ it is superseded by the easy observation that one can always pick a point in the plane (close to a vertex of H) that belongs to the interior of at least $|E(H)|/2n$ edges. We have no conjecture about the order of magnitude of the best possible bounds.

In view of 4.5, it should be easier to confirm the following conjecture which is true for $d \le 2$ [**BFL**].

4.7. CONJECTURE. Let K be a complete $(d+1)$-partite geometric $(d+1)$-hypergraph, $V(K) = V_1 \cup V_2 \cup \cdots \cup V_{d+1}$, and $|V_i| = k$ for every i. Then, for some $c_d > 0$, there exist $V_i' \subseteq V_i$, $|V_i'| \geq c_d k$ $(1 \leq i \leq d+1)$, such that all edges of the subhypergraph $K' \subseteq K$ induced by $V(K') = V_1' \cup V_2' \cup \cdots \cup V_{d+1}'$ have an interior point in common.

Obviously, this would be a generalization of Theorem 4.4.

5. Related problems

In spite of the fact that Ramsey's theorem became famous after its rediscovery and application to a geometric problem by Erdős and Szekeres [ES1, ES2], there are relatively few Ramsey-type results on geometric graphs and hypergraphs. In particular, we are unable to answer the following question strongly related to the problems discussed in §§2 and 3.

5.1. PROBLEM. What is the largest number $f = f(m)$ with the property that any set of m open segments in the plane contains f elements that are either pairwise crossing or pairwise disjoint?

Obviously, $f(m) = O(\sqrt{m})$ and $f(m) \geq c \log m$ by Ramsey's theorem.

It might be worth recalling the following one-dimensional result of this kind [Bi, R]. Given m segments on a line, we can always find $\lceil \sqrt{m} \rceil$ of them with pairwise disjoint interiors or having a point in common. For some generalizations, see [P]. A similar problem for abstract graphs has been solved by Sauer [Sa].

The connected components of the complement of the union of all edges of a geometric graph G are called the *faces* of G. We know very little about the facial structure of geometric graph. One of the few results is the following theorem whose proof is based on [PSS].

5.2. THEOREM [AHKMN]. *The number of sides of a single face of a geometric graph with n vertices cannot exceed $cn\alpha(n)$, where c is a constant and α denotes the (extremely slowly growing) inverse of Ackermann's function. Moreover, this bound is asymptotically tight (up to the exact value of c).*

Erdős, Lovász, Simmons, and Strauss [ELSS] have raised a new type of question by asking how many edges of a geometric graph can be crossed by a single line. According to their conjecture, for any geometric graph G with n vertices and m edges, there exists a straight line intersecting at least m^2/n^2 edges of G, which has been confirmed by Alon and Perles [AP] in many special cases. Similar questions can be asked for geometric hypergraphs.

5.3. PROBLEM. Determine or estimate the maximum number $f = f_{d,r}^k(m, n)$ such that, for any d-dimensional geometric r-hypergraph H, there exists a k-dimensional flat intersecting at least f edges of H.

In this brief survey we have attempted to review some recent problems and results indicating the emergence of a new trend in combinatorial geometry. To keep our exposition short, we did not discuss the vast literature about special geometric graphs, such as unit distance graphs (whose edges are unit segments), k-sets (geometric hypergraphs consisting of all $(d-1)$-dimensional simplices induced by a point set $V \subseteq \mathbb{R}^d$ such that the hyperplanes supporting them have exactly k points of V on one of their sides), etc. (See, e.g., [Er1, PA, ELSS, BFL, PSSz, ACEGSW].) Most of the problems mentioned above have obvious generalizations to graphs and hypergraphs embedded in surfaces.

REFERENCES

[AA] J. Akiyama and N. Alon, *Disjoint simplices and geometric hypergraphs*, Combinatorial Mathematics (G. S. Bloom et al., eds.), Ann. New York Acad. Sci., vol. 555, New York Acad. Sci., New York, 1989, pp. 1–3.

[ACEGSW] B. Aronov, B. Chazelle, H. Edelsbrunner, L. J. Guibas, M. Sharir, and R. Wenger, *Points and triangles in the plane and halving planes in space*, Proc. Sixth Annual Sympos. on Computational Geometry, ACM Press, New York, 1990, pp. 112–115.

[ACNS] M. Ajtai, V. Chvátal, M. M. Newborn, and E. Szemerédi, *Crossing-free subgraph*, Ann. Discrete Math. **12** (1982), 9–12.

[AEGKKPS] B. Aronov, P. Erdős, W. Goddard, D. J. Kleitman, M. Klugerman, J. Pach, and L. J. Schulman, *Crossing families*, Proc. Seventh Annual Sympos. on Computational Geometry, ACM Press, New York, 1991, pp. 351–356.

[AE] N. Alon and P. Erdős, *Disjoint edges in geometric graphs*, Discrete Comput. Geom. **4** (1989), 287–290.

[AHKMN] E. M. Arkin, D. Halperin, K. Kedem, J. S. B. Mitchell, and N. Naor, *Arrangements of segments that share endpoints; Single face results*, Proc. Seventh Annual. Sympos. on Computational Geometry, ACM Press, New York, 1991, pp. 324–333.

[AH] S. Avital and H. Hanani, *Graphs*, Gilyonot Lematematika 3 (1966), 2–8. (Hebrew)

[AP] N. Alon and M. Perles, *On the intersection of edges of a geometric graph by straight lines*, Discrete Math. **60** (1986), 75–90.

[Bá] I. Bárány, *A generalization of Carathéodory's theorem*, Discrete Math. **40** (1982), 141–152.

[BFL] I. Bárány, Z. Füredi, and L. Lovász, *On the number of halving planes*, Combinatorica **10** (1990), 175–183.

[BF] E. Boros and Z. Füredi, *The number of triangles covering the center of an n-set*, Geom. Dedicata **17** (1984), 69–77.

[Bi] A. Bielecki, *Problem 56*, Colloq. Math. **1** (1948), 333–334.

[Bo] K. Borsuk, *Drei Sätze über die n–dimensionale euklidische Sphäre*, Fund. Math. **20** (1933), 177–190.

[B] B. Bollobás, *Extremal graph theory*, Academic Press, London, 1978.

[CEGHSS] B. Chazelle, H. Edelsbrunner, L. J Guibas, J. E. Hershberger, R. Seidel, and M. Sharir, *Slimming down by adding; selecting heavily covered points*, Proc. Sixth Annual Sympos. on Computational Geometry, ACM Press, New York, 1990, pp. 116–127.

[CP] V. Capoyleas and J. Pach, *A Turán-type theorem on chords of a convex polygon*, J. Combin. Theory Ser. A (to appear).

[CW] B. Chazelle and E. Welzl, *Quasi-optimal range searching in spaces of finite VC-dimension*, Discrete Comput. Geom. **4** (1989), 467–490.

[C] J. H. Conway, *Open problem 42*, Theory of Graphs (P. Erdős and G. Katona, eds.), Akad. Kiadó, Budapest, 1968, p. 365.

[ELSS] P. Erdős, L. Lovász, A. Simmons, and E. G. Strauss, *Dissection graphs of planar point sets*, A Survey of Combinatorial Theory (J. N. Srivastava et al., eds.), North-Holland, Amsterdam, 1973, pp. 139–149.

[Er1] P. Erdős, *On sets of distances of n points*, Amer. Math. Monthly **53** (1946), 248–250.

[Er2] ____, *Extremal problems in graph theory*, Theory of Graphs and its Applications (M. Fiedler, ed.), Academic Press, New York, 1965, pp. 29–36.

[Er3] ____, *On extremal problems of graphs and generalized graphs*, Israel J. Math. **2** (1964), 183–190.

[ES1] P. Erdős and G. Szekeres, *A combinatorial problem in geometry*, Compositio Math. **2** (1935), 464–470.

[ES2] ____, *On some extremal problems in elementary geometry*, Ann. Univ. Sci. Budapest Eötvös Sect. Math. **3-4** (1961), 53–62.

[E] H. Edelsbrunner, *Algorithms in combinatorial geometry*, EATCS Monographs on Theoretical Comp. Sci., Springer-Verlag, Berlin-Heidelberg-New York-Paris-Tokyo, 1987.

[FPP] H. de Fraysseix, J. Pach, and R. Pollack, *Drawing a planar graph on a grid*, Combinatorica **10** (1990), 41–51.

[F] Z. Füredi, *Turán type problems*, Surveys in Combinatorics 1991, London Math. Soc. Lecture Note Series, Cambridge Univ. Press, 1991.

[GJ] M. K. Garey and D. S. Johnson, *Crossing number is NP-complete*, SIAM J. Algebraic Discrete Methods **4** (1983), 312–316.

[GMPP] P. Gritzmann, B. Mohar, J. Pach, and R. Pollack, *Problem E3341*, Amer. Math. Monthly **96** (1989), 642.

[GS] L. Guibas and M. Sharir, *Combinatorics and algorithms of arrangements*, New Trends in Discrete and Computational Geometry (J. Pach, ed.), Springer-Verlag (to appear).

[G] R. K. Guy, *Crossing numbers of graphs*, Graph Theory and Applications (Y. Alavi et al., eds.), Lecture Notes in Math., vol. 303, Springer-Verlag, Berlin and New York, 1972, pp. 111–124.

[HP] H. Hopf and E. Pannwitz, *Aufgabe No. 167*, Jber. Deutsch. Math. Verein. **43** (1934), 114.

[Kö] D. König, *Theorie der endlichen und unendlichen Graphen*, Leipzig, 1936; reprinted by Chelsea, New York, 1950.

[KST] T. Kővári, V. T. Sós, and P. Turán, *On a problem of K. Zarankiewicz*, Colloq. Math. **3** (1954), 50–57.

[Ku1] Y. S. Kupitz, *Extremal problems in combinatorial geometry*, Aarhus University Lecture Notes Series, no. 53, Aarhus University, Denmark, 1979.

[Ku2] ____, *On pairs of disjoint segments in convex position in the plane*, Ann. Discrete Math. **20** (1984), 203–208.

[K] D. J. Kleitman, *The crossing number of $K_{5,n}$*, J. Combin. Theory **9** (1970), 315–323.

[OP] P. O'Donnel and M. Perles, Lecture delivered at the DIMACS seminar on geometric graphs, Rutgers University, New Brunswick, NJ, 1990.

[P] J. Pach, *Decomposition of multiple packing and covering*, Diskrete Geometrie, 2. Kolloq. Inst. Math. Univ. Salzburg, 1980, pp. 169–178.

[PA] J. Pach and P. K. Agarwal, *Combinatorial geometry*, Tech. Report, Dept. Computer Science, Duke University, Durham, NC, 1991.

[PSSz] J. Pach, W. Steiger, and E. Szemerédi, *An upper bound on the number of planar k-sets*, Proc. 30th Annual IEEE Sympos. on Foundations of Computer Science, IEEE Comp. Soc. Press, Los Alamitos, 1989, pp. 72–79.

[PSS] R. Pollack, M. Sharir, and S. Sifrony, *Separating two simple polygons by a sequence of translations*, Discrete Comput. Geom. **3** (1988), 123–136.

[R] R. Rado, *Covering theorems for ordered sets*, Proc. London Math. Soc. (2) **50** (1948), 509–535.

[Si] M. Simonovits, *Extremal graph theory*, Selected Topics in Graph Theory 2 (L. W. Beineke and R. J. Wilson, eds.), Academic Press, London-New York, 1983, pp. 161–200.

[S] A. F. Sidorenko, *Asymptotic solution for a new class of forbidden r-graphs*, Combinatorica **9** (1989), 207–215.

[Sa] N. Sauer, *The largest number of edges of a graph such that not more than g intersect in a point or more than n are independent*, Combinatorial Mathematics and its Applications (D. J. A. Welsh, ed.), Academic Press, London-New York, 1971, 253–257.

[T1] P. Turán, *Applications of graph theory to geometry and potential theory*, Proc. Calgary Intern. Conf. on Combinatorial Structures and their Applications, Gordon and Breach, New York, 1970, pp. 423–434.

[T2] ——, *A note of welcome*, J. Graph Theory **1** (1977), 7–9.

[T3] ——, *Eine Extremalaufgabe aus der Graphentheorie*, Mat. Fiz. Lapok **48** (1941), 436–452. (Hungarian)

[Tv] H. Tverberg, *A generalization of Radon's theorem*, J. London Math. Soc. **41** (1966), 123–128.

[W] D. R. Woodall, *Thrackles and deadlock*, Combinatorial Mathematics and Its Applications (D. J. A. Welsh, ed.), Academic Press, London-New York, 1971, pp. 335–347.

[ZV] R. T. Živaljević and S. T. Vrećica, *The colored Tverberg problem and complexes of injective functions* (to appear).

COURANT INSTITUTE, NEW YORK UNIVERSITY, NEW YORK, NEW YORK 10012

MATHEMATICAL INSTITUTE, HUNGARIAN ACADEMY, H-1364 BUDAPEST, P.O.B. 127

DIMACS Series in Discrete Mathematics
and Theoretical Computer Science
Volume 6, 1991

Recent Progress on the Complexity
of the Decision Problem for the Reals

JAMES RENEGAR

This paper concerns recent progress on the computational complexity of decision methods and quantifier elimination methods for the first order theory of the reals. The paper begins with a quick introduction to the terminology, followed by a short survey of some complexity highlights. We then discuss ideas leading to the most (theoretically) efficient algorithms known. The discussion is necessarily simplistic, as a rigorous development of the algorithms forces one to consider a myriad of details.

This paper is similar to a talk the author gave at the DIMACS workshop on Algebraic Methods in Geometric Computations, held in May 1990. A complete development can be found in [**17**].

1. Some terminology

A *sentence* is an expression composed of certain ingredients. Letting $\mathbb{R}$ denote a real-closed field, the following is an example of a sentence:

$$(1.1) \quad (\exists x_1 \in \mathbb{R}^{n_1})(\forall x_2 \in \mathbb{R}^{n_2})[(g_1(x_1, x_2) > 0) \vee (g_2(x_1, x_2) = 0)]$$
$$\wedge (g_3(x_1, x_2) \neq 0).$$

The ingredients are: vectors of variables $(x_1$ and $x_2)$; the quantifiers $\exists$ and $\forall$; atomic predicates (e.g., $g_1(x_1, x_2) > 0$), which are real polynomial inequalities $(>, \geq, =, \neq, \leq, <)$; and a Boolean function holding the atomic predicates $([B_1 \vee B_2] \wedge B_3)$.

A sentence asserts something. The above sentence asserts that there exists $x_1 \in \mathbb{R}^{n_1}$ such that for all $x_2 \in \mathbb{R}^{n_2}$, (i) either $g_1(x_1, x_2) > 0$ or $g_2(x_1, x_2) = 0$, and (ii) $g_3(x_1, x_2) \neq 0$. Depending on the specific coefficients of the atomic predicate polynomials, this assertion is either true or false.

1991 *Mathematics Subject Classification*. Primary 03, 12, 14, 68.
This research was supported by NSF Grant No. DMS-8800835.

The set of all true sentences constitutes the first order theory of the reals. A *decision method* for the first order theory of the reals is an algorithm that, given any sentence, correctly determines if the sentence is true. Decision methods for the reals were first proven to exist by Tarski [18], who constructed one.

A sentence is a special case of a more general expression, called a *formula*. Here is an example of a formula:

$$
(1.2) \quad \begin{aligned} (\exists x_1 \in \mathbb{R}^{n_1})(\forall x_2 \in \mathbb{R}^{n_2})[(g_1(z, x_1, x_2) > 0) \vee (g_2(z, x_1, x_2) = 0)] \\ \wedge (g_3(z, x_1, x_2) \neq 0). \end{aligned}
$$

A formula has one thing that a sentence does not, namely, a vector $z \in \mathbb{R}^{n_0}$ of *free variables*. When specific values are substituted for the free variables, the formula becomes a sentence.

A vector $\bar{z} \in \mathbb{R}^{n_0}$ is a *solution* for the formula if the sentence obtained by substituting $\bar{z}$ is true.

Two formulae are *equivalent* if they have the same solutions.

A *quantifier elimination method* is an algorithm that, given any formula, computes an equivalent quantifier-free formula, i..e, for the above formula $(\exists x_1 \in \mathbb{R}^{n_1})(\forall x_2 \in \mathbb{R}^{n_2})P(z, x_1, x_2)$ such a method would compute an equivalent formula $Q(z)$ containing no quantified variables.

When a quantifier elimination method is applied to a sentence, it becomes a decision method. Thus, a quantifier elimination is in some sense more general than a decision method.

Tarski [18] actually constructed a quantifier elimination method.

Both (1.1) and (1.2) are said to be in *prenex* form, i.e., all quantifiers occur in front. More generally, a formula can be constructed from other formulae just as (1.1) was constructed from the atomic predicates, i.e., we can construct Boolean combinations of formulae, then quantify any or all of the free variables. If we quantify all free variables, then we obtain a sentence.

2. Some complexity highlights

We now present a brief survey of some complexity highlights for quantifier elimination methods, considering only formulae in prenex form. General bounds follow inductively. (If a formula is constructed from other formulae, first apply quantifier elimination to the innermost formulae, then to the innermost formulae of the resulting formula, etc.)

We consider the general formula

$$
(2.1) \quad (Q_1 x_1 \in \mathbb{R}^{n_1}) \cdots (Q_\omega x_\omega \in \mathbb{R}^{n_\omega}) P(z, x_1, \ldots, x_\omega)
$$

where $Q_1, \ldots, Q_\omega$ are quantifiers, assumed without loss of generality to alternate, i.e., Q_i is not the same as Q_{i+1}. Let m denote the number of distinct polynomials occurring among the atomic predicates, and let $d \geq 2$ be an upper bound on their degrees.

In the case of Turing machine computations where all polynomials coefficients are restricted to be integers, we let L denote the maximal bit length of the coefficients. In this context, we refer to the number of *bit operations* required by a quantifier elimination method. In the general and idealized case that the coefficients are real numbers, we rely on the computational model of Blum, Shub, and Smale [2] and refer to *arithmetic operations*, these essentially being field operations, including comparisons.

The sequential bit operation bounds that have appeared in the literature are all basically of the form

(2.2) $$(md)^E[L^{O(1)} + \text{Cost}]$$

where E is some exponent and *Cost* is the worst-case cost of evaluating the Boolean function holding the atomic predicates, i.e., worst-case over 0-1 vectors. The Cost term is generally relatively negligible compared to the factor $(md)^E$. Recent complexity bound improvements concern the exponent E.

The first reasonable upper bound for a quantifier elimination method was proven around 1973 by Collins [7]. He obtained $E = 2^{O(n)}$ where $n := n_0 + \cdots + n_\omega$. Collins' bound is thus *doubly exponential* in the number of variables. His method requires the formula coefficients to be integers, the number of arithmetic operations (not just bit operations) growing with the size of the integers. Collins' algorithm was not shown to parallelize, although enough is now known from work of Neff [15] that a parallel version probably could be developed. Collins' work has been enormously influential in the area.

The next major complexity breakthrough was made around 1985 by Grigor′ev [12], who developed a decision method for which $E \approx [O(n)]^{4\omega}$. Grigor′ev's bound is doubly exponential only in the number of quantifier alternations. Many interesting problems can be cast as sentences with only a few quantifier alternations. For these, Grigor′ev's result is obviously significant. Like Collins' quantifier elimination method, Grigor′ev's decision method requires integer coefficients and was not proven to completely parallelize.

Slightly incomplete ideas of Ben-Or, Kozen and Reif [1] (approximately 1986) were completed by Fitchas, Galligo, and Morgenstern [11] (approximately 1987) to construct a quantifier elimination method with arithmetic operation bound

(2.3) $$(md)^E \text{Cost}$$

where $E = 2^{O(n)}$. This provides an arithmetic operation analog of Collins' bit operation bound. When restricted to integer coefficients, the method also yields the Collins' bound if the arithmetic operations are carried out bit by bit. Moreover, the algorithm parallelizes. Assuming each arithmetic operation requires one time unit, the resulting time bound is

(2.4) $$[E \log(md)]^{O(1)} + \text{Time}(N)$$

if $(md)^E \ N$ parallel processors are used, where $\text{Time}(N)$ is the worst-case time required to evaluate the Boolean function holding the atomic predicates using N parallel processors. The analogous time bound for bit operations is also valid, namely, $[E \log(Lmd)]^{O(1)} + \text{Time}(N)$ if $(md)^E[L^{O(1)} + N]$ parallel processors are used.

In [17] (1989), the author introduced a new quantifier elimination method for which $E = \prod_{k=0}^{\omega} O(n_k)$. This was established for arithmetic operations and bit operations, i.e., (2.3) and (2.2). Regarding bit operations, the dependence of the bounds on L was shown to be very low, $L^{O(1)}$ in (2.2) being replaced by $L(\log L)(\log \log L)$, i.e., the best bound known for multiplying two L-bit integers. Moreover, the method was shown to parallelize, resulting in the arithmetic operation time bound (2.4) if $(md)^E N$ parallel processors are used and the bit operation time bound $\log(L)[E \log(md)]^{O(1)} + \text{Time}(N)$ if $(md)^E[L^2 + N]$ parallel processors are used.

Independently and simultaneously, Heintz, Roy, and Solernó [14] developed a quantifier elimination method for which $E = O(n)^{O(\omega)}$, both for arithmetic and bit operations. Their method also completely parallelizes.

The various bounds are best understood by realizing that quantifier elimination methods typically work by passing through a formula from back to front. First the vector x_ω is focused on, then the vector $x_{\omega-1}$, and so on. The work arising from each vector results in a factor for E. For Collins' quantifier elimination method, the factor corresponding to x_k is $2^{O(n_k)}$ (note that $2^{O(n)} = 2^{O(n_0)} \cdots 2^{O(n_\omega)}$). For the method introduced by the author, the factor is $O(n_k)$. The factor corresponding to Grigor'ev's decision method is approximately $O(n)^4$ independently of the number of variables in x_k. In that method a vector with few variables can potentially create as large of a factor as one with many variables. Similarly, the factor corresponding to the quantifier elimination method of Heintz, Roy, and Solernó is $O(n)^{O(1)}$ independently of x_k.

Many interesting formulae have blocks of variables of various sizes. For example, sentences asserting continuity generally have $(\forall \varepsilon)(\exists \delta)$, along with large blocks of variables. The exponent $E = \prod O(n_k)$ is especially relevant for such sentences, as the contribution to E from the smaller blocks is modest.

For the record, the quantifier elimination method in [17] produces a quantifier-free formula of the form

$$\bigvee_{i=1}^{I} \bigwedge_{j=1}^{J_i} (h_{ij}(z) \Delta_{ij} 0)$$

where $I \leq (md)^E$, $J_i \leq (md)^{E/n_0}$, $E = \prod_{k=0}^{\omega} O(n_k)$, the degree of h_{ij} is at most $(md)^{E/n_0}$ and the Δ_{ij} are standard relations $(\geq, >, =, \neq, <, \leq)$. If the coefficients of the original formula are integers of bit length at most L, the coefficients of the polynomials h_{ij} will be integers of bit length at most $(L + n_0)(md)^{E/n_0}$.

Results of Weispfenning [20], and Davenport and Heintz [9], show that the double exponential dependence on ω of the above bound on the degrees of the polynomials h_{ij} cannot be improved in the worst case.

Fisher and Rabin [10] proved an exponential worst-case lower bound for decision methods. However, the lower bound is exponential only in the number of quantifier alternations and is only singly exponential in that. A tremendous gap remains between the known upper and lower bounds for decision methods.

In closing this section we mention that work of Canny [4–6] has been especially influential in this area in recent years, both for the techniques he has developed and employed and for the connections he has established between the area and robotics. Work of Vorobjov [13] has also been very influential. Recent work of Pedersen [16] is especially relevant.

3. Discussion of ideas behind the algorithms

3.1. We now discuss some of the main ideas behind the algorithms with the best complexity bounds. Our discussion is rather loose. At times it is not quite correct, i.e., *some of the assertions that follow are simply not quite true.* However, the discussion is *close in spirit* to a rigorous development. Only by hiding technical details can we present a comprehensible discussion that can be read in a reasonable amount of time. The reader can find all of the details in [17].

3.2. We take as our starting point a fundamental result due to Ben-Or, Kozen, and Reif regarding univariate polynomials.

Let $g_1, \ldots, g_m : \mathbb{R} \to \mathbb{R}$ be real univariate polynomials. To each $x \in \mathbb{R}$ there is associated a sign vector $\sigma(x) \in \mathbb{R}^m$ whose i th coordinate is defined as follows:

$$\sigma_i(x) := \begin{cases} 1 & \text{if } g_i(x) > 0, \\ 0 & \text{if } g_i(x) = 0, \\ -1 & \text{if } g_i(x) < 0. \end{cases}$$

The sign vectors of $g_1, \ldots, g_m$ are the vectors $\sigma(x)$ obtained as x varies over $\mathbb{R}$. If each of the univariate polynomials g_i is of degree at most d, then the set of sign vectors has at most $2md + 1$ elements because there are at most md distinct roots and $md + 1$ open intervals defined by them.

The entire set of sign vectors of $g_1, \ldots, g_m$ can be constructed efficiently by algorithms based on Sturm sequence computations. In fact, they can be constructed quickly in parallel (i.e., in polylogarithmic time), as was shown by Ken-Or, Kozen, and Reif [1]. We take this fundamental result as a given and do not elaborate except to mention that the ideas developed in [1] can be found throughout the real decision method literature that followed its appearance. We refer to the algorithm of Ben-Or, Kozen, and Reif as the BKR algorithm.

A simple corollary is the fact that there exists an efficient decision method for sentences with exactly one variable (and hence one quantifier), i.e., an efficient decision method for univariate sentences in the *existential theory of the reals*. This is because it is trivial to determine if a sentence with one quantifier is true if one knows the sign vectors of the atomic predicate polynomials.

3.3. Our first step in designing general decision and quantifier elimination methods is to provide a method for efficiently computing all sign vectors of multivariate polynomials. The approach is to efficiently reduce the set of multivariate polynomials whose sign vectors are to be determined to a set of univariate polynomials with the same sign vectors; then the BKR algorithm can be applied.

There are various ways to efficiently reduce multivariate polynomials to appropriate univariate ones. One approach is based on the effective Nullstellensatz (Brownawell [3]). Such an approach was used by Heintz, Roy, and Solernó [14]. Another approach is based on so-called *u-resultants*. This is the approach that we discuss; it has led to the best complexity bounds.

For the moment we set aside the problem of efficiently computing the sign vectors for a set of multivariate polynomials and present a brief, but not quite accurate, discussion of *u*-resultants. Renegar [17] presents a completely accurate, but lengthy, discussion.

3.4. The *u*-resultant is an algebraic construction that allows one to reduce the problem of finding the zeros of a system of multivariate polynomials to the problem of finding the zeros of a univariate polynomial.

Let $f: \mathbb{R}^n \to \mathbb{R}^n$ be a system of n polynomials in n variables and let $F: \mathbb{R}^{n+1} \to \mathbb{R}^n$ denote the homogenization of f, i.e., the terms of the ith coordinate polynomial F_i of F are obtained by multiplying the terms of f_i by the appropriate power of an additional variable x_{n+1} so as to make all terms have the same degree as that of the highest degree term in f_i. Then $F^{-1}(0)$, the zero set of F, consists of lines through the origin.

Assume that $F^{-1}(0)$ consists of only finitely many such *zero lines*. If one knows all of these zero lines, then it is a trivial matter to compute all of the zeros of f. Thus, the problem of finding the zeros of f reduces to the problem of finding the zero lines of F.

From the coefficients of F, one can construct the *u*-resultant $R: \mathbb{R}^{n+1} \to \mathbb{R}$ of F. This is a single homogeneous polynomial with the same domain as F, i.e., $\mathbb{R}^{n+1}$. If F has infinitely many zero lines, then the *u*-resultant is identically zero. However, if F has only finitely many zero lines then the zero set of the *u*-resultant, $R^{-1}(0)$, consists of those hyperplanes in $\mathbb{R}^{n+1}$ that contain the origin and are orthogonal to the zero lines of F; so each zero line is associated with a hyperplane in $R^{-1}(0)$. It is this property that allows one to reduce the problem of finding the zero lines of F to the problem of finding the zeros of a univariate polynomial, as we now show.

Consider an arbitrary point x (other than the origin) on one of the hyperplanes of $R^{-1}(0)$. From calculus, the gradient of R at x, $\nabla R(x)$, is perpendicular to the hyperplane and, hence, lies in a zero line of F. Thus, if we could merely get a point on each of the hyperplanes of $R^{-1}(0)$, and then compute the gradient of R at those points we would have all of the zero lines of F. (We are glossing over subtleties like hyperplanes occurring with multiplicities, in which case the gradient will be the zero vector.)

The fact that the hyperplanes are n-dimensional makes finding points on them relatively easier than finding one-dimensional zero lines of F directly.

Consider a line $\{t\alpha + \beta \,;\, t \in \mathbb{R}\}$ that probes $\mathbb{R}^{n+1}$ in search of hyperplanes; here, α and β are fixed vectors in $\mathbb{R}^{n+1}$. Now consider the restriction of the u-resultant to this line, that is, consider the univariate polynomial $t \to R(t\alpha + \beta)$. Then t' is a zero of this univariate polynomial iff $t'\alpha + \beta$ is a point on one of the hyperplanes. Assuming, for simplicity, that the probe line is in general position relative to the hyperplanes, so that it intersects each of the hyperplanes, there is thus a corresponding zero of the univariate polynomial for each of the hyperplanes. (In a complete development one needs a method of choosing α and β to guarantee that the probe line is indeed in general position.)

We have reduced the problem of finding all of the zero lines of F to the problem of finding the zeros of a univariate polynomial, assuming F has only finitely many zero lines. In short, letting

$$p(t) := R(t\alpha + \beta) \qquad p : \mathbb{R} \to \mathbb{R},$$
$$q(t) := \nabla R(t\alpha + \beta) \qquad q : \mathbb{R} \to \mathbb{R}^{n+1},$$

where α and β are fixed vectors in $\mathbb{R}^{n+1}$, we have that $p(t') = 0$ if and only if $q(t')$ is a zero line of F.

Of course, one can carry out the construction of the polynomials p and q efficiently if one can construct the u-resultant efficiently. Constructing the u-resultant essentially amounts to computing the determinants of matrices. As it is well known how to compute determinants efficiently, the polynomials p and q can be computed efficiently. Again, we refer to [**17**] for details.

The u-resultant has been around for a long time, e.g., one can find it discussed in earlier editions of Van der Waerden [**19**] in the chapter on elimination theory. (Unfortunately, that chapter was eliminated in later editions, but there is rumor that it will be included again in a forthcoming edition.)

3.5. Now we simplify again, but not in a serious way. For our interests, the last section can be summarized as follows: $p(t') = 0$ iff $q(t')$ is a zero line of F, where F is the homogenization of $f : \mathbb{R}^n \to \mathbb{R}^n$. Rather than this, we will pretend the last section can be summarized in another way: $p(t') = 0$

iff $q(t')$ is a zero of f. In other words, we pretend the polynomials p and q given us the zeros of f directly rather than give us something that has to be dehomogenized. Our intent here is simply to make the following discussion easier to follow. We want to present the discussion in terms of the geometry of the atomic predicate polynomials occurring in a sentence or formula rather than in terms of the geometry of their homogenizations, but by doing so we are not quite honest (although we are close to being honest).

3.6. Now we show the relevance of u-resultants for the problem of computing the sign vectors of a set of multivariate polynomials. We begin by considering a more structured problem, namely, given a polynomial system $f : \mathbb{R}^n \to \mathbb{R}$ with only finitely many zeros, and given polynomials $g_1, \ldots, g_m :$ $\mathbb{R}^n \to \mathbb{R}$, compute the sign vectors of $g_1, \ldots, g_m$, which occur at the zeros of f.

The more structured problem is easily solved. Letting $p(t)$ and $q(t)$ be as in the last subsection, we can apply the BKR algorithm to compute all sign vectors of the univariate polynomials

$$p(t), g_1(q(t)), \ldots, g_m(q(t)).$$

We can then search through the sign vectors and find those whose first coordinate is zero. Clearly, the remaining coordinates of such a sign vector yield a sign vector for $g_1, \ldots, g_m$ at a zero of f; moreover, all sign vectors of $g_1, \ldots, g_m$ at zeros of f are obtained in this way.

This motivates an approach to computing *all* sign vectors of $g_1, \ldots, g_m$. First construct systems $f : \mathbb{R}^n \to \mathbb{R}^n$ at whose zeros all of the sign vectors of $g_1, \ldots, g_m$ occur, i.e., for each sign vector at least one of the systems has a zero at which that is indeed the sign vector. Then proceed as above.

3.7. We are now motivated to consider the following problem: Given polynomials $g_1, \ldots, g_m : \mathbb{R}^n \to \mathbb{R}$ construct systems $f : \mathbb{R}^n \to \mathbb{R}^n$ at whose zeros all of the sign vectors occur. We require each of the systems to have only finitely many zeros; otherwise, the polynomials $p(t)$ and $q(t)$ will be identically zero as the u-resultant will be identically zero; if $p(t)$ and $q(t)$ are identically zero, the preceding approach founders.

The polynomials g_i partition $\mathbb{R}^n$ into maximal connected components where two points are in the same connected component only if they have the same sign vector. This is shown schematically in Figure 1. The zero sets of the polynomials determine the boundaries of the connected components; in the figure the components are 0-, 1-, and 2-dimensional. We refer to the partition as the *connected sign partition for* $g_1, \ldots, g_m$.

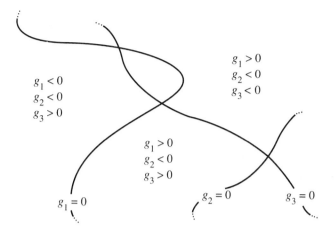

$g_1 > 0$
$g_2 < 0$
$g_3 < 0$

$g_1 < 0$
$g_2 < 0$
$g_3 > 0$

$g_1 > 0$
$g_2 < 0$
$g_3 > 0$

$g_1 = 0$

$g_2 = 0$

$g_3 = 0$

FIGURE 1

To construct systems $f: \mathbb{R}^n \to \mathbb{R}^n$ at whose zeros all of the sign vectors occur, it certainly suffices to construct systems $f: \mathbb{R}^n \to \mathbb{R}^n$ which have zeros in all of the connected components.

We again change focus slightly. Consider the following problem: Given polynomials $g_1, \ldots, g_m: \mathbb{R}^n \to \mathbb{R}$ construct systems $f: \mathbb{R}^n \to \mathbb{R}^n$ with zeros in the *closures* of all of the connected components. As is indicated in §3.8, for our purposes it actually suffices to solve this apparently easier problem. More specifically, given arbitrary polynomials $g_1, \ldots, g_m: \mathbb{R}^n \to \mathbb{R}^n$, it suffices to be able to compute nonvanishing pairs (p, q), $p: \mathbb{R} \to \mathbb{R}$, $q: \mathbb{R} \to \mathbb{R}^n$ with the property that for each component of the connected sign partition for $g_1, \ldots, g_m$, there exists t' for which one of the pairs (p, q) satisfies $p(t') = 0$ and $q(t')$ is in the closure of the component.

(The following discussion is somewhat technical and represents the hardest parts of [17]; the reader may wish to skip ahead to the §3.8. For the reader who does skip ahead, we remark that in the following discussion we construct pairs (p, q) with the property stated in the last sentence of the previous paragraph. The construction naturally leads to the following indexing of the pairs: $(\overline{p}_A, \overline{q}_A)$, where $A \subseteq \{1, \ldots, m\}$ and $\#A \leq n$, i.e., the number of elements in A does not exceed n.)

A naive solution to the problem of constructing systems with zeros in all of the closures can be obtained via calculus. Noting that the closure of each connected component has a point closest to the origin (i.e., a minimizer of $\sum_j x_j^2$ restricted to the closure) standard arguments regarding necessary conditions for optimality lead to the following: For each connected component there exists a subset $A \subseteq \{1, \ldots, m\}$ such that the system $f_A: \mathbb{R}^n \to \mathbb{R}^n$ defined by

$$(3.7.1) \qquad f_A(x) := \nabla[\det(M_A^T(x)M_A(x)) + \sum_{i \in A}(g_i(x))^2]$$

has a zero in the closure of the component; ∇ denotes *the gradient of* and $M_A(x)$ is the matrix whose rows are the polynomials $\nabla g_i(x)$, for $i \in A$, and the identity polynomial $x \mapsto x$. (We do not digress to discuss necessary conditions for optimality.)

There are, undoubtedly, numerous elementary ways to construct systems that have zeros in the closures of each of the connected components, each way having shortcomings for our purposes. The rather strange looking systems (3.7.1) we focus on have some especially useful properties. We motivate discussion of these useful properties by discussing the systems' shortcomings.

One glaring shortcoming of the systems f_A is simply that there are too many of them, i.e., 2^m. If for each system f_A we formed the corresponding polynomials $p_A(t)$ and $q_A(t)$ and applied the BKR algorithm to the univariate polynomials $g_1(q_A(t)), \ldots, g_m(q_A(t))$, we would end up with an algorithm for the existential theory of the reals requiring at least 2^m operations. However, the number of atomic predicate polynomials does not appear exponentially in the best complexity bounds for decision methods, so use of all of the systems f_A is certainly not the way to proceed.

Another shortcoming of the systems f_A is that for some polynomials $g_1, \ldots, g_m$ they may have infinitely many zeros. Consequently, the corresponding u-resultants and, hence, the corresponding polynomials p_A and q_A, may be identically zero, in which case we get no information from the sign vectors of the univariate polynomials $g_1(q_A(t)), \ldots, g_m(q_A(t))$, i.e., our approach flounders.

A way to circumvent these shortcomings is to slightly alter our approach to computing all of the sign vectors of $g_1, \ldots, g_m$, the alteration depending heavily on special properties of u-resultants and the systems f_A. Briefly, as these ideas will be developed more fully in the following paragraphs, we carefully perturb the polynomials $g_1, \ldots, g_m$, construct the exactly analogous systems f_A for the perturbed polynomials, construct the pairs (p_A, q_A) from these, and from these construct pairs $(\overline{p}_A, \overline{q}_A)$, which are something like limits of the pairs (p_A, q_A) as the perturbation goes to zero; the true limits of the pairs (p_A, q_A) may be identically zero, but the pairs $(\overline{p}_A, \overline{q}_A)$ will not be. The *limit* paris $(\overline{p}_A, \overline{q}_A)$ have the property that for each connected component of the sign partition of $g_1, \ldots, g_m$, there exists $A \subseteq \{1, \ldots, m\}$, $\#A \leq n$, such that for some t', $\overline{p}_A(t') = 0$ and $\overline{q}_A(t')$ is in the closure of the connected component. Note that this circumvents both of the previously mentioned shortcomings! Even though the systems f_A constructed directly from $g_1, \ldots, g_m$ may have infinitely many zeros so that we cannot construct useful pairs (p_A, q_A) from their u-resultants, we are still able to construct useful pairs indirectly! Moreover, rather than having 2^m pairs to contend with, we have $\binom{m}{n} = m^{O(n)}$ pairs to contend with, a major improvement.

Now we elaborate a bit. The perturbed polynomials we use are defined as follows:

$$g_i(x, \delta) := (1 - \delta)g_i(x) + \delta \left(1 + \sum_j i^j x_j^d\right)$$

where d is the maximal degree occurring among the unperturbed polynomials; the additional variable δ is a perturbation parameter; when $\delta = 0$ we
have $x \mapsto g_i(x, \delta) = g_i(x)$.

Viewing δ as fixed, consider the polynomials

$$x \mapsto g_1(x, \delta), \ldots, x \mapsto g_m(x, \delta).$$

Just as we defined the systems $f_A(x)$ from the polynomials $g_1, \ldots, g_m$, we
can define the systems $x \mapsto f_A(x, \delta)$; simply substitute $g_i(x, \delta)$ for $g_i(x)$
in (3.7.1), the gradients referring to derivatives only in the x-variables.

Continuing to view δ as fixed, we construct the pairs of polynomials $t \mapsto$
$p_A(t, \delta)$, $t \mapsto q_A(t, \delta)$ from the u-resultants of the systems $x \mapsto f_A(x, \delta)$.
The careful choice of the perturbed polynomials and the definition of the
systems $x \mapsto f_A(x, \delta)$ allow one to prove (something similar to) the following: for each sufficiently small $\delta > 0$ and for each component of the
connected sign partition of $x \mapsto g_1(x, \delta), \ldots, x \mapsto g_m(x, \delta)$ there exists
$A \subseteq \{1, \ldots, m\}$, $\#A \le n$, and there exists $t' \in \mathbb{R}$ such that $p_A(t') = 0$ and
$q_A(t')$ is in the closure of the connected component.

We would like the same conclusion to be true for $\delta = 0$. However, it may
not be true for $\delta = 0$ because the systems $x \mapsto f_A(x, 0)$ may have infinitely
many zeros causing the corresponding pairs $t \mapsto p_A(t, 0)$, $t \mapsto q_A(t, 0)$ to
be identically zero. This problem can be handled as follows.

Tracing through definitions (all of which can be found in [17]), one finds
the pairs $p_A(t, \delta)$, $q_A(t, \delta)$ to be polynomials in δ as well as in t. Expanding in powers of δ,

(3.7.2) $$p_A(t, \delta) = \sum_i \delta^i p_A^{[i]}(t),$$

(3.7.3) $$q_A(t, \delta) = \sum_i \delta^i q_A^{[i]}(t),$$

let $\overline{p}_A(t) := p_A^{[i']}(t)$ where the polynomial $p_A^{[i]}(t)$ is identically zero for all
$i < i'$, but not for i'; similarly, let $\overline{q}_A(t) := q_A^{[i'']}(t)$ where the polynomial
system $q_A^{[i]}(t)$ is identically zero for all $i < i''$, but nor for i''.

Relying on the definition of the perturbed polynomials, now considering the perturbation parameter $\delta > 0$ to tend to zero, one can prove that
(something very similar to) the following is true: For each component of
the sign partition of the original polynomials $g_1, \ldots, g_m$, there exists $A \subseteq$
$\{1, \ldots, m\}$, $\#A \le n$, and there exists t' such that $\overline{p}_A(t') = 0$ and $\overline{q}_A(t')$
is in the closure of the connected component.

The idea of combining u-resultants and perturbations can be found in Canny [6], which focuses on the problem of solving degenerate systems of polynomial equations $f(x) = 0$ where $f: \mathbb{C}^n \to \mathbb{C}^n$.

3.8. Recall that our real goal is to construct all of the sign vectors of $g_1, \ldots, g_m$. If in the conclusion of the next-to-last paragraph the words "the closure of" did not appear, we would be in a position to accomplish the goal: simply apply the BKR algorithm to the univariate polynomials $t \mapsto g_1(\overline{q}_A(t)), \ldots, t \mapsto g_m(\overline{q}_A(t))$ for all $A \subseteq \{1, \ldots, m\}$ satisfying $\#A \leq n$.

However, the words "the closure of" certainly do appear; but this problem is easily remedied. From the polynomials $g_1, \ldots, g_m: \mathbb{R}^n \to \mathbb{R}$ one can easily construct polynomials $\tilde{g}_1, \ldots, \tilde{g}_m: \mathbb{R}^{n+1} \to \mathbb{R}$ with the property that for each connected component C_1 in the sign partition of $g_1, \ldots, g_m$, there exists a connected component C_2 in the sign partition of $\tilde{g}_1, \ldots, \tilde{g}_{\tilde{m}}$ such that C_1 is the projection of the closure of C_2 onto $\mathbb{R}^n$. (We do not provide details here but, as an example, note that the positive real numbers are the projection of the closed semialgebraic set $\{(x, y); xy \geq 1, x \geq 0, y \geq 0\}$.) Consequently, if we construct the analogous pairs $\overline{p}_A: \mathbb{R} \to \mathbb{R}$, $\overline{q}_A: \mathbb{R} \to \mathbb{R}^{n+1}$ for $\tilde{g}_1, \ldots, \tilde{g}_{\tilde{m}}$, and discard the last coordinate polynomial of $\overline{q}_A$, we obtain appropriate pairs for $g_1, \ldots, g_m$. Henceforth, we use the notation $(\overline{p}_A, \overline{q}_A)$ to denote pairs for $g_1, \ldots, g_m$ obtained in this way, so we no longer need to refer to the closure of the components.

3.9. The pairs $(\overline{p}_A, \overline{q}_A)$ constructed for $g_1, \ldots, g_m$ have another property to be highlighted; the coefficients of the pairs are polynomials in the coefficients of $g_1, \ldots, g_m$. If one fixes m, n, and d and considers the vector space of tuples $g_1, \ldots, g_m$ of m polynomials $g_i: \mathbb{R}^n \to \mathbb{R}$ of degree at most d (identifying polynomials with their coefficients), one can replace the univariate polynomial pairs $(\overline{p}_A(t), \overline{q}_A(t))$ for fixed $g_1, \ldots, g_m$, with pairs $(\overline{p}_A(g_1, \ldots, g_m, t), q_A(g_1, \ldots, g_m, t))$ for variable $g_1, \ldots, g_m$; to obtain the corresponding univariate polynomials for particular $g_1, \ldots, g_m$, just substitute the coefficients of $g_1, \ldots, g_m$. In particular, *construction of the pairs can be accomplished by straight line programs*, with input being the coefficients of $g_1, \ldots, g_m$. This is very important in obtaining the best complexity bounds.

(To be more accurate, to obtain the pairs as straight line programs one needs to include all pairs $(p_A^{[i]}(t), q_A^{[i]}(t))$ where $p_A^{[i]}$ and $q_A^{[i]}$ are as in (3.7.2) and (3.7.3) rather than simply $\overline{p}_A = p_A^{[i']}$ and $\overline{q}_A = \overline{q}_A^{[i'']}$ although for simplicity we ignore this in what follows.)

Something similar to the above pairs are constructed in Heintz, Roy, and Solerno [14] by using the effective Nullstellensatz rather than u-resultants. However, the construction there is not by straight line programs.

3.10. To simplify what follows, we consider the following to be established: we can construct a single pair (rather than a reasonably small set of pairs) of polynomials $p(g_1, \ldots, g_m, t)$, $q(g_1, \ldots, g_m, t)$, with the property that for any fixed $g_1, \ldots, g_m : \mathbb{R}^n \to \mathbb{R}$ of degree at most d and any component of the connected sign partition of $g_1, \ldots, g_m$, there exists t' such that $p(g_1, \ldots, g_m, t') = 0$ and $q(g_1, \ldots, g_m, t')$ is in the component.

Whenever we write $p(t)$, $q(t)$, we are viewing $g_1, \ldots, g_m$ as fixed as will be clear from context.

3.11. We now have an efficient decision method for the existential theory of the reals, i.e., a method for deciding sentences with only one quantifier. Simply apply the BKR algorithm to the univariate polynomials

$$g_1(q(t)), \ldots, g_m(q(t))$$

where $g_1, \ldots, g_m$ are the atomic predicate polynomials in the sentence, thus constructing all sign vectors for $g_1, \ldots, g_m$. From these, the truth or falsity of the sentence is easily determined.

3.12. We are now in position to describe a quantifier elimination method, which works inductively.

First, assume we wish to obtain a quantifier-free formula equivalent to a formula $(\exists x)P(z, x)$ with a single quantifier; here, z and x are vectors of variables, z being the free variables. The variables z can be viewed as the coefficients of the atomic predicate polynomials; different values for z yield different sentences. Viewing z as fixed, thereby obtaining a sentence with a single quantifier, we could apply the decision method of §3.11 to determine if the sentence is true.

From this viewpoint, the decision method for the existential theory of the reals is a method for deciding if sentences corresponding to different values of z are true; input to the algorithm is simply specific values for z. *Unrolling* the algorithm, which requires a close examination of the BKR algorithm, one obtains an algebraic decision tree with input z as depicted in Figure 2.

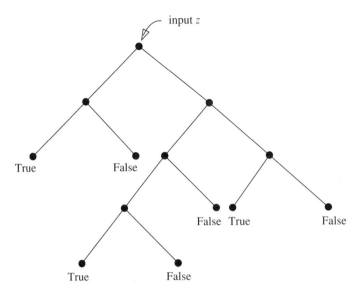

FIGURE 2

Each node of the tree is associated with a polynomial in z. Upon input z, the polynomial at the initial node is evaluated, and the value is compared to zero. Depending on the comparison, one branches to the left or to the right, thereby determining the next node and polynomial evaluation to be made, and so on until a leaf is reached.

Each leaf is assigned the value *True* or *False*. All inputs z, arriving at a leaf marked True, correspond to sentences that are true. Similarly, all inputs z, arriving at a leaf marked False, correspond to sentences that are false.

Here is a quantifier elimination method for the sentence: First, unroll the decision method to obtain an algebraic decision tree with input z. Next, determine all paths from the root that lead to leaves marked True. For each of these paths p, the decision method follows that path if and only if the corresponding sequence of polynomial inequalities and equalities are satisfied, i.e., z must satisfy the appropriate conjunction of polynomial equalities and inequalities

$$\bigwedge_{i \in I_p} (h_i(z) \Delta_i 0),$$

where I_p is an index set for path p and the Δ_i are standard relations ($\geq$, $>$, $=$, $\neq$, $<$, $\leq$). Taking the disjunction over all paths p leading to leaves marked True, one obtains a quantifier-free formula

$$\bigvee_p \bigwedge_{i \in I_p} (h_i(z) \Delta_i 0),$$

which is equivalent to $(\exists x) P(z, x)$.

3.13. The problem with this quantifier elimination method for formulae with a single quantifier is that there are too many paths in the tree; naively unrolling the decision method with input z leads to a bad complexity bound. However, this can be avoided with close examination of the BKR algorithm.

One can prove that most paths in the decision tree are not followed for any input z, although which paths are not followed is a function of the particular formula.

Rather than unroll the decision method naively, we can use the decision method to unroll itself. First, assuming the polynomial associated with the root node is $h_0(z)$ and one branches to the left if and only if $h_0(0) < 0$, we determine if the sentence $(\exists z)(h_0(z) < 0)$ is true. If it is false, we know there is no need to investigate paths beginning with the edge to the left of the initial node; so we can *prune* it (and all paths containing it). Continuing in the obvious manner, pruning as we unroll the tree, we do indeed obtain a fairly efficient quantifier elimination method for formulae with a single quantifier.

3.14. To obtain a general quantifier elimination method, we can simply proceed inductively on the blocks of variables in a formula

$$(Q_1 x_1) \cdots (Q_\omega x_\omega) P(z, x_1, \ldots, x_\omega).$$

We first eliminate the block x_ω from the formula

$$(Q_\omega x_\omega) P(z, x_1, \ldots, x_\omega)$$

to obtain an equivalent quantifier-free formula $P'(z, x_1, \ldots, x_{\omega-1})$. We then eliminate the block $x_{\omega-1}$ from the formula

$$(Q_{\omega-1} x_{\omega-1}) P'(z, x_1, \ldots, x_{\omega-1})$$

to obtain an equivalent quantifier-free formula $P''(z_1, x_1, \ldots, x_{\omega-2})$, and so on, until we have finally eliminated all of the quantified variables in the original formula.

This quantifier elimination method is fairly efficient. In terms of the notation introduced in §2, it readily yields a complexity bound with $E = O(n)^{O(\omega)}$, where n is the total number of variables in the formula. This is the bound obtained by Heintz, Roy, and Solernó [14] by a somewhat different approach relying on the effective Nullstellensatz, rather than u-resultants.

In the following sections, we describe further ideas that lead to a method with the best complexity bound, i.e., $E = \prod O(n_k)$. The speed-up is strongly dependent on the fact that the pairs (p, q) can be computed by straight-line programs as we discussed earlier.

3.15. Our decision method for the existential theory of the reals was essentially obtained by replacing the block of variables by a single variable, that is, the sentence $(\exists x)P(x)$ with atomic predicate polynomials $g_1(x), \ldots, g_m(x)$ was essentially reduced to the equivalent univariate sentence $(\exists t)P(t)$ obtained by substituting $g_i(q(t))$ for $g_i(x)$.

It is natural to attempt to extend this idea to sentences involving many quantifiers, reducing each block of variables to a single variable. We discuss a way to do this for sentences involving two quantifiers. The generalization to arbitrary sentences if fairly straightforward; complete details can be found in [17].

Consider a sentence $(\exists y)(\forall x)P(y, z)$ with atomic predicate polynomials $g_1(y, x), \ldots, g_m(y, x)$. To each fixed y, there corresponds the set of sign vectors of the polynomials in x,

$$x \mapsto g_1(y, x), \ldots, x \mapsto g_m(y, x).$$

Varying y, we obtain a family of sets of sign vectors.

It is easy to see that if we knew the entire family then we could decide if the sentence is true. With this as motivation, we proceed.

3.16. Consider the special case of a sentence with two quantifiers $(\exists y)(\forall x)$ $P(y, x)$, for which x consists of a single variable; to distinguish this special case, we write $(\exists y)(\forall t)P(y, t)$. Momentarily, we reduce a general sentence with two quantifiers to the special case.

For the special case, to each y there corresponds the sign vectors of the univariate polynomials $t \mapsto g_1(y, t), \ldots, t \mapsto g_m(y, t)$. We aim to construct the family of all such sets of sign vectors obtained as y varies over its domain.

The polynomials $g_1, \ldots, g_m$ partition (y, t)-space as is shown schematically in Figure 1, i.e., the connected sign partition of $g_1, \ldots, g_m$. When we focused on sentences with a single quantifier, we aimed at obtaining a point in each of the connected components of the partition. Now our goal is different.

For fixed y, the univariate polynomials $t \mapsto g_1(y, t), \ldots, t \mapsto g_m(y, t)$ are simply the restrictions of the polynomials $g_1, \ldots, g_m$ to the line through $(y, 0)$ that is parallel to the t-axis. It is easily seen that the sign vectors of the univariate polynomials are completely determined by which components of the connected sign partition of $g_1, \ldots, g_m$ the line passes through. Now viewing y as varying continuously, the set of sign vectors of the univariate polynomials $t \mapsto g_1(y, t), \ldots, t \mapsto g_m(y, t)$ can change only when y passes through one of the points indicated in Figure 3.

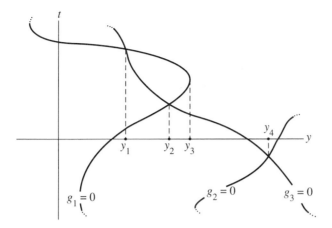

FIGURE 3

We thus see that y-space is partitioned into maximal connected components with the property that if y' and y'' are in the same component, then the set of sign vectors for the univariate polynomials $t \mapsto g_1(y', t), \ldots, t \mapsto g_m(y', t)$ is the same as for the univariate polynomials $t \mapsto g_1(y'', t), \ldots, t \mapsto g_m(y'', t)$. Hence, if for each component of this partition of y-space we could construct a point y' in it, then we could apply the BKR algorithm to the resulting sets of univariate polynomials $t \mapsto g_1(y', t), \ldots, t \mapsto g_m(y', t)$ to compute the entire family of sets of sign vectors that we desire, and from those determine if the sentence is true.

3.17. We now construct the partition of y-space, that is, we present a method for computing polynomials $h_1(y), \ldots, h_l(y)$ with the property that if y' and y'' are in the same component of the connected sign partition of $h_1, \ldots, h_l$, then the set of sign vectors for the univariate polynomials $t \mapsto g_1(y, t), \ldots, t \mapsto g_m(y, t)$ is the same for $y = y'$ and $y = y''$. The method figures prominently in Collins' work [7] and is very well known. (We over simplify the method to avoid technicalities, but our oversimplification is close in spirit to the true method.)

The breakpoints in Figure 3 fall into two categories. The first category consists of points that are the projection onto y-space of intersections of the form $\{(y, t); g_i(y, t) = 0\} \cap \{(y, t); g_j(y, t) = 0\}$ for some $i \neq j$. The points y_1, y_2, and y_4, in the figure, all fall into this category.

The second category consists of points that are the projection onto y-space of intersections of the form $\{(y, t); g_i(y, t) = 0\} \cap \{(y, t); \frac{d}{dt} g_i(y, t) = 0\}$. The point y_3 in the figure falls into this category.

A point y is in the first category only if the Sylvester resultant of the univariate polynomials $t \mapsto g_i(y, t)$ and $t \mapsto g_j(y, t)$ is zero at y for some $i \neq j$; the Sylvester resultant is a polynomial in the coefficients of the univariate polynomials, which here means that it is a polynomial in y.

Similarly, a point y is in the second category only if the Sylvester resultant of the univariate polynomials $t \mapsto g_i(y, t)$ and $t \mapsto \frac{d}{dt} g_i(y, t)$ is zero at y for some i.

Collecting the Sylvester resultant polynomials, we obtain the desired polynomials $h_1(y), \ldots, h_l(y)$, which appropriately partition y-space. (Again, to be honest, the construction of appropriate polynomials h_i is a bit more complicated.)

3.18. Polynomials analogous to the h_i can be readily obtained for general sentences $(\exists y)(\forall x) P(y, x)$ involving two quantifiers, where x is not required to consist of a single variable.

First, viewing y as the *coefficients* of the atomic predicate polynomials $g_1, \ldots, g_m$, from the polynomials $x \mapsto g_1(y, x), \ldots, x \mapsto g_m(y, x)$ we construct the $p(y, t)$, $q(y, t)$; for each component of the connected sign partition of $x \mapsto g_1(y, x), \ldots, x \mapsto g_m(y, x)$ there exists t' such that $p(y, t') = 0$ and $q(y, t')$ is in the component; the pair $p(y, t)$, $q(y, t)$ are polynomials in y as well as in t.

Then we apply the preceding Sylvester matrix determinant computations to the univariate polynomials

$$t \mapsto g_1(y, q(y, t)), \ldots, t \mapsto g_m(y, q(y, t)),$$

thereby obtaining polynomials $h_1(y), \ldots, h_l(y)$ with the following property: If y' and y'' are in the same component of the connected sign partition of $h_1, \ldots, h_l$, then the set of sign vectors for the polynomials $x \mapsto g_1(y, x), \ldots, x \mapsto g_m(y, x)$ is the same for $y = y'$ and $y = y''$.

This is where it is crucial in designing the most efficient decision and quantifier-elimination methods that the pair $p(y, t)$, $q(y, t)$ be constructed by straight-line program, that is, the same pair works for all y (actually, recalling we have suppressed the subscripts A, the same reasonably "small" set of pairs suffices for all y). The number l of polynomials $h_1, \ldots, h_l$ does not depend on the number of variables in y. Methods whose construction of similar pairs requires branching on the coefficients y, lead to polynomials $h_1, \ldots, h_l$ where l depends exponentially on the number of variables in y. If our l did depend exponentially on the number of variables in y, the approach we are now discussing would again lead to a complexity bound with $E = O(n)^{O(\omega)}$ rather than with $E = \prod_k O(n_k)$.

3.19. For an arbitrary formula $(\exists y)(\forall y) P(y, x)$, we have now described a method for constructing $h_1, \ldots, h_l$, which partition y-space in a useful way; if for each connected component of this partition we could construct a point y' in it, then we could apply the BKR algorithm to the resulting sets of univariate polynomials $t \mapsto g_1(y', q(y', t)), \ldots, t \mapsto g_m(y', q(y', t))$ to compute the entire family of sets of sign vectors that we desire, i.e., the family of sets of sign vectors for $x \mapsto g_1(y, x), \ldots, x \mapsto g_m(y, x)$ obtained as y varies.

Motivated by this, from the polynomials $h_1, \ldots, h_l$, we construct a corre-

sponding pair $\overline{p}(s)$, $\overline{q}(s)$; for each component of the connected sign partition of $h_1, \ldots, h_l$, there exists s' such that $\overline{p}(s') = 0$ and $\overline{q}(s')$ is in the
component. Moreover, we already know that for each y and each compo-
nent of the connected sign partition of $x \mapsto g_1(y, x), \ldots, x \mapsto g_m(y, x)$,
there exists t' such that $p(y, t') = 0$ and $q(y, t')$ is in the component.

Composing these facts, we have that for each set of sign vectors for polynomials of the form

$$x \mapsto g_1(y, x), \ldots, x \mapsto g_m(y, x),$$

there exists s' such that $\overline{p}(s') = 0$ and the set is identical to the set of sign
vectors for the polynomials

$$x \mapsto g_1(\overline{q}(s'), x), \ldots, x \mapsto g_m(\overline{q}(s'), x).$$

Moreover, for each sign vector in this set, there exists t' such that $p(\overline{q}(s'), t')$
$= 0$; and the sign vector is identical to that for the vector

$$g_1(\overline{q}(s'), q(\overline{q}(s'), t')), \ldots, g_m(\overline{q}(s'), q(\overline{q}(s'), t')),$$

i.e., has the same positive, zero, and negative coordinates.

3.20. In essence, we have now reduced the sentence $(\exists y)(\forall x)P(y, x)$ to
the bivariate sentence $(\exists s)(\forall t)P(s, t)$, obtained by replacing the atomic predicate polynomials $g_i(y, x)$ with the polynomials $g_i(\overline{q}(s), q(\overline{q}(s), t))$. Directly generalizing this approach one can obtain a procedure for replacing
each block of variables in an arbitrary sentence by a single variable. Then
simply applying the quantifier elimination method of §3.14 to the resulting
sentence in many fewer variables, one obtains a decision method with a complexity bound for which $E = \prod O(n_k)^{O(1)}$.

However, this approach is not good for quantifier elimination because the
block of free variables cannot be replaced by a single variable. Consequently,
only replacing each block of quantified variables by a single variable and then
applying the quantifier elimination method of §3.14, one obtains a complexity
bound with $E = (n_0)^{O(\omega)} \prod_{k=1}^{\omega} O(n_k)^{O(1)}$, where n_0 is the number of free
variables. Moreover, even for the decision problem alone, I see no way of
refining the above approach to obtain a bound with $E = \prod O(n_k)$, rather
than with $E = \prod O(n_k)^{O(1)}$.

3.21. The key to obtaining the most efficient decision and quantifier elimination methods is to also rely on the first polynomial in the pair $(\overline{p}, \overline{q})$;
note that it did not appear in the approach of the previous subsection.

The reliance is via Thom's lemma (cf. Coste and Roy [8]), which asserts
that if $p(t)$ is a univariate polynomial of degree d, then no two of the sign
vectors of the set of derivative polynomials $p, p', \ldots, p^{(d-1)}$ for which the
first coordinate is zero are identical, i.e., if $p(t') = 0 = p(t'')$ then the sign
of $p^{(i)}(t')$ differs from that of $p^{(i)}(t'')$ for some i. The point of this is that

the sign vector of the polynomial and its derivatives at a root provides a signature for the root that allows it to be distinguished from the other roots.

Again, consider an arbitrary sentence $(\exists y)(\forall x)P(y, x)$ with two quantifiers, and let (p, q), $(\overline{p}, \overline{q})$ be as before. Consider the set of sign vectors of the system of bivariate polynomials,

$$(3.20.1) \qquad \frac{d^i}{ds^i}\overline{p}(s) \qquad i = 0, \ldots, D,$$

$$g_i(\overline{q}(s), q(\overline{q}(s), t)) \qquad i = 1, \ldots, m,$$

where D is at least as large as the degree of $\overline{p}$. As we already know from earlier sections, the entire set of sign vectors for these polynomials can be computed efficiently.

We claim that from this set of sign vectors one can easily construct the family of sets of sign vectors for the polynomials $x \mapsto g_1(y, x), \ldots, x \mapsto g_m(y, x)$ obtained as y varies. To see this first recall two observations recorded in §3.19: (i) For each component of the connected sign partition of $h_1, \ldots, h_l$, there exists s' such that $\overline{p}(s') = 0$ and $\overline{q}(s')$ is in the component; (ii) For each y, and each component of the connected sign partition of $x \mapsto g_1(y, x), \ldots, x \mapsto g_m(y, x)$, there exists t' such that $p(y, t') = 0$ and $q(y, t')$ is in the component.

Proceed as follows. First, partition the set of sign vectors for the set of polynomials (3.20.1), according to the coordinates corresponding to the polynomials $\frac{d^i}{ds^i}\overline{p}(s)$, $i = 0, \ldots, D$, grouping those vectors together for which the coordinates are identical and discarding all vectors for which the first of the coordinates is not zero. By Thom's lemma, for each group there exists a root s' of $\overline{p}$ such that the sign vectors in the group are precisely those occurring at points of the form (s', t). Hence, by (i) and (ii) above, for each y there exists one of the groups such that sign vectors of $x \mapsto g_1(y, x), \ldots, x \mapsto g_m(y, x)$ are identical to the sign vectors obtained by truncating the sign vectors in the group, where the truncation leaves only the coordinates corresponding to the polynomials

$$g_1(\overline{q}(s), q(\overline{q}(s), t)), \ldots, g_m(\overline{q}(s), q(\overline{q}(s, t)).$$

Conversely, for each of the groups, the truncated sign vectors are precisely the sign vectors of

$$x \mapsto g_1(y, x), \ldots, x \mapsto g_m(y, x)$$

for some y.

We thereby obtain another approach to determining if a sentence with two quantifiers is true. This approach directly reduces the general decision problem to the problem of determining signs of multi-variate polynomials and allows us to avoid unrolling any algorithms.

3.22. The above procedure generalizes, in a straightforward manner to arbitrary sentences,

$$(Q_1 x_1) \cdots (Q_\omega x_\omega) P(x_1, \ldots, x_\omega).$$

First viewing $(x_1, \ldots, x_{\omega-1})$ as coefficients of the atomic predicate polynomials, one constructs a pair $(p^{[\omega-1]}, q^{[\omega-1]})$ and polynomials $h_i^{[\omega-1]}$, which partition $(x_1, \ldots, x_{\omega-1})$-space. Then, viewing $(x_1, \ldots, x_{\omega-2})$ as coefficients of the polynomials $h_i^{[\omega-1]}$, one constructs a pair $(p^{[\omega-2]}, q^{[\omega-2]})$ and polynomials $h_i^{[\omega-2]}$, which partition $(x_1, \ldots, x_{\omega-2})$-space, and so on.

Next, one computes the sign vectors of the polynomials

$$\frac{d^i}{dt_1^i} p^{[1]}(t_1) \qquad i = 0, \ldots, D$$

$$\frac{d^i}{dt_2^i} p^{[2]}(q^{[1]}(t_1), t_2) \qquad i = 0, \ldots, D$$

$$\vdots$$

$$\frac{d^i}{dt_{\omega-1}^i} p^{[\omega-1]}(q^{[1]}(t_1), q^{[2]}(q^{[1]}(t_1), t_2), \ldots,$$

$$q^{[\omega-2]}(q^{[1]}(t_1), q^{[2]}(q^{[1]}(t_1), t_2), \ldots, t_{\omega-2}), t_{\omega-1}) \qquad i = 0, \ldots, D$$

$$g_i(q^{[1]}(t_1), q^{[2]}(q^{[1]}(t_1), t_2), \ldots,$$

$$q^{[\omega-1]}(q^{[1]}(t_1), q^{[2]}(q^{[1]}(t_1), t_2), \ldots, t_{\omega-1}), t_\omega)) \qquad i = 1, \ldots, m.$$

These are the polynomials obtained by first substituting $q^{[\omega]}(x_1, \ldots, x_{\omega-1}, t_\omega)$ for x_ω in the polynomials $(d^i/dt_j)p(x_1, \ldots, x_{j-1}, t_j)$ and $g_i(x_1, \ldots, x_\omega)$. Then, one substitutes $q^{[\omega-1]}(x_1, \ldots, x_{\omega-2}, t_{\omega-2})$ for $x_{\omega-2}$ in the resulting polynomials, and so on. From these sign vectors, one is able to determine if the original sentence is true. (See [17] for the details.)

3.23. The principal advantage to the above decision method, in contrast to the one obtained in §3.20 by replacing each block of variables with a single variable and then applying the quantifier elimination method of §3.14, is that the above method does not require any algorithm to be unrolled into an algebraic decision true; the quantifier elimination method of §3.14 was obtained by repeated unrolling, once each time a block of quantified variables was eliminated.

To obtain a quantifier elimination method for an arbitrary formula

$$(Q_1 x_1) \cdots (Q_\omega x_\omega) P(z, x_1, \ldots, x_\omega)$$

with free variables z, we can unroll our new decision method into an algebraic decision tree in z, *pruning* branches (as in §3.13) as we unroll. Hence, to obtain a quantifier elimination method, we need only unroll once as opposed to once for each block of quantified variables. In this manner, one is led to a quantifier elimination method with complexity bound for which $E = \prod_{k=0}^{\omega} O(n_k)^{O(1)}$. With considerable attention to detail, the method can be refined to provide a complexity bound for which $E = \prod O(n_k)$, the best available; once again, we remark that the refinements are in [17].

REFERENCES

1. M. Ben-Or, D. Kozen and J. Reif, *The complexity of elementary algebra and geometry*, J. Comput. System Sci. **32** (1986), 251–264.
2. L. Blum, M. Shub and S. Smale, *On a theory of computation and complexity over the real numbers: NP-completeness, recursive functions and universal machines*, Bull. Amer. Math. Soc. (N.S.) **21** (1989), 1–46.
3. D. Brownawell, *Bounds for the degrees of the Nullstellensatz*, Ann. of Math. (2) **126** (1987), 577–591.
4. J. Canny, *The complexity of robot motion planning*, MIT Press, Cambridge, MA, 1989.
5. ____, *Some algebraic and geometric computations in PSPACE*, Proc. 20th Annual ACM Sympos. on the Theory of Computing, Association for Computing Machinery, NY, 1988, pp. 460–467.
6. ____, *Generalized characteristic polynomials*, J. Symbolic Comput. **9** (1990), 241–250.
7. G. Collins, *Quantifier elimination for real closed fields by cylindrical algebraic decomposition*, Lectures Notes in Comput. Sci., vol. 33, Springer-Verlag, New York, Berlin, and Heidelberg, 1975, pp. 515–532.
8. M. Coste and M. F. Roy, *Thom's lemma, the coding of real algebraic numbers and the computation of the topology of semi-algebraic sets*, J. Symbolic Comput. **5** (1988), 121–129.
9. J. Davenport and J. Heintz, *Real quantifier elimination is doubly exponential*, J. Symbolic Comput. **5** (1988), 29–35.
10. M. Fischer and M. Rabin, *Super-exponential complexity of Presburger arithmetic*, Complexity of Computations, SIAM-AMS Proc., vol. 7, Amer. Math. Soc., Providence, RI, 1974, pp. 27–41.
11. N. Fitchas, A. Galligo and J. Morgenstern, *Algorithmes rapides en sequentiel et en parallel pour l'élimination de quantificateurs en geométrie élémtaire*, Séminaire Structures Ordonnées, U. E. R. de Math. Univ. Pairs VII, 1987.
12. D. Yu. Grigor'ev, *The complexity of deciding Tarski algebra*, J. Symbolic Comput. **5** (1988), 65–108.
13. D. Yu. Grigor'ev and N. N. Vorobjov, Jr., *Solving systems of polynomial inequalities in subexponential time*, J. Symbolic Comput. **5** (1988), 37–64.
14. J. Heintz, M.-F. Roy and P,. Solernó, *Sur la complexite du principe de Tarski-Seidenberg*, Bull. Soc. Math. France **118** (1990), 101–126.
15. A. C. Neff, *Specified precision polynomial root isolation is in NC*, J. Comput. Systems Sci. (to appear).
16. P. Pedersen, *Counting real zeros*, Ph. D. Thesis, Courant Inst. of Math. Sci., New York Univ., 1991.
17. J. Renegar, *On the computational complexity and geometry of the first-order theory of the reals*, J. Symbolic Comput. (to appear).
18. A. Tarski, *A decision method for elementary algebra and geometry*, Univ. of California Press, Berkeley, CA, 1951.
19. B. L. Van der Waerden, *Modern algebra*, Vol. II, Ungar, 1950.
20. V. Weispfenning, *The complexity of linear problems in fields*, J. Symbolic Comput. **5** (1988), 3–27.

SCHOOL OF OPERATIONS RESEARCH AND INDUSTRIAL ENGINEERING, COLLEGE OF ENGINEERING, CORNELL UNIVERSITY, ITHACA, NEW YORK 14853-3801

DIMACS Series in Discrete Mathematics
and Theoretical Computer Science
Volume **6**, 1991

Sweeping Arrangements of Curves

JACK SNOEYINK AND JOHN HERSHBERGER

ABSTRACT. We consider arrangements of curves that intersect pairwise in at most k points. We show that a curve can sweep any such arrangement and maintain the k-intersection property if and only if k equals 1 or 2. We apply this result to an eclectic set of problems: finding Boolean formulae for polygons with curved edges, counting triangles and digons in arrangements of pseudocircles, and finding extension curves for arrangements. We also discuss implementing the sweep.

1. Introduction

When is it important that lines in an arrangement are straight, or that circles in an arrangement are circular? Of course, it depends on the questions one asks. Many questions, such as "What is the minimum number of triangles in an arrangement of lines," do not depend on straightness, but on the fact that two lines intersect in a single point [25]. One can answer such questions for more general arrangements of pseudolines. Branko Grünbaum, in a lecture entitled "The importance of being straight," points out that because there are arrangements of pseudolines that cannot be stretched to lines, there are questions in which straightness is crucial. He says that we cannot yet answer these because "most of our tools and methods are general (or vague and imprecise?) enough to apply to the case of pseudolines" [20].

In this paper, we look at sweeping arrangements of curves with intersection restrictions. Before proceeding, let us define *curves*, *arrangements*, and *sweeping*, and look at reasons to study these objects.

The curves that we consider in this paper lie in the Euclidean plane or on the sphere, are smooth, have no self-intersections, and are endless (either closed or bi-infinite). Any two curves intersect in a finite number of points, at which they cross.

A set of curves Γ has the *k-intersection property* if every two of them intersect in at most k points. If any two curves of Γ intersect in exactly k points, then Γ has the *exact k-intersection property*.

1991 *Mathematics Subject Classification.* Primary 51H15; Secondary 57N05, 68U05.

This topological or combinatorial restriction on the intersection of curves is different from the restrictions used in the field of computer graphics. Computer-aided design systems usually place algebraic restrictions on curves; for example, they may require that all curves be lines, conic sections, or cubic plane curves (the components of cubic splines). Natural families of algebraic curves satisfy the k-intersection property for some k, however: lines have the one-intersection property, vertical parabolas are two-intersecting curves, general conics are four-intersecting curves, and cubic plane curves are nine-intersecting. The topological restriction is more general in the sense that if a property holds or an algorithm works for k-intersecting curves, then it will apply to any family of algebraic curves with the k-intersection property. (Unfortunately, our positive results extend only to curves with the two-intersection property; we cannot say much about cubic plane curves.)

When calculating arrangements using finite precision arithmetic, one usually cannot preserve straightness. Greene and Yao [18] and Milenkovic [28] have given algorithms that do preserve the one-intersection property for line arrangements.

Finally, Sharir and other researchers in computational geometry have used intersection restrictions in applying the theory of Davenport-Schinzel sequences to curves [1, 11, 23, 33, 34]. This theory has been an important tool in the analysis of algorithms that deal with algebraic curves.

A finite set of curves Γ partitions a surface into three types of maximal connected regions: *vertices* are regions that are contained in two or more curves of Γ, *edges* are contained in only one curve, and *faces* are connected regions contained in no curves of Γ. We call this partition the *arrangement* of Γ. An arrangement is *simple* if no three curves share a common point. We deal primarily with simple arrangements in this paper; we will note where our statements apply to nonsimple arrangements. Figure 1 illustrates these definitions.

The names given to the sets of the partition suggest that the arrangement of Γ is a graph embedded in the plane. It makes an unusual graph—some edges are infinite rays, and many edges can connect a pair of vertices. If a curve of Γ does not intersect any other curve, we have an edge with no vertices. With these differences in mind, however, it should not cause confusion to think of the arrangment as a planar graph.

In the past, arrangements of lines and planes were studied in recreational mathematics and because of their relationships to configurations of points and to certain convex polytopes. Grünbaum collected many results and conjectures on arrangements of lines and curves in the plane in his 1972 monograph [22]. Other early results are contained in [19] and [21]. Grünbaum's terminology differs slightly from ours. He discusses "arrangements of pseudolines," "arrangements of curves," and "weak arrangements of curves"; we call them arrangements of curves with the exact one-, exact two-, and two-intersection properties, respectively. We shall call one-intersecting curves

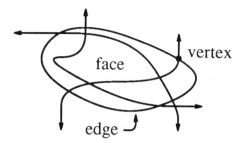

FIGURE 1. A simple arrangement

pseudolines and two-intersecting curves *pseudocircles*. This means that two pseudolines (or two pseudocircles) are not required to intersect in one point (or two points).

Researchers in computational geometry have found numerous applications for line arrangements in algorithms for geometry and graphics (see [7]). They have also considered arrangements of curves with intersection conditions. Edelsbrunner et al. [8] apply Davenport-Schinzel sequences to prove generalizations of the horizon theorem for arrangements of lines [5, 10] and to construct such arrangements incrementally in nearly quadratic time. McKenna and O'Rourke [27] independently proved and used the horizon theorem for the case of pseudocircles.

Sweeping is important both as a paradigm for developing graphical and geometric algorithms and as tool for use in mathematical proofs. The underlying idea is to determine properties of a collection of objects in a space of dimension d by looking at a series of consecutive $(d-1)$-dimensional slices. Sweeping converts a static problem into a dynamic problem of lower dimension.

As examples of sweep algorithms in the literature, consider the problem of finding the intersections of n lines or segments in the plane. Shamos and Hoey [32] showed how to detect an intersection in $O(n \log n)$ time by sweeping the plane with a line. Bentley and Ottman [2] extended their ideas and developed a practical algorithm to report all K intersections between n segments in $O((n + K) \log n)$ time. Mairson and Stolfi [26] used a sweeping line to report all K intersections between two sets of segments, each of which has no self-intersections, in $O(n \log n + K)$ time. They also applied their techniques to segments of x-monotone curves.

If a straight sweeps the plane to find the intersection points of a set of segments, then it encounters the points in sorted order in the direction perpendicular to the sweep. In such a sweep, the logarithmic factors in the time complexity seem inescapable. For infinite lines, Edelsbrunner and Guibas [9] showed that one can avoid the logarithmic factor by sweeping the plane with a pseudoline. Their algorithm runs in optimal $O(n^2)$ time. Chazelle and Edelsbrunner [4] recently developed an algorithm to report all K segment intersections in $O(n \log n + K)$ time, which also sweeps the plane with a pseudoline.

In the next section, we will define precisely what it means to sweep arrangements of curves that have the k-intersection property, but let us first think about what condition we would want on the sweeping curve c. In designing algorithms, we would like the curves of our arrangement to have few intersections with c, since we must keep track of each intersection. In proofs dealing with a family of k-intersecting curves, we would like c to fit into the family.

In §3 we prove the sweeping theorem, which says that we can always sweep an arrangement of pseudolines or pseudocircles starting from any curve in the arrangement, and cannot always sweep k-intersecting curves for $k \geq 3$.

THEOREM 3.1 (Sweeping theorem). *Let Γ be a finite set of bi-infinite curves (and closed curves, if $k > 1$) in the plane or sphere that have the k-intersection property. Let c be a curve of Γ. If $k \in \{1, 2\}$, then we can sweep Γ starting with c and maintain the k-intersection property. If $k > 2$, then arrangements exist that cannot be swept.*

In the next section, we define local operations by which a sweeping curve can advance. For pseudolines $(k = 1)$ and pseudocircles $(k = 2)$ we can always apply one of the local operations. For $k > 2$ we indicate how to construct an arrangement that cannot be swept by a given curve.

In §4 we present two applications of the sweep theorem. We use the pseudoline case to extend the work of Dobkin et al. [6]; we find a short Boolean formula to describe a polygon with curved edges. We use the pseudocircle case to find a relationship between the minimum number of digons and triangles in an arrangement of 2-intersecting curves. This is related to Grünbaum's conjecture 3.7 [22].

Section 5 defines another type of sweep in the plane: sweeping a double wedge. We use this sweep to prove an extension theorem for arrangements that includes Levi's lemma [25] as a special case.

THEOREM 5.1 (Extension theorem). *Let Γ be a finite set of k-intersecting curves and P be a set of $k+1$ points, not all on the same curve. If $k \in \{1, 2\}$, then there is a curve that contains the points of P and has the k-intersection property with respect to Γ. If $k > 2$, then arrangements and point sets exist such that any curve through the points violates the k-intersection property.*

Section 6 points out that we can easily implement the local operations used in the proof of the sweeping theorem when we know the arrangement. When we do not, we can apply the ideas of Edelsbrunner and Guibas [9] to sweep a set of pseudolines using linear space. We leave sweeping pseudocircles as an open problem.

The operations that we define in §2 can also sweep arrangements of $(k > 2)$-intersecting curves if we do not require the sweep to have the k-intersection property. But is there some intersection property that is maintained by such a sweep? For algorithmic purposes, it would be satisfactory to have a function of k (and not of n) bound the number of points of intersection the sweep has with any curve.

2. Definitions for sweeping

In the beginning of this section, we define sweeping as a continuous process; later we will see that we can carry it out in discrete steps. We also define notation that we use in the proofs of §3.

Let c be an endless curve in the sphere or Euclidean plane with an orientation, which we depict as left to right. Since c is smooth and has no self-intersections, it divides the plane or sphere into two connected components: one with boundary oriented clockwise, the other counterclockwise. We say that the component with the counterclockwise orientation is *above* c, and the other component is *below*. A curve γ *lies above* c if the component above c contains γ. When we sweep, we handle these two components separately.

To sweep the component of the plane or sphere above c, we want to move c continuously to infinity if the component is unbounded, or to a point if it is bounded. Formally, we say that a family of curves indexed by the nonnegative reals, $\mathscr{C} = \{c_\alpha\}_{\alpha \geq 0}$, *covers the component above c* if

- the curve $c_0 = c$,
- every point above c (except one, if the component above c is bounded) lies on exactly one curve c_α, and
- the curve c_β is above c_α whenever $\beta > \alpha$.

To say that we can *sweep the component above c* means that a family of curves $\mathscr{C}$ exists that covers the component above c.

Suppose c is a member of a set of k-intersecting curves, Γ. We can *sweep the arrangement of Γ above c* if there is a family of curves $\mathscr{C}$ that covers the component above c, and the set $\Gamma \cup \mathscr{C}$ has the k-intersection property. Finally, we say that we can *sweep Γ starting from c* if we can sweep above c and below c.

Though we employ this continuous definition to prove negative results in §3.3, it is unwieldy for proving positive results. In the literature, sweeping is investigated as a discrete process by identifying events where the intersection between the sweep and the plane changes significantly. One can usually interpolate between two consecutive curves of a discrete sweep in a natural way to obtain a continuous family that covers the plane or sphere.

In our case, the oriented sweeping curve c intersects the other curves of Γ in some order; this order changes when c passes a vertex of the arrangement or changes the number of its intersections with some curve. If we choose these changes as our events and find the curves of the sweep where the events occur, then we obtain a discrete set of curves. We can interpolate between two consecutive curves with the help of the Schönflies theorem [31] as follows: Map adjacent curves c and c' to parallel lines. Map the curve segments between c and c' to nonintersecting line segments in the strip between the lines. The Schönflies theorem implies that we can carry out this mapping by a homeomorphism of the plane or sphere to itself. Under the inverse mapping, the parallel lines in the strip become a family of curves that interpolate between c and c' and intersect the curves of the arrangement in a consistent order. By interpolating between each pair of adjacent curves, we can extend the curves from a discrete sweep to a continuous family that covers the plane.

We will define local operations that allow us to advance the sweeping curve past an intersection point of two curves and to add or remove curves from the set of curves intersected by the sweep, without violating the k-intersection property. But first, let us give names to some of the things that appear along the sweeping curve.

We can represent the portion of the arrangement Γ that lies on and above the sweep c as a graph $G = \langle V, E \rangle$. (We assume that $c \in \Gamma$.) Let A denote the component of the plane or sphere on and above c. The vertices V of G are the points in A where two curves intersect. The edges E are the edges of the arrangement that are contained in A. To avoid having edges that are not adjacent to vertices, we introduce an artificial vertex on any curve that has no intersection points. We denote the edges in E by $\langle u, v \rangle$, where $u, v \in V \cup \{\infty\}$—ambiguity caused by more than one edge having the same endpoints will be resolved in context. An edge $\langle u, \infty \rangle$ is an infinite ray.

Assume that we are sweeping above c. A point p on a curve γ is *visible from an edge e* of c if there is a face above c whose boundary contains both p and the edge e. In particular, p *is visible from c* if p lies on a face immediately above c.

Suppose that m of the vertices of G appear on the sweeping curve c. We number these vertices with the integers $1, \ldots, m$ in order of their appearance along c. In this and the next section, we use the names i, j, k, and l for vertices on c, and the names u, v, and w for vertices above the sweep.

Figure 2 illustrates some other notation that we will use. We define a function $\gamma()$ from vertices on c to curves that cause them: if the intersection of the curve $\alpha \in \Gamma$ with c forms the vertex i, then we define $\gamma(i) = \alpha$. Some edges of the graph G lie on the curve c—we give these edges special status and denote them with square brackets, for example, $[j, j+1]$, instead of angle brackets. Thus, if we write $\langle i, j \rangle$, we mean the edge from vertex i to vertex j along the curve $\gamma(i) = \gamma(j)$ and not an edge along the curve c.

We also define terms to describe some special configurations in G. A *hump* is an edge $\langle i, j \rangle$ of G with both i and j on the sweeping curve c. The hump $\langle i, j \rangle$ is *empty* if it contains no other curves. Clearly, for an empty hump, $|j - i| = 1$. The hump pictured is not empty. A *triangle* is a pair of edges, $\langle k, v \rangle$ and $\langle l, v \rangle$, with two distinct vertices k and l, both

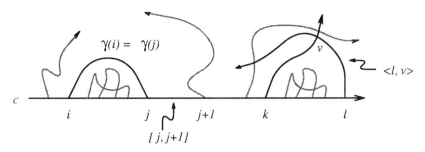

FIGURE 2. A hump $\langle i, j \rangle$ and a triangle $\Delta(k, v, l)$ in G.

on the sweep and one vertex v above the sweep—we denote this triangle by $\Delta(k, v, l)$. The triangle $\Delta(k, v, l)$ is *empty* if it contains no other curves. For an empty triangle, $|l - k| = 1$.

With this notation, we can now define the local operations that move c forward while preserving the k-intersection property. The operations, shown in Figure 3, are

1. *Taking a loop.* Let γ be a curve that intersects the sweep in at most $k - 2$ points. If an edge e of γ is visible from an edge $[i, i+1]$ of c, then there is a path from a point on e to a point on $[i, i+1]$ that intersects no other curves. The sweep c can advance along this path and intersect the edge e in two places without adding any other intersections—preserving the k-intersection property.

2. *Passing an empty triangle.* If $\Delta(i, v, i+1)$ is an empty triangle, then c can move past the vertex v, interchanging the order of $\gamma(i)$ and $\gamma(i+1)$ along the sweep.

3. *Passing an empty hump.* If the edge $\langle i, i+1 \rangle$ is an empty hump, then

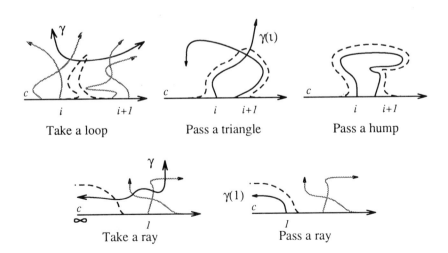

Take a loop   Pass a triangle   Pass a hump

Take a ray   Pass a ray

FIGURE 3. Operations by which the sweep progresses

c can advance past it and reduce the number of intersections with $\gamma(i)$ by two.

4. *Taking the first ray.* If a bi-infinite curve γ intersects the sweep in fewer than $k - 1$ points and is visible as an infinite ray from edge $[\infty, 1]$ of c, then c can move forward and introduce one intersection point with this ray; γ becomes $\gamma(1)$. We can define *taking the last ray* in a similar fashion.

5. *Passing the first ray.* If the leftmost curve $\gamma(1)$ is an infinite ray with no intersections above the sweep c, then c can move past the ray and lose one intersection point with $\gamma(1)$.

If we can apply an operation, we say the sweep can make progress. A discrete sweep terminates when either there are no curves above the sweep or the only curves above the sweep are nonintersecting rays. In these situations, one can continuously sweep above c and maintain the k-intersection property without difficulty.

In a nonsimple arrangement, we may encounter a number of triangles with a common vertex v above the sweep. We can handle this either by passing all the triangles simultaneously or by perturbing the curves (either actually or conceptually) to form a simple arrangement. Since our theorems will not restrict how the operations are applied, either method works.

3. The sweeping theorem

Using the definition of sweeping from the previous section, we now prove

THEOREM 3.1 (Sweeping theorem). *Let Γ be a finite set of bi-infinite curves (and closed curves, if $k > 1$) in the plane or sphere that have the k-intersection property. Let c be a curve of Γ. If $k \in \{1, 2\}$, then we can sweep Γ starting with c and maintain the k-intersection property. If $k > 2$, then arrangements exist that cannot be swept.*

We establish the pseudoline and pseudocircle cases, $k = 1$ and $k = 2$, by showing that certain local operations can always make progress. In §3.3 we present some unsweepable arrangements with $k > 2$.

3.1. Sweeping pseudolines. To prove the pseudoline case of the sweep theorem, we show that the sweep can always make progress using three of the operations defined in §2.

LEMMA 3.1. *Any arrangement of bi-infinite curves Γ having the one-intersection property can be swept starting with any curve $\gamma \in \Gamma$ using three operations: passing a triangle, passing the first ray, and taking the first ray.*

PROOF. We will see that the sweep can either pass the first ray, take the first ray, or pass some triangle.

Consider the edge $\langle 1, u \rangle$ of the curve $\gamma(1)$ that intersects the sweep first. If $u = \infty$, then we can pass $\gamma(1)$. Otherwise the vertex u comes from the intersection of a curve α_1 with $\gamma(1)$.

Suppose that the curve α_1 does not intersect the sweep and that the curves $\{c, \gamma(1), \alpha_1, \dots, \alpha_m\}$ border an unbounded face in counterclockwise order, as illustrated in Figure 4. We now show by induction that α_m does not intersect the sweep.

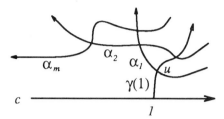

FIGURE 4. α_1 does not intersect the sweep.

Assume that these curves are oriented in the counterclockwise direction. Then the curve α_{i-1} divides α_i into two pieces, one to the left and one to the right. If α_{i-1} does not intersect the sweeping curve c, then it separates the right piece of α_i from c. The left piece of α_i must remain in the region bounded by $\{c, \gamma(1), \alpha_1, \ldots, \alpha_{i-1}\}$; since vertex 1 is the first vertex on c, the left piece also cannot intersect c. Therefore, if α_1 does not intersect c, then neither does α_m, and c can take the first ray.

Otherwise, α_1 intersects the sweep at a point l.

Let us call a triple (i, v, j) a *half-triangle* if $\langle i, v \rangle$ is an edge and $\gamma(i)$ and $\gamma(j)$ intersect at v. No curve crosses the edge $\langle i, v \rangle$ of (i, v, j), so any curve visible in the interval (i, j) crosses c between i and j and $\gamma(j)$ between v and as illustrated in Figure 5. By induction on the size of the interval (i, j), we show that every half-triangle contains an empty triangle.

In the base case, $|i - j| = 1$ and $\Delta(i, v, j)$ is an empty triangle. Otherwise a curve β intersects $\gamma(j)$ at w such that $\langle j, w \rangle$ is an edge. Since β is visible from j, it intersects c at $k \in (i, j)$. But now (j, w, k) is a smaller half-triangle, which contains an empty triangle by the induction hypothesis.

Since α_1 intersects the sweep at l, the triple $(1, u, l)$ is a half triangle. The sweep can make progress by passing the empty triangle contained in $(1, u, l)$. This establishes the lemma.   □

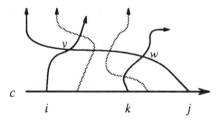

FIGURE 5. An empty triangle in a half-triangle.

3.2. Sweeping pseudocircles. To establish the pseudocircle $(k = 2)$ case of the sweep theorem, we use operations that change an even number of intersections. Once, again, we show that the sweep can always advance.

LEMMA 3.2. *Any arrangement of curves* Γ *with the two-intersection prop-erty can be swept starting from any curve* $\gamma \in \Gamma$ *by using three operations: passing a triangle, passing a hump, and taking a loop.*

We first prove that a bi-infinite curve can sweep an arrangement of two-intersecting curves in the Euclidean plane. We assume that a counterexample exists and then derive a contradiction. At the end of this section, we show that a closed sweeping curve can sweep a sphere or the Euclidean plane.

For each arrangement in the plane with a bi-infinite sweeping curve, we can form a graph G of the portion on or above the sweep. We define the *size* of an arrangement with a sweeping curve to be the pair $(|V|, n)$; that is, the number of vertices on and above the sweep followed by the number of curves. Now, choose a set of two-intersecting curves Γ, including an oriented bi-infinite sweeping curve c, such that c cannot make progress and the size of the arrangement is lexicographically minimum. We prove a sequence of lemmas about the structure of this arrangement and wind up with a contradiction. This contradiction implies that a bi-infinite sweeping curve can always sweep an arrangement of two-intersecting curves.

LEMMA 3.3. *Removing any curve* γ *from the arrangement* Γ *allows the sweep to progress.*

PROOF. By minimality of the arrangement. □

LEMMA 3.4. *All curves in* Γ *intersect the sweeping curve c.*

PROOF. Since the sweep cannot make progress by taking a new curve, any curve γ that does not intersect c is not visible from c. But removing such a curve γ does not change c's ability to make progress. Therefore, all curves in the smallest counterexample intersect the sweeping curve c. □

LEMMA 3.5. *We can choose a minimum counterexample such that all of the curves of* Γ *are rays to infinity below the sweep c.*

PROOF. Map the bi-infinite sweeping curve c to a line c' by a contin-uous mapping of the plane onto itself, cut all curves below c' and extend them perpendicularly to infinity, then apply the inverse mapping. When this procedure is applied to a minimum counterexample, neither the arrangement above the sweep nor the size is affected. Thus, it remains a minimum coun-terexample. □

From now on, we assume that our unsweepable arrangement, Γ, is chosen in accordance with Lemma 3.5.

LEMMA 3.6. *The arrangement* Γ *cannot contain a hump or a triangle.*

PROOF. The arrangement certainly cannot contain an empty triangle or hump; otherwise, the sweep could make progress. Suppose, however, that Γ contains vertices i and j on the sweep, with $i < j$, which are the ends of a hump or a triangle. Since none of the curves that intersect the sweep between $i + 1$ and $j - 1$ intersect the hump or triangle, the smaller arrangement $\Gamma' = \{\gamma(i + 1), \gamma(i + 2), \ldots, \gamma(j - 1)\}$ lies completely inside the hump or triangle. But, by minimality, the sweep can make progress in the arrangement Γ'. This contradicts the fact that Γ is unsweepable and establishes the lemma.   □

Let us pause a moment and consider what these first four lemmas tell us. We have an arrangement Γ in which the sweep cannot progress. If we remove any curve γ, then some operation applies. Lemmas 3.4 and 3.6 tell us that, after removing γ, there must be an empty hump or triangle where there was none before. We also have a notational convenience from Lemma 3.4; every curve in Γ can be denoted $\gamma(i)$ for some vertex i on the sweep.

Next, we consider configurations in which removing a curve leaves a hump or triangle. By finding contradictions to Lemma 3.6 or to the minimality of Γ, we can show that most configurations cannot occur.

LEMMA 3.7. *For any vertex j on the sweep, the removal of the curve $\gamma(j)$ cannot leave an empty hump or empty triangle with vertices i and k satisfying $i < j < k$.*

PROOF. By case analysis; we sketch the idea: Lemma 3.6 implies that the curve $\gamma(j)$ intersects the triangle or hump that is formed if $\gamma(j)$ is removed. That intersection point forms a new empty triangle. To break up the new triangle, $\gamma(j)$ must reenter it—but in doing so, the curve $\gamma(j)$ creates another triangle that cannot be avoided.   □

LEMMA 3.8. *The removal of a curve $\gamma(k)$ cannot leave a hump.*

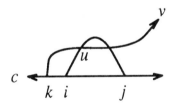

FIGURE 6. $\gamma(k)$ cannot leave a hump.

PROOF. Suppose that the arrangement of $\Gamma - \{\gamma(k)\}$ contains the hump $\langle i, j \rangle$. Consider the arrangement Γ' with only this hump removed: $\Gamma' = \Gamma - \{\gamma(i)\}$. By Lemma 3.3, the sweep can make progress in Γ'. But $\gamma(i)$ intersects only $\gamma(k)$; therefore, in Γ', the curve $\gamma(k)$ contributes an edge to an empty hump or triangle—let $\langle k, v \rangle$ be that edge. In the original arrangement Γ, however, $\gamma(i)$ intersects $\gamma(k)$ at some vertex u between k and v as shown in Figure 6. Thus, $\Delta(i, u, k)$ is a triangle in Γ, contradicting Lemma 3.6.   □

Because the removal of a curve γ does not allow us to move past a hump (Lemma 3.8) or take an edge (Lemma 3.4), it must leave an empty triangle in the arrangement $\Gamma - \{\gamma\}$, which we denote $\Delta(i, v, j)$. Since γ cannot intersect the sweep between i and j (Lemma 3.7), we can assume that $j = i + 1$. The next lemma restricts the way γ can break up this triangle, $\Delta(i, v, i + 1)$.

LEMMA 3.9. *Suppose that the arrangement $\Gamma - \{\gamma\}$ has an empty triangle $\Delta(i, v, i + 1)$. Then, in the original arrangement Γ, the face above the edge $[i, i + 1]$ has v as a vertex.*

PROOF. Suppose that v is not a vertex of the face above the edge $[i, i+1]$. Then the curve γ must *hide* v from that portion of the sweep by intersecting both legs, $\langle i, v \rangle$ and $\langle i + 1, v \rangle$, of the triangle $\Delta(i, v, i + 1)$. Lemma 3.7 implies that γ intersects both legs once or both legs twice; we consider these possibilities in two separate cases.

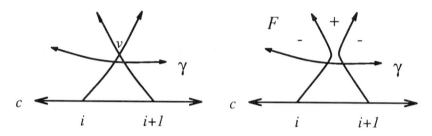

FIGURE 7. The curve γ intersects the legs once. We untangle the triangle.

CASE 1. The curve γ intersects the legs $\langle i, v \rangle$ and $\langle i + 1, v \rangle$ once each, as shown in Figure 7. We can form a new arrangement Γ' without the vertex v by untangling the crossing of curves $\gamma(i)$ and $\gamma(i+1)$ as pictured—essentially, we switch triangle legs $\langle i, v \rangle$ and $\langle i + 1, v \rangle$. Since removal of γ leaves an empty triangle $\Delta(i, v, i + 1)$ and since γ intersects each leg once, the arrangement Γ' still has the two-intersection property. It also has one fewer vertex above the sweep, so the sweep can make progress.

How can the sweep advance? The curve that we originally removed, γ, still intersects both $\gamma(i)$ and $\gamma(i+1)$, so none of these three curves is a hump. Thus, some face adjacent to the sweep must have become an empty triangle in Γ'. Only the faces that were touching v are affected by the untangling—two faces, marked by minus signs in Figure 7, lose a vertex, one gains a vertex, and the face bounded by $[i, i + 1]$ is unchanged. Since the configuration is symmetric, we can assume that the face F, bounded by the curves $\gamma(i)$ and γ, is an empty triangle $\Delta(j - 1, u, j)$ in Γ'.

Figure 8 shows the two ways in which the face F can become a triangle when the vertex v is untangled. Notice that the curve $\gamma(j) = \gamma$ and that u is the intersection of γ and leg $\langle i, v \rangle$. If $j < i$, then both $\langle j, u \rangle$ and $\langle i, u \rangle$ are edges in the original arrangement Γ and $\Delta(j, u, i)$ is a triangle—contradicting Lemma 3.6.

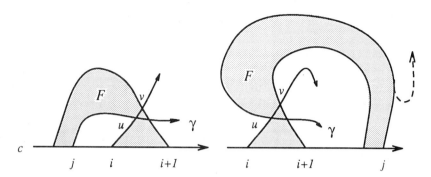

FIGURE 8. Two ways that F can become triangle after untangling vertex v. In the first, we have $\Delta(j, u, i)$. In the second, we can remove the vertex j.

So suppose that $j > i$. No curve intersects the edges $\langle i, u \rangle$ or $\langle j, u \rangle$ in the original arrangement Γ, so if the sweep could jump past these edges in a single operation, then the remaining arrangement, $\Gamma - \{\gamma(i), \gamma(i+1), \ldots, \gamma(j)\}$, would be smaller than the smallest counterexample and thus be sweepable. Moreover, these remaining curves do not affect the ability to sweep past edges $\langle i, u \rangle$ and $\langle j, u \rangle$. By minimality, there are no curves $\gamma(m)$ with $m < i$ or $j < m$. Now we can direct $\gamma(j)$ to infinity without causing any more intersections or allowing c to make progress, as indicated by the dotted line in Figure 8. Thus, we delete the vertex j. Since the other end of the curve $\gamma(j)$ does not leave the region bounded by $\langle i, u \rangle$ and $\langle j, u \rangle$, it still intersects the sweep. But this gives a smaller unsweepable arrangement—violating the minimality of Γ. This contradiction proves that Case 1 cannot occur.

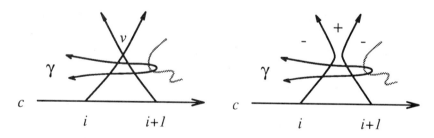

FIGURE 9. The curve γ intersects the legs twice. (The grey curve may or may not exist.)

CASE 2. The curve γ intersects each of the legs $\langle i, v \rangle$ and $\langle i+1, v \rangle$ twice. Without loss of generality, the curve γ hits leg $\langle i, v \rangle$ before leg $\langle i+1, v \rangle$, as shown in Figure 9. Again, we eliminate v by untangling it. We form the arrangement Γ' in which the sweep must make progress by finding an empty triangle.

As in Case 1, two faces that lose the vertex v gain no new vertices—one on the right and one on the left, as shown in Figure 9. By choosing our arrangement in accordance with Lemma 3.5 (so that curves go to infinity after crossing the sweep), we know that the curve γ cannot intersect the sweep between its two intersection points with leg $\langle i, v \rangle$. Since the face on the right has a portion of γ on its boundary, it cannot become a triangle. Thus, the face on the left becomes an empty triangle.

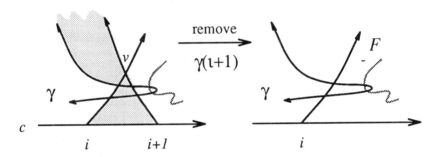

FIGURE 10. If untangling vertex v forms a triangle, discard $\gamma(i + 1)$.

Here we know enough about the structure of Γ to completely eliminate a curve; this is a trickier procedure than what we have done so far. Suppose that we remove the curve $\gamma(i+1)$ from Γ as in Figure 10. Only the face F, which is to the right of both $\gamma(i)$ and γ, and adjacent to the vertex v, can become an empty triangle—the other faces keep the same number of vertices. But we argued above that the middle of γ does not intersect the sweep c, so no triangles are formed when we remove $\gamma(i + 1)$. This contradicts the minimality of our counterexample and proves that the face above $[i, i + 1]$ has the point v as a vertex. $\square$

We have shown that removing any curve γ from the minimum counterexample Γ must leave an empty triangle and that γ must intersect only one leg of that triangle. If removing γ leaves $\Delta(i, v, i + 1)$, then we say that the edge $[i, i + 1]$ is *responsible* for γ. Notice that each edge of the sweep is responsible for at most one curve. (See Figure 11.)

In order to have an edge of the sweep responsible for every one of the n curves in the set Γ, we need at least $n + 1$ vertices on the sweep. Thus, there is a curve that intersects the sweep twice. We call a curve a *loop with respect to c*, or, more simply, a *loop*, if it intersects c twice and is connected in the above component of the plane. The points of intersection with c are called the *endpoints* of the loop.

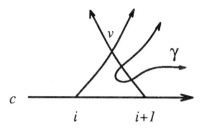

FIGURE 11. $[i, i+1]$ is responsible for γ.

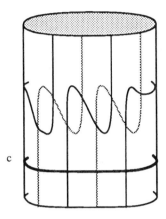

FIGURE 12. Cannot be swept.

The presence of a loop is critical to our argument. If we were sweeping an infinite cylinder, we would not always have a loop; in fact, there are unsweepable arrangements of curves that have the two-intersection property and have no loops. Figure 12 shows an arrangement in which the sweeping circle, c, cannot make progress. The existence of a loop is a property of the plane that prevents the sweep from becoming stuck.

We can define a partial order on loops. Let γ and γ' be loops with endpoints $i < j$ and $i' < j'$. We say that γ is *inside* γ' if $i' < i < j < j'$. It is not hard to see that the inside relation is acyclic and gives a partial order. We prove one technical lemma before we get to the final lemma for the bi-infinite case.

LEMMA 3.10. *Let the curve γ have endpoints $i < j$ and γ' have $i' < j'$. The following are equivalent*:

(1) *The curve γ is inside γ'.*

(2) *In the arrangement $\{\gamma, \gamma', c\}$ there is an unbounded face above the sweep. Either γ does not bound the face, or γ is a loop with respect to γ'.*

(3) *The endpoints satisfy $i' < i < j'$ and γ intersects γ' in zero or two points above the sweep.*

PROOF. We sketch a proof that $1 \Rightarrow 2 \Rightarrow 3 \Rightarrow 1$.

$1 \Rightarrow 2$. Suppose that γ bounds the face but is not a loop with respect to γ'. Then one or both of its endpoints do not lie between i' and j'.

$2 \Rightarrow 3$. No portion of γ that bounds the infinite face intersects c by the condition of 2. Thus the vertex i must lie between i' and j'. If γ does not bound the face, then it does not intersect the curve γ', since γ' does bound the face. If γ is a loop, then it intersects γ' twice.

$3 \Rightarrow 1$. Follows from the Jordan curve theorem.   □

Finally, we prove the last lemma for the bi-infinite case.

LEMMA 3.11. *There exists a curve γ whose removal does not leave a triangle.*

PROOF. In the plane, there is at least one loop. Choose γ to be a loop with no loop inside it—if there are many innermost loops, then choose one arbitrarily. Let the endpoints of γ be k and l, with $k < l$.

Suppose that the removal of γ leaves the empty triangle $\Delta(i, v, j)$, with $i = j \pm 1$. By Lemma 3.9, the curve γ intersects only one leg. Without loss of generality, γ intersects leg $\langle j, v \rangle$, and Lemma 3.7 implies that it does so twice. By the way we chose γ, the curve containing this leg, curve $\gamma(j)$, cannot be inside γ. Thus, Lemma 3.10 (3), tells us that either $j < k$ or $j > l$. Without loss of generality, $j < k$.

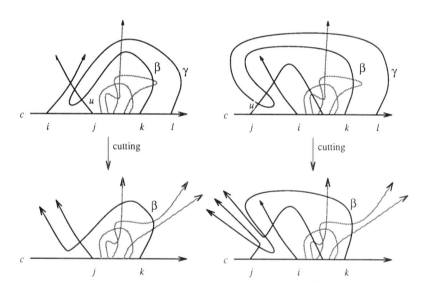

FIGURE 13. The cases for a loop intersecting a triangle.

Let u be the intersection point of γ and $\gamma(j)$ that is nearest to the vertex j in the original arrangement Γ—equivalently, let $\langle j, u \rangle$ be an edge of $\gamma(j)$. Let β denote the portion of the curve γ between vertices k and u. There are two cases, both portrayed in Figure 13, depending on whether the empty triangle $\Delta(i, v, j)$ left after removing γ has $i = j + 1$ or $i = j - 1$.

In both cases, the edge $\langle j, u \rangle$ and the curve β, together, partition the plane above the sweep into a bounded and an unbounded component. We perform radical surgery in the unbounded component to obtain a smaller counterexample. Discard all portions of the arrangement that lie in the unbounded component of the plane above the sweep. Any curves that are cut by this process are cut where they intersect β. We replace the discarded portions by rays to infinity, following the procedure used in the proof of Lemma 3.5. Let Γ' denote the resulting arrangement. The cutting process may increase the number of curves, but it does not give more intersections. In fact, we eliminated at least the vertex l; and, therefore, the sweep can make progress in Γ'.

But how can the sweep make progress? Which operation can apply? Suppose that Γ' contains a hump or triangle with vertices m and $m+1$ on the sweep. Since the surgery leaves no vertices in the unbounded component, we have $j \leq m < m+1 \leq k$. But this means that the hump or triangle was contained in the bounded component above the sweep and was not affected by the cutting process. Thus, it exists in the original arrangement Γ—a contradiction.

Therefore, we must have a curve that does not intersect the sweep in Γ'. Since all curves in Γ do intersect the sweep by Lemma 3.4, there must be some curve α that has a portion in the bounded component and is cut during the surgery. Curve α must enter and leave the bounded component without intersecting the sweep c or the edge $\langle j, u \rangle$. Thus, α crosses β twice and, as a result, α is a loop with respect to γ above the sweep. However, this is equivalent to α being inside γ, by Lemma 3.10 (2). This contradicts the choice of γ.

Therefore, the sweep cannot make progress in Γ'. This contradiction proves that there is some curve whose removal does not leave an empty triangle. □

Lemma 3.3 states that if we remove one curve from the smallest counterexample, then some operation applies. Lemmas 3.4 and 3.8, however, say that the operations of taking a curve or passing a hump never apply after removing one curve. Lemma 3.11 proves the existence of a curve whose removal does not leave a triangle. This contradiction proves that any arrangement of bi-infinite, two-intersecting curves curves in the Euclidean plane can be swept by a curve that maintains the two-intersection property.

Having established the bi-infinite case, we can use it to prove that a closed curve can always make progress. To sweep a sphere with a closed curve c, simply find a point that lies only on c and remove this point from the plane. Sweeping this punctured sphere is equivalent to sweeping the Euclidean plane with a bi-infinite curve.

To sweep an arrangement in the Euclidean plane with a closed curve, we first show how to embed it in an arrangement of two-intersecting curves on the sphere. The following lemma shows that we can map the plane into a region of a sphere bounded by a two-intersecting curve ξ, and then close off the bi-infinite curves outside of ξ to form an arrangement on the sphere with the two-intersection property.

LEMMA 3.12. *The bounded faces of an arrangement of endless plane curves having the two-intersection property can be embedded into a region bounded by a curve in an arrangement of closed two-intersecting curves on the sphere.*

PROOF. Draw an auxiliary curve ξ around the arrangement Γ so that all the vertices and closed curves of Γ are in the interior of ξ and modify the infinite rays so ξ intersects each ray once. Take a solid half-ball H and map ξ to the rim of H so that the interior of the curve ξ is mapped to the rounded surface of the half-ball. Every bi-infinite curve of Γ has two points of intersection with ξ; connect these two points by a chord along the flat surface of the half-ball. These chords turn the bi-infinite curves of Γ into closed curves; we must show that any two closed curves intersect in at most two points.

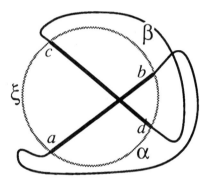

FIGURE 14. Closing off curves with chords.

Suppose two chords, $\overline{ab}$ and $\overline{cd}$, intersect. Let α denote the closed curve including chord $\overline{ab}$, and let β denote the closed curve including chord $\overline{cd}$ as illustrated in Figure 14. The nonchordal portion of β goes from the interior of the closed curve α to the exterior, so it intersects α an odd number of times. But, by the two-intersection property, it must do so only once. Therefore, closing off the curves of Γ with these chords does not violate the two-intersection property. $\square$

Now, given an arrangement in the Euclidean plane and a closed sweeping curve c, embed the arrangement on the sphere according to Lemma 3.12. To sweep the interior of c, simply sweep on the sphere. To sweep the exterior of c, sweep on the sphere until c reaches the auxiliary curve ξ. Then return to the plane, cutting c where it meets ξ and extending it to infinity. Finish the sweep in the plane with this bi-infinite sweeping curve. Thus, a closed sweeping curve can always progress and Lemma 3.2 is established.

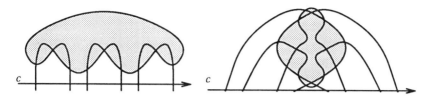

FIGURE 15. Unsweepable arrangements of three-intersecting curves.

3.3. Arrangements that cannot be swept. There are arrangements with the
three-intersection property that cannot be swept.

Consider the arrangements in Figure 15. In both of them, the sweeping
curve c intersects every curve twice. When the sweep first passes over one
of the vertices on the boundary of the shaded region, it intersects one or
two curves, each in two additional places—in both instances violating the
three-intersection property. Thus, these arrangements are unsweepable.

It is not difficult to construct arrangements with the $(k > 3)$-intersection
property that have copies of the arrangements of Figure 15 stopping progress
in triangles or humps. This establishes the final case of the sweeping theorem.

4. Applications of sweeping

In this section, we apply the sweeping theorem to solve two problems: finding Boolean formulae for polygons with curved edges and counting triangles
and digons in arrangements of exactly two-intersecting curves.

4.1. Boolean formulae for polygons. The lines supporting the edges of a
simple polygon define half-spaces in the plane; we can describe the set of
points in the interior of the polygon by a Boolean formula on these half-
spaces. For example, a convex polygon is the AND of the half-spaces defined
by its edges. Peterson [30] showed that every polygon with n edges can be
represented by a monotone formula that uses each half-space once.

Any bi-infinite curve divides the plane into two half-spaces; we can attempt
to find similar formulae for polygons with curved edges if these edges are
pieces of bi-infinite curves. Define a *curved segment* to be a simply connected
portion of a curve. We call the points that bound a segment *vertices* in this
section; segments with only one endpoint we call *rays*. A *polygonal chain*
is a sequence of curved segments $\{s_1, s_2, \ldots, s_n\}$ such that s_i and s_{i+1}
share a common vertex, either s_1 and s_n are rays (and the chain is *open*)
or they share a common endpoint (and the chain is *closed*), and no other
intersections between segments occur. A vertex v of a segment s divides a

bi-infinite curve into two pieces; we call the piece not containing the segment the *extension of s through* v.

Dobkin et al. [6] gave a simple proof that every polygon with n edges has a *Peterson-style formula*—a monotone Boolean formula that uses each edge's half-space once. Their proof depends on the fact that every polygonal chain has a *splitting vertex*—a vertex v such that the (straight-line) extensions of the incident edges through v do not intersect the chain. They find a Peterson-style formula recursively by splitting the chain at v, finding subformulae for the two subchains, and combining the subformulae with an AND if v is convex or an OR if v is concave.

We can use the sweep theorem to show that polygonal chains, whose segments are portions of pseudolines, also have a splitting vertex; this will imply that they also have Peterson-style formulae.

THEOREM 4.1. *In any polygonal chain whose segments are connected portions of distinct pseudolines, there is a vertex v such that the extensions of the incident segments through v do not intersect the polygonal chain.*

PROOF. Let $S = \{s_1, s_2, \ldots, s_n\}$ be a polygonal chain satisfying the hypothesis.

If S is closed, sweep its arrangement of pseudolines starting from a bi-infinite curve that intersects no curves. Stop when the sweeping curve c first crosses the chain S. Since all segments of S are bounded, the sweep first intersects S by passing a triangle and not by taking or passing the first ray. Hence, the intersection point crossed is a vertex v of the chain, as shown in Figure 16a. The incident segments both intersect c, so their extensions through v lie wholly on the swept side of c. The chain S, however, lies on the unswept side, except for v. Therefore, the extensions cannot intersect S.

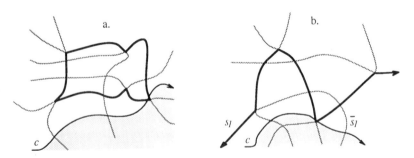

FIGURE 16. Sweeping closed and open polygonal chains.

If S is open, let $\bar{s}_1$ be the ray that is the extension of s_1 through its vertex. Assume that the rays $s_1, \bar{s}_1$, and s_n appear counterclockwise in this order on the line at infinity; if they are reversed, consider the mirror arrangement. Sweep the arrangement beginning with a bi-infinite curve c that intersects only $\bar{s}_1$ and stop when c intersects the chain S, as shown in Figure 16b.

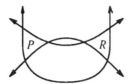

FIGURE 17. P has no formula.

Recall that we sweep by passing triangles and taking and passing the first (and not the last) ray. Since the sweep c intersects $\bar{s}_1$, it cannot cross the ray s_1; ray s_1 blocks c from crossing s_n. Furthermore, c cannot pass $\bar{s}_1$ without intersecting the chain S since $s_1 \in S$. Therefore, c first intersects S by passing a triangle, and the extensions of the segments incident to the triangle's apex are separated from the chain S as in the previous case. This proves the theorem. □

Thus, by the previous discussion, polygons bounded by portions of pseudolines have Peterson-style formulae. This result cannot be extended to polygons bounded by portions of pseudocircles. In Figure 17, the polygons R is contained in the same half-spaces as the polygon P—to represent P requires auxiliary curves and variables that separate R from P.

4.2. Triangles and digons in arrangements of curves with the exactly two-intersection property. We can model lines and pseudolines in the projective plane by great circles on a sphere with opposite points identified. In *Arrangements and spreads*, Grünbaum notes that if we cease to identify opposite points of the sphere, we obtain an arrangement of exactly two-intersecting curves with no digons (two-sided faces) [**22**]. He investigates minimum and maximum number of triangles in such arrangements and asks what is the relationship between the number of triangles and number of digons.

We can prove

LEMMA 4.1. *In an arrangement of closed curves with the exact two-intersection property, let p_i denote the number of i-sided faces. Then $2p_2 + 3p_3 \geq 4n$.*

PROOF. Assume the arrangement is on the sphere. We will find four digons or triangles on each curve.

Choose a curve c. Map c and one of the hemispheres it defines to a disk in the plane by the Schönflies theorem [**31**]. Extend the curves to infinity outside the disk.

Sweep the disk starting with c. Since c already intersects all of the curves, it advances by passing a triangle or hump (a digon). Cut c inside the hump or triangle and extend the ends to infinity without crossing any other curves. We can still sweep the disk with this bi-infinite curve, so there is another digon or triangle. Similarly, we find two digons or triangles in the other hemisphere defined by c.

Since we find each digon at most two times and each triangle at most three times, we have $2p_2 + 3p_3 \geq 4n$. $\square$

Grünbaum conjectures [22, Conjecture 3.7] that every digon-free arrangement of exactly two-intersecting curves has at least $2n - 4$ triangles. Specializing our result proves that there are at least $4n/3$ triangles—we suspect that his conjecture is closer to the truth.

5. Sweeping wedges

In the previous section our sweeping curves moved forward to sweep the plane. In this section, we see how to sweep by rotation about a point. We use this sweep to prove the extension theorem, which extends arrangements of one- and two-intersecting curves by adding new curves that pass through specified points.

THEOREM 5.1 (Extension theorem). *Let Γ be a finite set of k-intersecting curves, and P be a set of $k+1$ points, not all on the same curve. If $k \in \{1, 2\}$, then there is a curve that contains the points of P and has the k-intersection property with respect to Γ. If $k > 2$ then arrangements and point sets exist such that any curve through the points violates the k-intersection property.*

Levi proved the exact $k = 1$ case of the extension theorem in 1926 [25]. Levi's extension lemma, as his result is known, has become an important tool for generalizing properties of arrangements of lines to arrangements of pseudolines. For example, Goodman and Pollack use it, in concert with their "circular sequences" [7, 17], in a series of papers that prove duals of Radon's and Helly's theorems for pseudolines [16], prove that all arrangements of eight pseudolines are stretchable [13], and define a duality for configurations of points and pseudolines [12, 15] to establish a conjecture of Burr, Grünbaum, and Sloane [3]. They also show that no extension lemma can hold for planes in space [14].

As we mentioned in §4.2, Grünbaum [22] is interested in arrangements of exactly two-intersecting curves. He ways, "Another open and seemingly hard problem is to find the right analogue for (digon-free) arrangements of curves of Levi's extension lemma. It is well possible that an appropriate result in this direction would lead to solutions of some of the other problems mentioned." The pseudocircle case of the extension theorem finds a two-intersecting curve, which is not necessarily an exactly two-intersecting curve—we do not guarantee that our new curve intersects all other curves.

We establish both the pseudoline and pseudocircle cases of the extension theorem by showing that a bi-infinite curve can sweep a double wedge in the Euclidean plane. Let us define such a sweep formally before we give the proofs.

Let c and c' be oriented bi-infinite curves that intersect only at a point p. Curves c and c' divide the plane into four regions bounded by curve segments, as illustrated in Figure 18. We call the region where both boundary

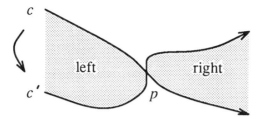

FIGURE 18. A double wedge.

segments are oriented towards p the *left wedge*, and call the region where both are oriented away from p the *right wedge*. The left and right wedges together comprise the *double wedge*, denoted wedge(c, c'). The double wedge wedge(c, c') contains the curves c and c'.

The segment of c bounding the left wedge is the *left half* of c, that bounding the right wedge is the *right half*. We assume that the left wedge is below c and the right wedge is above, according to the definitions of below and above given in §2. We want to rotate c continuously, moving downward on the left and upward on the right until c reaches c'.

A family of curves, $\mathscr{C} = \{c_\alpha\}_{0 \le \alpha \le 1}$, indexed by reals from the interval $[0, 1]$, *covers the double wedge* wedge(c, c') if

• the curves $c_0 = c$ and $c_1 = c'$,

• every point in wedge(c, c') lies on exactly one curve c_α except the point p, which lies on every curve in wedge(c, c'), and

• for all $0 \le \alpha < \beta < \alpha' \le 1$, the curve c_β lies in wedge$(c_\alpha, c_{\alpha'})$.

Suppose that the curves c and c' form a wedge wedge(c, c') and belong to a set of curves Γ that have the k-intersection property. We say that c can *sweep the double wedge* wedge(c, c') if there is a family of curves $\mathscr{C}$ that covers wedge(c, c'), and the set $\Gamma \cap \mathscr{C}$ has the k-intersection property. As before, we assume that the sweeping curve c is part of the arrangement.

We use the local operations from §2; for example, passing triangles and humps and taking loops. Recall that the operation of taking a loop added two intersection points to a curve that was visible from the sweep—to apply this operation to sweep a wedge instead of a half-plane, we modify the definition of visibility slightly: A point q on a curve γ is visible from an interval I of the sweep c if q and some point $r \in I$ both lie on the boundary of a common face contained in the double wedge wedge(c, c').

We can view the sweeping process in another way. If we remove the point p from the plane, the resulting surface is topologically equivalent to a cylinder. The sweeping curve c and its destination c' each become two bi-infinite curves. Both pieces of c are moving in the same direction (counterclockwise in Figure 19). These two pieces together are not allowed to intersect any curve in more than k points; the sweep ends when the pieces of c reach the pieces of c'.

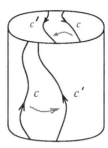

FIGURE 19. *c* on a cylinder.

Even though we have defined sweeping as a continuous operation, we again want to perform the sweep in discrete steps. We use the local operations from §2–modifying the *taking* operations to use the new definition of visibility. We prove the extension theorem in the next three subsections.

5.1. Extension lemma for pseudolines. To prove the extension theorem for exactly one-intersecting curves, it is not necessary to use a sweep method. Levi originally proved his lemma by arguing that a curve that connected two points and had the minimum number of intersections with other curves had the exact one-intersection property. See [**25**] or, for a proof in English [**22, Theorem 3.4**]. Our proof, which sweeps a double wedge, handles the (nonexactly) one-intersecting curves case and foreshadows the proof the pseudocircle case.

Suppose we have an arrangement of pseudolines and points p and q not on the same pseudoline. If no curve passes through p then we add a bi-infinite curve that intersects no other curves and then sweep with it until we encounter p. Now, if only one curve passes through p, duplicate it, reverse its orientation and perturb it so that the original and the copy intersect only at p and the point q lies in the double wedge of these curves. Otherwise, q lies in some region bounded by two curves through p. Orient these two curves so that q lies in their double wedge. Lemma 5.1 shows that we can sweep the double wedge containing q and maintain the one-intersection property. This establishes the $k = 1$ case of the extension theorem.

LEMMA 5.1. *Given a finite set of pseudolines* Γ *that includes two curves, c and c', defining a double wedge, the curve c can sweep* wedge(c, c') *using the operations of passing a triangle and taking or passing the first or last ray.*

PROOF. First apply operations that advance the right half of the sweep c. If c can no longer make progress in the right wedge, then it will be able to advance in the left wedge.

Assume c cannot advance on the right—no operation applies in the right wedge. We perform surgery: Erase everything in the plane except c and the curve segments in the right wedge, and extend these segments to bi-infinite

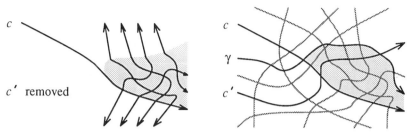

FIGURE 20. Surgery on the right wedge finds curve γ.

curves without adding any intersections. The only effect of this surgery that is visible from the right half of c is that c' disappears. Lemma 3.1 says that c can make progress in the reduced arrangement by passing triangles and passing and taking the last ray. Since no operation applied in the original arrangement and the curves visible from c did not change, c cannot pass a triangle or ray after the surgery. Therefore c can now take a curve γ that intersected the left half of c before the surgery, as shown in Figure 20.

Look at the right wedge of wedge(c, γ). Since γ is visible from c as a ray in this wedge, every pseudoline that intersects the right half of c also intersects γ above c.

Returning to wedge(c, c'), perform the same surgery as above on the left wedge. Lemma 3.1 says that c can advance downward in the reduced arrangement by passing triangles and passing and taking the first ray. As before, if we can pass a triangle or a ray, we can also pass it in the original arrangement. If we can take the ray of a curve α, then α intersects γ below the sweep c. Therefore, α cannot intersect the right half of c—if it did, it would intersect γ above c as well. This means that c can take α in the original arrangement.

Therefore, c can advance on the left to sweep the double wedge. □

This complete the proof of the pseudoline case.

5.2. Extension theorem for pseudocircles.

In this section, we prove the pseudocircle case of the extension theorem: that given three points in an arrangement of two-intersecting curves, one can find a curve through the points that satisfies the two-intersection property. We begin by reducing the problem to that of sweeping a double wedge defined by two bi-infinite curves in an arrangement of closed curves. Then Lemma 5.2 proves that such a sweep is always possible and establishes the theorem.

Let p, q, and r be the three points given by the hypothesis of the extension theorem; for simplicity, let us assume that they do not lie on any curves of the arrangement. We first show that it is sufficient to consider only closed curves: By Lemma 3.12, we can map any plane arrangement into a region

$\mathscr{R}$ of a sphere that is bounded by a pseudocircle and close off all bi-infinite curves without violating the two-intersection property. Any extension pseudocircle that passes through p, q, and r either remains in $\mathscr{R}$—and, thus, maps to a closed pseudocircle in the plane—or intersects the boundary of $\mathscr{R}$ in two points—and, thus, maps to a bi-infinite pseudocircle in the plane. Therefore, it is enough to consider closed curves on a sphere.

Next, we reduce the problem to sweeping a double wedge in the plane: If we remove the point r from the sphere, we obtain a plane containing points p and q and closed pseudocircles. As in the pseudoline case, Theorem 3.1 implies that we can sweep this plane, starting with a bi-infinite curve that initially intersects no curves, until we encounter the point p. Then we duplicate the sweep, perturb it, and orient it so that the original and the copy intersect only at p and the point q lies in their double wedge. The following lemma says that we can sweep this double wedge with a bi-infinite curve until we encounter q. Then the resulting curve passes through the pivot point p, the point q, and the point r at infinity—since it has the two-intersection property, it establishes the extension theorem for pseudocircles. The rest of this section is devoted to the proof of Lemma 5.2.

LEMMA 5.2. *Let Γ be a finite set of pseudocircles, all closed except two, c and c', which define a double wedge. The curve c can sweep $\operatorname{wedge}(c, c')$ using the operations of passing a triangle, passing a hump, and taking a loop.*

The proof of this lemma is a long investigation into the structures that could stop the advance of the sweep in an arrangement of closed two-intersecting curves. The next two paragraphs give a brief outline of the three things, crescents, their shadows, and their nesting properties, that will be important in the proof.

First, we call a curve a *crescent* if it is visible from one-half of the sweep and intersects the other half in two points. We refer to crescents that intersect the left (right) half of c as above (below) crescents. Figure 21 shows an arrangement with two crescents, one above and one below. Lemma 5.3 will use the sweep theorem, Theorem 3.1, to prove that the sweep can advance if there are no crescents.

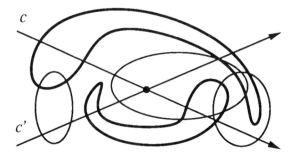

FIGURE 21. The two darker curves are crescents.

If the side of the sweep that intersected a crescent could advance and drop γ, then the other side of the sweep could advance by taking γ and both sides would have made progress. Thus, we will define our second concept: the *shadow* of a crescent will be a set of curve segments that must be swept before the crescent can be passed and dropped from the sweep. After also defining *shadow regions* of the plane and *shadow intervals* of the sweep, we will show that the sweep can advance in the shadow region of a crescent A unless some other crescent B is visible but does not intersect the sweep within the shadow interval of A—we say that B *stops* A. Third, but examined concurrently with the second, the *nesting properties* of crescents and curves bounding the shadow regins will be used to show that the stops relation is acyclic. Once the stops relation is proved acyclic, there must be some crescent not stopped by any other crescent and the sweep can make progress in its shadow.

The next lemma proves that crescents are important.

LEMMA 5.3. *If an arrangement does not have both above and below crescents, then the sweep can make progress.*

PROOF. Suppose the arrangement has no above crescents; the proof for below crescents is similar. Perform surgery on the right wedge: erase everything from the plane except c and the curve segments in the right wedge, then extend these segments to bi-infinite curves without adding intersections. (See Figure 20.)

Lemma 3.2 says that c can now advance by passing a hump or triangle or by taking a loop. If a passing operation applies in the reduced arrangement, then it also applies in the original. Furthermore, no curve visible from the right half of the sweep intersects the left half, so any loop that can be taken in the reduced arrangement can also be taken in the original. Therefore, c can make progress in sweeping the double wedge.   □

Assume for now that γ is a below crescent—it intersects the right half of the sweep c and is visible from a point p in the left half of c. The *hump* of γ is the portion of γ above the right half of c. If the sweep c could pass the hump of γ on the right, then c could subsequently take the loop of γ on the left. In a moment, we will iteratively define the *shadow of* γ in an attempt to capture the set of curves (or curve segments) that prevent c from passing the hump of γ. Here we outline the motivation for these definitions: surgery and the sweep theorem.

Initially, the first shadow region is the hump of γ. Suppose we performed surgery by erasing all curve segments outside of this shadow region except for c. Then, the sweep theorem for pseudocircles says that the sweep c can make progress by passing a hump or triangle or by taking a curve that is visible inside the shadow region. Only one of these operations could apply in this reduced arrangement but not in the original: the operation of taking a curve α that has a point p_α visible inside the hump, but whose intersections with c were removed by the surgery. If this curve α is not a crescent, then

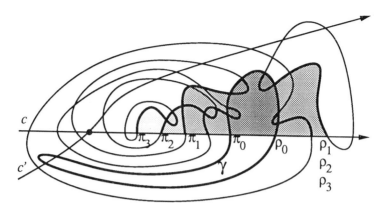

FIGURE 22. The curves that contribute segments to the shadow of γ.

it intersects the right half of c in one or two points. We expand the shadow region to include a connected portion of α from the visible point p_α to one of these intersection points. Surgery on this expanded shadow region will again find an operation by which the sweep can make progress, but it will not be the operation of taking α. Eventually we will either discover a crescent or we run out of curves to take and discover that the sweep can make progress in the original arrangement.

After two more paragraphs of definitions and notation, we will iteratively define the *shadows* of γ, denoted $\mathrm{Sh}_0(\gamma)$, $\mathrm{Sh}_1(\gamma)$, $\mathrm{Sh}_2(\gamma)$, ..., as sets of curve segments with an endpoint on the right half of c. These curve segments are drawn darkly in Figure 22. We frequently drop the subscript or the curve (γ) when we make statements that apply to all shadows or when the specific shadow can be determined from the context.

Given a shadow Sh, its *shadow interval* $\overline{\mathrm{Sh}}$ is the smallest interval of the sweep c that contains all the intersection points $\{c \cap s : s \in \mathrm{Sh}\}$. The *inner curve* of Sh is the curve that contains the endpoint of $\overline{\mathrm{Sh}}$ closest to the pivot point $c \cap c'$ of the double wedge; the *outer curve* of Sh is the curve that contains the endpoint farthest from the pivot point. The *shadow region* of Sh is the largest simply connected region bounded by the curve segments of Sh and by the interval $\overline{\mathrm{Sh}}$. In Figure 22, the inner curve of shadow $\mathrm{Sh}_i(\gamma)$ is π_i, the outer curve is ρ_i, and the shadow interval $\overline{\mathrm{Sh}}_i(\gamma)$ is the portion of c from π_i to ρ_i. The shadow regions are shaded.

We say that a curve α is *eligible in shadow* Sh if three conditions are satisfied: α contains a point p_α visible from the shadow interval $\overline{\mathrm{Sh}}$, α intersects the right half of c, and α does not intersect the interval $\overline{\mathrm{Sh}}$. We will show, in Corollary 5.5, that if both points of the intersection $\alpha \cap c$ lie on the right half of c then they both preceed or both follow the interval Sh on the sweep c. Thus, we can unambiguously define the intersection point of α and the right half of c that is nearest the interval $\overline{\mathrm{Sh}}$ to be the *anchor* of α, denoted a_α. See Figure 23.

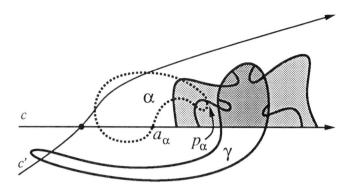

FIGURE 23. α is eligible in shadow $Sh_1(\gamma)$.

Finally, here is the constructive definition of the shadows $Sh_0(\gamma)$, $Sh_1(\gamma)$, $Sh_2(\gamma)$, ... We drop the (γ) in this paragraph. Let Sh_0 be the hump of γ. Given Sh_i, we form the shadow Sh_{i+1} by adding, for each curve α eligible in Sh_i, the longest segment above c from α's anchor a_α to a point p_α visible from $\overline{Sh}_i$. If no curves are eligible, then we end the construction procedure and call the last shadow constructed the *maximal shadow*. Since a finite number of curves intersect c, the construction eventually terminates. (You may note that we have made no reference to c', the other curve defining the double wedge. Corollary 5.6 will show that c' is never visible in the shadow of a crescent.)

These definitions have been made for a below crescent, which intersects the right half of the sweep. For an above crescent, simply interchange right and left.

Lemma 5.4 will prove that the crescents and eligible curves visible from a shadow interval $\overline{Sh}$ must *nest* with the shadow's inner and outer curves; these nesting properties are crucial to our proofs. Let us denote the inner curves of $Sh_0(\gamma)$, $Sh_1(\gamma)$, ... by $\pi_0, \pi_1, ...$ and the outer curves by $\rho_0, \rho_1, ...$ Then, as illustrated in Figure 22, $\gamma = \pi_0 = \rho_0$. Corollary 5.5 will imply that no curve other than γ is both an inner and outer curve. Thus, we define nesting differently for γ than for inner and outer curves with indices greater than zero.

We say that a curve α *nests inside* an inner curve π_i, with $i > 0$, if both points of the intersection $\alpha \cap c$ lie inside the closed curve π_i. A curve α *nests outside of* an outer curve ρ_i, with $i > 0$, if both points of $\alpha \cap c$ lie outside the closed curve ρ_i. A curve α *nests inside* the crescent γ if both points of $\alpha \cap c$ lie in the interval of c between the hump of γ and p, a point from which γ is visible below c. The curve α *nests outside* γ if both points of $\alpha \cap c$ lie outside of the hump of γ and outside the interval from the hump to p. In Figure 24 the curve α, which happens to be a crescent, nests

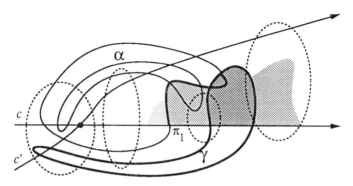

FIGURE 24. α nests inside π_1 and both nest inside γ.

inside the inner curve π_1 and both curves nest inside γ. The next lemma establishes the nesting properties of crescents and eligible curves.

LEMMA 5.4. *Let α be an eligible curve or crescent that is visible from $\overline{\mathrm{Sh}}_i(\gamma)$ but not from $\overline{\mathrm{Sh}}_{i-1}(\gamma)$. Then α nests inside the inner curves $\pi_0, \pi_1, \ldots, \pi_i$ or outside of the outer curves $\rho_0, \rho_1, \ldots, \rho_i$. If $i > 0$, then α nests inside if and only if it is visible between the sweep pivot point $c \cap c'$ and the hump of γ.*

PROOF. We first prove by induction that the inner curve π_i nests inside π_j for all $i > j$ and that the outer curve ρ_i nests outside of ρ_j for all $i > j$. Then we notice that the lemma holds for other eligible curves and crescents.

For the base case, if a curve α is visible from $\overline{\mathrm{Sh}}_0$, but does not intersect $\overline{\mathrm{Sh}}_0$, then α intersects the hump of γ in two points above the sweep. When α crosses the sweep, α cannot intersect γ further nor obstruct the visibility of γ from some point p on the left half of the sweep. Thus, both or neither of the points of $\alpha \cap c$ lie between p and the hump of γ—α nests inside γ if one of these points is between p and the hump of γ; otherwise, α nests outside of γ. Setting $\alpha = \pi_1$ and $\alpha = \rho_1$ establishes the base cases for inner and outer curves.

Now, assume that $i > 1$ and that the nesting properties of inner curve π_i and outer curve ρ_i have been established. The interval difference $\overline{\mathrm{Sh}}_i \backslash \overline{\mathrm{Sh}}_{i-1}$ may consist of two intervals, one closer to and one farther from the pivot point than the hump of γ. Assume that α is an eligible curve visible from the interval closer to the pivot point.

Let β denote the curve segment of π_i contained in the shadow $\mathrm{Sh}_i(\gamma)$; the endpoint p_β is visible from the interval $\overline{\mathrm{Sh}}_{i-1}(\gamma)$. (See the darkened portion of π_1 in Figure 24, for example.) Since α neither obstructs this visibility nor intersects $\overline{\mathrm{Sh}}_i(\gamma)$, the curve α intersects β twice and can only intersect c inside π_i. Therefore, α nests inside π_i. Because nesting inside is transitive for inner curves, α nests inside $\pi_0, \pi_1, \ldots, \pi_i$. Similarly, if α was visible from the farther interval, α would nest outside of $\rho_0, \rho_1, \ldots, \rho_i$.

Now, consider $\alpha = \pi_{i+1}$; we know from the previous paragraph that π_{i+1} nests inside $\pi_0 = \gamma$ or outside of $\rho_0 = \gamma$. But, because π_{i+1} is an inner curve, one of the intersection points $\pi_{i+1} \cap c$ is between the hump of γ and the pivot point. Therefore, π_{i+1} nests inside $\pi_0, \pi_1, \ldots, \pi_i$. A similar argument proves that ρ_{i+1} nest outside of $\rho_0, \rho_1, \ldots, \rho_i$ and establishes the inductive assumption.

Notice that we used only the fact that α was visible from, but did not intersect, a certain shadow interval. Thus, the lemma holds for crescents and other eligible curves as well.   □

Corollaries to this lemma prove that the definitions of anchors and inner and outer curves are unambiguous, that the wedge boundary c' never appears in a shadow, and that curves that do appear are crescents or are properly anchored.

COROLLARY 5.5. *If a curve α that is eligible in* $\mathrm{Sh}(\gamma)$ *intersects the right half of c in two points, then both preceed or both follow the interval* $\overline{\mathrm{Sh}}(\gamma)$ *along the sweep.*

PROOF. Follows from the fact that the eligible curve α either nests inside or nest outside of γ.   □

COROLLARY 5.6. *The wedge boundary c' is never visible in the shadow of a crescent.*

PROOF. For every crescent γ, the wedge pivot point $c \cap c'$ is between the hump of γ and a point from which γ is visible. The point at infinity is outside this interval. Thus, c' does not nest with γ.   □

COROLLARY 5.7. *Let α be a curve with a point p that is visible from the shadow interval* $\overline{\mathrm{Sh}}(\gamma)$ *of the maximal shadow of γ. Either α is a crescent, does not intersect c, or connects p to* $\overline{\mathrm{Sh}}(\gamma)$ *within the shadow region of* $\mathrm{Sh}(\gamma)$.

PROOF. Assume that α does intersect the sweep c. As in the proof of Lemma 5.4, we consider cases depending on the first shadow in which p is visible.

Suppose that the point p is first visible from $p' \in \overline{\mathrm{Sh}}_0(\gamma)$—from a point in the hump of γ. Since α and γ intersect in only two points, if α intersects the sweep within the hump, that is, if α intersects $\overline{\mathrm{Sh}}_0(\gamma)$, then p is connected in $\overline{\mathrm{Sh}}_0(\gamma)$ within the shadow region of $\mathrm{Sh}_0(\gamma)$. If α intersects the sweep only outside of the hump, then α is a crescent or is eligible and p will be properly anchored in the shadow region $\mathrm{Sh}_1(\gamma)$.

Otherwise, the point p is first visible from a point $p' \in \overline{\mathrm{Sh}}_i(\gamma)$ with $i > 0$—meaning that p' lies in one of the (at most two) intervals of the difference $\overline{\mathrm{Sh}}_i(\gamma) \backslash \overline{\mathrm{Sh}}_{i-1}(\gamma)$. Assume that p' lies in the interval closer to the pivot point; then p' is between the endpoint of the inner curve π_i and a point $q' \in \overline{\mathrm{Sh}}_{i-1}(\gamma)$ from which π_i is visible, as depicted in Figure 25.

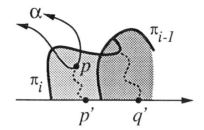

FIGURE 25. Visibility of α and π_i.

We have assumed that α does intersect the sweep. Since the curve α does not obscure the visibility of π_i from q', the curve α either joins p to $\overline{\mathrm{Sh}}_i(\gamma)$ within the shadow region of $\mathrm{Sh}_i(\gamma)$ or intersects π_i twice and intersects the sweep on the other side of π_i from p. In the latter case, Lemma 5.4 implies that α nests inside π_i and must be a crescent or an eligible curve in shadow $\mathrm{Sh}_i(\gamma)$. If α is eligible, then it will be anchored in shadow $\mathrm{Sh}_{i+1}(\gamma)$.

If p' lies in the other interval of $\overline{\mathrm{Sh}}_i(\gamma) \backslash \overline{\mathrm{Sh}}_{i-1}(\gamma)$, then a symmetric argument can be made using the outer curve ρ_i. □

We have already mentioned that the only event that can prevent the sweep from advancing in the shadow of one crescent is the visibility of another crescent. We say that an above or below crescent A *stops* below or above crescent B, denoted $A \oslash B$, if A is visible from the shadow interval $\overline{\mathrm{Sh}}(B)$. In Figure 24, for example, $\alpha \oslash \gamma$.

LEMMA 5.8. *If no crescent stops a crescent γ, then the sweep can advance in the shadow region of $\mathrm{Sh}(\gamma)$.*

PROOF. Suppose that no operation applies in $\mathrm{Sh}(\gamma)$. As hinted before, we perform surgery: erase every curve outside the shadow region except c and extend all curve segments to infinity without adding intersections.

Sweep Lemma 3.2 says that c can make progress in the reduced arrangement. If c could advance by passing a hump or triangle, then it could do so in the original arrangement. Therefore, it can take a loop of a curve α that is not anchored in the shadow region. Since α is not a crescent, Corollary 5.7 implies that α does not intersect c. Thus, c can take α in the original arrangement. □

Because of this lemma, if we show that some crescent is not stopped by any other crescent, then we know that the sweep can make progress. In the remainder of this section we will use the nesting properties of the inner and outer curves to prove the stronger fact that the stops relation $\oslash$ is acyclic. In Lemma 5.9 we prove that there are no cycles of length two by deriving a contradiction to the nesting lemma, Lemma 5.4. Once that is out of the way, it is less difficult to prove the relation acyclic by examining the structure of a cycle using the smallest number of crescents, and again deriving a contradiction to the nesting lemma.

LEMMA 5.9. *There are no two crescents A and B such that $A \oslash B$ and $B \oslash A$.*

PROOF. Suppose $A \oslash B$ and $B \oslash A$; we will derive a contradiction to the nesting lemma. By the definition of the $\oslash$ relation, the crescent B is visible from a point $p_B \in c$ that lies in the shadow interval of A and A is visible from a point p_A in the shadow interval of B.

First, we prove that p_B does not lie in the hump of A. By Lemma 5.4, we know that B either nests inside or nests outside of A and that A nests inside or outside of B. We cannot have p_B inside the hump of A; otherwise, the two intersection points of A with c would lie on opposite sides of p_B, which would contradict the fact that A nests inside or outside B. Similarly, p_A is not inside the hump of B.

Second, we prove that we can insure that B nests inside A and A nests outside of B. Let us assume that B nests inside A, as shown in Figure 26. Since B is not visible in the hump of A, the inner curve π_1 of $\mathrm{Sh}_1(A)$ exists and B nests inside π_1. But π_1 forces A to nest outside of B on the side of the sweep that B intersects. Moreover, if B nests outside of A, a similar argument using the first outer curve of $\mathrm{Sh}_1(A)$ shows that A nests inside B and the rôles of A and B can be reversed.

Third, we look at the inner curves of A's shadows and the outer curves of B's shadows. Let B first be visible from $\overline{\mathrm{Sh}}_i(A)$, then Lemma 5.4 implies that B nests inside the inner curves of $\mathrm{Sh}_0(A)$, $\mathrm{Sh}_1(A)$, ..., $\mathrm{Sh}_i(A)$, which we denote $\{\pi_0 = A, \pi_1, \ldots, \pi_i\}$. Similarly, let A first be visible from $\overline{\mathrm{Sh}}_j(B)$; the crescent A nests outside the outer curves of B's shadows, $\{\rho_0 = B, \rho_1, \ldots, \rho_j\}$. (Note that $i > 0$ and $j > 0$.) For all $1 \le k \le i$, define j_k to be the smallest index such that point of $\pi_k \cap c$ lies in the interval $\overline{\mathrm{Sh}}_{j_k}(B)$. The fact that B nests inside π_k, which nests inside A, implies that π_k intersects the sweep c exactly once between the hump of B and the visibility of A. Since the shadow interval $\overline{\mathrm{Sh}}_j(B)$ includes this intersection point, j_k is well defined and satisfies $1 \le j_k \le j$.

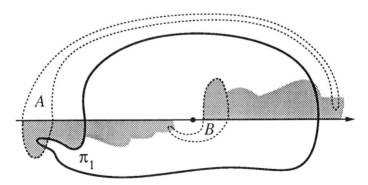

FIGURE 26. Proving that if B nests inside A, then A nests outside of B.

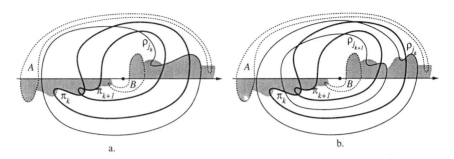

FIGURE 27. Two cases for proving that π_k is inside ρ_{j_k} below the sweep.

Fourth, we prove by induction from $k = i$ down to $k = 1$ that π_k is contained inside the curve ρ_{j_k} below the sweep, as depicted in Figure 27. We will first establish the general case and then verify the base case. Our inductive assumption is that, below the sweep, π_{k+1} is contained in $\rho_{j_{k+1}}$ as shown in Figure 27. We have two cases to consider—either $j_k = j_{k+1}$ or $j_k > j_{k+1}$.

Suppose $j_k = j_{k+1}$, as shown in Figure 27a, and look at the two points $\pi_k \cap c$. The rightmost point is in the shadow interval $\text{Sh}_{j_k}(B)$ by definition of j_k, so it is inside ρ_{j_k}. The leftmost point bounds the shadow interval $\text{Sh}_k(A)$; the curve π_{k+1} is visible in this shadow and $\rho_{j_{k+1}} = \rho_{j_k}$ encloses all visible points of π_{k+1}, so the leftmost point is also inside ρ_{j_k}. This implies that π_k intersects ρ_{j_k} an even number of times below the sweep. But π_k must intersect ρ_{j_k} at least once above the sweep because it enters the portion of the shadow $\text{Sh}_{j_k}(B)$ and does not obscure the visibility of ρ_{j_k}. Thus, π_k is contained in ρ_{j_k} below the sweep.

Otherwise, $j_k > j_{k+1}$, as shown in Figure 27b. We again look at the two points $\pi_k \cap c$. As before, the rightmost point is in the shadow interval $\text{Sh}_{j_k}(B)$, so it is inside ρ_{j_k}. Also as before, the leftmost point is inside $\rho_{j_{k+1}}$, which nests inside ρ_{j_k}. And, just as before, π_k intersects ρ_{j_k} at least once above the sweep. Thus π_k is contained in ρ_{j_k} below the sweep.

The base case of the induction is nearly the same as the $j_k > j_{k+1}$ case with $B = \rho_0$ playing the rôle of $\rho_{j_{k+1}}$.

By induction, we have proved that the leading curve π_1 is not visible outside of ρ_{j_1}. But A nests outside of ρ_{j_1} and π_1 is visible in A. This contradiction proves that there are no two-cycles.   □

We define one final relation before we investigate longer cycles in the $\varnothing$ relation. Given two above or below crescents, A and B, we say that A *overshadows* B if the maximal shadow interval $\overline{\text{Sh}}(A)$ contains both intersection

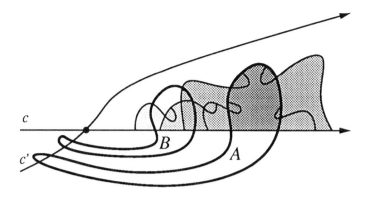

FIGURE 28. A overshadows B.

points of $B \cap c$. (See Figure 28.) The next lemma implies that if crescent A overshadows B, then any crescent visible in the shadow of B is also visible in the shadow of A—in other words, any crescent that stops B also stops A.

LEMMA 5.10. *If A overshadows B, then the maximal shadow interval $\overline{\mathrm{Sh}}(A)$ contains the maximal shadow interval $\overline{\mathrm{Sh}}(B)$.*

PROOF. We prove by induction that shadow intervals $\overline{\mathrm{Sh}}_0(B)$, $\overline{\mathrm{Sh}}_1(B)$, ... are contained in $\overline{\mathrm{Sh}}(A)$. For the base case, $\overline{\mathrm{Sh}}_0(B) \subseteq \overline{\mathrm{Sh}}(A)$ by the definition of the overshadows relation.

Let α be a curve that is eligible in $\mathrm{Sh}_{i-1}(B)$. The curve α cannot be a crescent and must intersect c, so Corollary 5.7 implies that α does not leave the shadow region of $\mathrm{Sh}(A)$. Thus, $\overline{\mathrm{Sh}}_i(B) \subseteq \overline{\mathrm{Sh}}(A)$.   □

COROLLARY 5.11. *Let $\gamma_1 \oslash \gamma_2 \oslash \cdots \oslash \gamma_k \oslash \gamma_1$ be a minimum length cycle in the $\oslash$ relation. If $i \neq j$, then γ_i does not overshadow γ_j.*

PROOF. Suppose γ_i overshadows γ_j. Then $\gamma_{j-1} \oslash \gamma_j$ and $\overline{\mathrm{Sh}}(\gamma_i) \supset \overline{\mathrm{Sh}}(\gamma_j)$ imply that $\gamma_{j-1} \oslash \gamma_i$. But this relation shortcuts the cycle and contradicts its minimality.   □

Now we can prove the final lemma.

LEMMA 5.12. *The $\oslash$ relation is acyclic.*

PROOF. We will prove this by contradiction. Let $\gamma_1 \oslash \gamma_2 \oslash \cdots \oslash \gamma_k \oslash \gamma_1$ be a cycle with minimum length. Assume, for the moment, that γ_2 nests inside γ_1, as shown in Figure 29. We can prove by induction that each successive crescent also nests inside the previous. But then, by transitivity of *nesting inside*, we find that γ_1 nests inside itself, which is false.

We have already proved that there are no cycles of length two, so there are at least three distinct curves in the minimum length cycle. (Actually four, since above and below crescents will alternate.) The general induction step would take each successive triple of crescents γ_{i-1}, γ_i, and γ_{i+1} and show

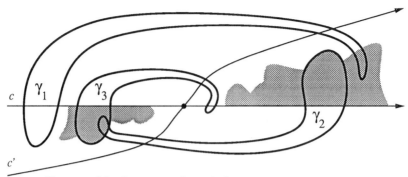

FIGURE 29. Cresents of a minimum length cycle nest.

that if γ_i nests inside γ_{i-1} then γ_{i+1} nests inside γ_i. We just illustrate it with the base case.

Consider where the interval of the hump of γ_3, that is, $\mathrm{Sh}_0(\gamma_3)$, can lie. Lemma 5.11 says that γ_1 does not overshadow γ_3 and γ_3 does not overshadow γ_1. The former implies that the intersection points of γ_3 and c cannot both lie in $\overline{\mathrm{Sh}}(\gamma_1)$. The latter, with the fact that γ_2 is visible from $\overline{\mathrm{Sh}}(\gamma_3)$, implies that γ_3 cannot intersect c at all before $\overline{\mathrm{Sh}}(\gamma_1)$. So at least one point of $\gamma_3 \cap c$ follows $\overline{\mathrm{Sh}}(\gamma_1)$. Since γ_3 does nest with γ_2 by Lemma 5.4, γ_3 nests inside of γ_2. (By transitivity of nesting, γ_3 also nests inside γ_1.)

A similar induction can show that if γ_2 nests outside of γ_1, then each successive crescent also nests outside of the previous. Again, the transitivity of *nesting outside of* implies that γ_1 nests outside of itself, which is false. These contradictions imply that there are no cycles in the $\varnothing$ relation. $\square$

This lemma, combined with Lemma 5.8, shows that we can make progress even when crescents exist. Thus, the proof of Lemma 5.2 is complete.

5.3. Arrangements with no extension curves. Figure 30 shows that when we specify $k + 2$ points in pseudoline $(k = 1)$ or pseudocircle $(k = 2)$ arrangements, an extension curve may not exist.

In Figure 30a, we want to draw a new curve through the three points, so one

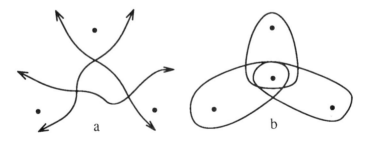

FIGURE 30. No extension curve.

point lies between the other two. But then one of the pseudolines separates the middle point from the adjacent points—the new curve intersects this pseudoline at least twice.

In Figure 30b, we can assume that the new curve through these four points is closed. The center point then has two adjacent points and one opposite point on this closed curve. One of the pseudocircles separates the center and its opposite from their adjacent points. The new curve intersects this pseudocircle in at least four points. Since the pseudocircles of Figure 30b also satisfy the three-intersection property, the above argument proves that not all arrangements of three-intersecting curves with four points have an extension curve. We show how to convert this to an arrangement of bi-infinite curves in Lemma 5.13.

LEMMA 5.13. *For every* $k \geq 3$, *there is an arrangement of k-intersecting curves and* $k + 1$ *points such that any curve through these points violates the k-intersection property.*

PROOF. If $k > 3$ is even, consider arrangements of four curves constructed after the scheme in Figure 31. The figures on the left show arrangements with the four- and six-intersection property. Those on the right indicate how to form arrangements for other even k—curve segments are drawn to represent the open curves that surround their perimeter.

We can connect one pair of points by a (curved) segment that crosses curves two times. Any other segment that connects two points crosses the curves at least four times. To connect all $k + 1$ points requires $4k + 2$ crossings, or an average of $k + 1/2$ crossings per curve. Thus, any extension curve through the points intersects one of the curves in more than k points.

Since the extension must cross each curve an even number of times, it intersects one of the curves in $k + 2$ points. Thus, the arrangements of Figure 31 also show $(k + 1)$-intersecting curves with no extension curves.

Each closed curve of the preceding constructions can be replaced by three nested bi-infinite curves, shown in Figure 32, each defining a half plane containing the interior of the closed curve. A bi-infinite extension curve can extend infinitely in only two of these three half planes, so we can close off one of the curves without crossing the extension curve. The previous analysis then applies to these closed curves.   □

6. Algorithms for sweeping curves

If we know the graph that represents the arrangement, then we can easily implement the local operations defined in §2. The only complication could be dealing with separate connected components of the graph. If we sweep pseudolines or pseudocircles, Lemmas 5.1 and 5.2 say that we can apply any operation to make progress—we may even be able to apply operations in parallel.

Since an arrangement of n curves can have quadratic size, it is important

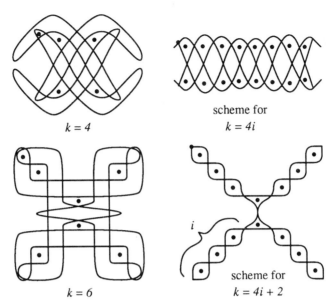

FIGURE 31. Arrangements with no extension curves for even k.

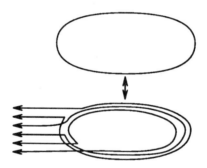

FIGURE 32. Bi-infinite curves.

to ask whether one can perform the sweep when the arrangement is not known explicitly. Edelsbrunner and Guibas showed that this could be done efficiently for line arrangements [9]. Let us look at their ideas and try to use them to sweep arrangements of pseudolines and pseudocircles.

6.1. Implementing a pseudoline sweep. When sweeping an arrangement of lines, Edelsbrunner and Guibas can start with a curve that intersects every line—they do not need to use taking or dropping operations. Thus, their problem is to recognize empty triangles. For this purpose they keep two data structures that they call upper and lower horizon trees.

The horizon trees could also be called *envelope trees* because for any interval $[i, i+1]$ on the sweep, one tree encodes the lower envelope of the lines that intersect the sweep before the interval, specifically the lines $\gamma(1), \dots,$

$\gamma(i)$, and the other encodes the upper envelope of the lines $\gamma(i+1), \ldots, \gamma(n)$. (Unfortunately, the upper horizon tree encodes the lower envelopes.) In this section, we want to concentrate on the envelope properties, so we refer to the trees as lower and upper envelope trees.

Any empty triangle that occurs in both envelope trees is truly empty. The lower envelope tree certifies that no line cuts it from above and the upper that no line cuts it from below. Since we can pass triangles in any order, we need only one tree: one can prove that in the upper envelope tree (lower horizon tree) the uppermost or first triangle cannot be cut by a line above it. Thus, if we always pass the upper triangle in the upper envelope tree, we can sweep the plane. Overmars and Welzl noticed this fact in the dual [29].

For lines, and also for pseudolines, the envelope trees have linear size—once a curve crosses into the envelope, it cannot leave. The initial trees are easy to construct. Simply add the curves in increasing order along the sweep to build the lower envelope tree, and in decreasing order to build the upper envelope tree. The time to update the trees can be related to the horizon complexity of the lines of the arrangement (thus the original names) so it amortizes to constant time per triangle.

We can now modify the method of Edelsbrunner and Guibas to use a single envelope tree to sweep pseudolines in linear space and in time proportional to the size of the arrangement. Initially, we need to know the order along the sweep c of the curves that intersect c. We also assume that we know the ordering around the line at infinity. With this information, we can take rays until we reach a curve that already intersects the sweep—say it is γ intersecting the sweep at i. That is, $\gamma = \gamma(i)$.

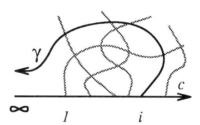

FIGURE 33. Sweeping curves $\gamma(1), \ldots, \gamma(i)$.

We find the upper envelope tree (lower horizon tree) for the curves $\Gamma' = \{\gamma(1), \ldots, \gamma(i)\}$. The first triangle in this tree is an empty triangle in the arrangement Γ'. Every curve that is visible in the interval $[\infty, i]$ of the sweep c must intersect c at or before i to avoid intersecting γ twice; therefore, the first triangle is an empty triangle in Γ and c can pass it. The sweep c can continue to advance until c drops the curve γ. If the sweep is not yet complete, then c returns to taking curves.

6.2. Open problems for $(k > 1)$**-intersecting curves.** Applying these ideas to sweep pseudocircles seems more difficult. There are two obvious complications: Since pseudocircles can be closed, we may have to sweep several connected components of the graph of their arrangement. The fact that each curve can intersect the sweep twice is certain to complicate the description if not the algorithm.

A special case that avoids these two complications is the case in which the sweep intersects every pseudocircle once. Since passing a hump and taking a loop change two intersections, Lemma 3.2 says that an arrangement of such curves can be swept by passing triangles. An algorithm for this case might have application to computing skewed projections [24, 27].

There are complications even in this case. The envelope trees can have size $\Omega(n^2)$, so we do not want to store them explicitly. We need some structure that can store them compactly and allow efficient updates when two curves change order. We leave this as an open problem.

Another open problem involves sweeping arrangements of k-intersecting curves, for $k > 2$. We can sweep any arrangement by first applying our local operations until we get stuck, then taking curves, violating the k-intersection property, until other operations apply. In an extreme case, if the sweep takes all visible edges, then clearly it can make progress. What intersection property does the sweeping curve satisfy if we sweep with this procedure? How can such a procedure be implemented if the arrangement is not known explicitly? These questions are important for practical applications to sweeping curves.

Acknowledgment

We thank the referee for a careful reading and helpful suggestions and Digital Equipment Corporation for their support of this research.

REFERENCES

1. P. K. Agarwal, M. Sharir, and P. Shor, *Sharp upper and lower bounds on the length of general Davenport-Schinzel sequences*, J. Combin. Theory Ser. A **52** (1989), 228–274.
2. J. L. Bentley and T. A. Ottman, *Algorithms for reporting and counting geometric intersections*, IEEE Trans. Comput. **28** (1979), 643–647.
3. S. A. Burr, B. Grünbaum, and N. J. A. Sloane, *The orchard problem*, Geom. Dedicata **2** (1974), 397–424.
4. B. Chazelle and H. Edelsbrunner, *An optimal algorithm for intersecting line segments in the plane*, Proc. 29th IEEE Sympos. on Foundations of Computer Sci., 1988, pp. 590–600.
5. B. Chazelle, L. J. Guibas, and D. T. Lee, *The power of geometric duality*, BIT **25** (1985), 76–90.
6. D. Dobkin, L. Guibas, J. Hershberger, and J. Snoeyink, *An efficient algorithm for finding the CSG representation of a simple polygon*, Proc. SIGGRAPH '88, Comput. Graphics **22** (1988), 31–40.
7. H. Edelsbrunner, *Algorithms in combinatorial geometry*, Springer-Verlag, Berlin, 1987.
8. H. Edelsbrunner, L. Guibas, J. Pach, R. Pollack, R. Seidel, and M. Sharir, *Arrangements of curves in the plane—topology, combinatorics, and algorithms*, Proc. 15th Internat. Colloq. on Automata, Languages and Programming, 1988, Springer-Verlag, Berlin.
9. H. Edelsbrunner and L. J. Guibas, *Topologically sweeping an arrangement*, Proc. 18th Annual ACM Sympos. on Theory of Computing, 1986, pp. 389–403.

10. H. Edelsbrunner, J. O'Rourke, and R. Seidel, *Constructing arrangements of lines and hyperplanes with applications*, SIAM J. Comput. **15** (1986), 341–363.
11. H. Edelsbrunner, I. Pach, J. T. Schwartz, and M. Sharir, *On the lower envelope of bivariate functions and its applications*, Proc. 28th IEEE Sympos. on Foundations of Computer Science, 1987, pp. 27–37.
12. J. E. Goodman, *Proof of a conjecture of Burr, Grünbaum, and Sloane*, Discrete Math. **32** (1980), 27–35.
13. J. E. Goodman and R. Pollack, *Proof of Grünbaum's conjecture of the stretchability of certain arrangements of pseudolines*, J. Combin. Theory Ser. A **29** (1980), 385–390.
14. ____, *Three points do not determine a (pseudo-) plane*, J. Combin. Theory Ser. A **31** (1981), 215–218.
15. ____, *A theorem of ordered duality*, Geom. Dedicata **12** (1982), 63–74.
16. ____, *Helly-type theorems for pseudoline arrangements in P^2*, J. Combin. Theory Ser. A **32** (1984), 1–19.
17. ____, *Semispaces of configurations, cell complexes of arrangements*, J. Combin. Theory Ser. A **37** (1984), pp. 257–293.
18. D. H. Greene and F. F. Yao, *Finite-resolution computational geometry*, Proc. 27th IEEE Sympos. on Foundations of Computer Science, 1986, pp. 143–152.
19. B. Grünbaum, *Convex polytopes*, Wiley, London, 1967.
20. ____, *The importance of being straight*, Proc. Twelfth Biennial Seminar of the Canad. Math. Congress on Time Series and Stochastic Processes; Convexity and Combinatorics, 1970, pp. 243–254.
21. ____, *Arrangements of hyperplanes*, Proc. Second Louisiana Conf. on Combinatorics, Graph Theory and Computing, 1971, pp. 41–106.
22. ____, *Arrangements and spreads*, CBMS Regional Conf. Ser. in Math., no. 10, Amer. Math. Soc., Providence, RI, 1972.
23. S. Hart and M. Sharir, *Nonlinearity of Davenport-Schinzel sequences and of genralized path compression schemes*, Combinatorica **6** (1986), 151–177.
24. J. W. Jaromczyk and M. Kowaluk, *Skewed projections with an application to line stabbing in R^3*, Proc. Fourth Annual ACM Sympos. on Computational Geometry, 1988, pp. 362–370.
25. F. Levi, *Die Teilung der projektiven Ebene durch Gerade oder Pseudogerade*, Ber. Math. Phys. Kl. Sächs. Akad. Wiss. Leipzig. **78** (1926), 256–267.
26. H. G. Mairson and J. Stolfi, *Reporting line segment intersections*, Theoretical Foundations of Computer Graphics and CAD (R. Earnshaw, ed.) NATO Adv. Sci. Inst. Ser. F: Comput. Systems Sci, no. F40, Springer-Verlag, Berlin and New York, 1988.
27. M. McKenna and J. O'Rourke, *Arrangements of lines in 3-space: A data structure with applications*, Proc. Fourth Annual ACM Sympos. on Computational Geometry, 1988, pp. 371–380.
28. V. J. Milenkovic, *Verifiable implementations of geometric algorithms using finite precision arithmetic*, Ph.D. thesis, Carnegie-Mellon University, Pittsburg, PA, 1988.
29. M. H. Overmars and E. Welzl, *New methods for computing visibility graphs*, Proc. Fourth Annual ACM Sympos. on Computational Geometry, 1988, pp. 164–171.
30. D. P. Peterson, *Halfspace representation of extrusions, solids of revolution, and pyramids*, SANDIA Report SAND84-0572, Sandia National Laboratories, Albuquerque, NM 1984.
31. D. Rolfsen, *Knots and Links*, Publish or Perish, Berkeley, CA, 1976.
32. M. I. Shamos and D. Hoey, *Geometric intersection problems*, Proc. 17th IEEE Sympos. on Foundations of Comput. Science, Houston, TX, 1976, pp. 208–215.
33. M. Sharir, R. Cole, K. Kedem, D. Leven, R. Pollack, and S. Sifrony, *Geometric applications of Davenport-Schinzel sequences*, Proc. 27th IEEE Sympos. on Foundations of Computer Science, 1986, pp. 77–86.
34. A. Wiernik and M. Sharir, *Planar realization of non-linear Davenport-Schinzel sequences by segments*, Discrete Comput. Geom. **3** (1988), 15–47.

DEPARTMENT OF COMPUTER SCIENCE, UNIVERSITY OF BRITISH COLUMBIA, 6356 AGRICULTURAL ROAD, VANCOUVER, BRITISH COLUMBIA, V6T 1W5, CANADA

DEC SYSTEMS RESEARCH CENTER, 130 LYTTON AVENUE, PALO ALTO, CALIFORNIA 94301

DIMACS Series in Discrete Mathematics
and Theoretical Computer Science
Volume **6**, 1991

On Geometric Permutations
and the Katchalski-Lewis Conjecture
on Partial Transversals for Translates

HELGE TVERBERG

1. Introduction

Let F be a family of disjoint convex sets in the plane, and let L be a line meeting all sets in F. Then L is said to be a transversal (or 1-transversal) for F. L meets the sets in some order and this induces two permutations of them. We identify these two permutations and call the ensuing object a geometric permutation of F.

The study of transversal problems goes back to 1935 (see [18]), and the Danzer-Grünbaum-Klee article [2] gives a survey of what was known in 1963. The field seems to have blossomed in recent years, and a survey paper [4] has just been written by Goodman, Pollack, and Wenger. It should be noted that even from the beginning the topic was more general than the description above indicates. The plane can be replaced by R^n, lines by m-flats, disjointness may not be required, and convexity may be replaced by connectedness. Our framework will be as described, however.

The study of geometric permutations is much more recent, and the definition of the concept seems to stem from the paper [9] by Katchalski, Lewis, and Liu. This concept has also been generalized to higher dimensions but we do not go into that here.

The main purpose of this paper is (hopefully) to start the proof of a conjecture by Katchalski and Lewis from 1981, (see [8]). It deals with the case when the sets in F are translates of one compact set K, and one knows that every three sets in F have a transversal. The conjecture states that F will then have a large partial transversal, in the sense that some line will meet all but at most two sets in F. A short formulation of this would be: $T(3) \Rightarrow T - 2$. They make an important step toward this by proving that

---

1991 *Mathematics Subject Classification.* Primary 52A10, 52A37.

$T(3) \Rightarrow T - 603$. This result can be easily strengthened if one combines their approach with results and methods from Eckhoff [3] and Tverberg [17]. One gets something like $T(3) \Rightarrow T - 108$. We do not elaborate upon this, however, but aim straight for $T - 2$.

Our approach to the Katchalski-Lewis conjecture is mainly the same as that of the proof of Grünbaum's conjecture $(T(5) \Rightarrow T - 0)$, in [17]. There we started by proving that, given a counterexample, one can be fabricated consisting of just six sets. Then came a study of the small counterexample. In the present study, we first prove a reduction to 49 sets (which is probably too much, ≈ 20 is more likely). The study of the small counterexample in [17] required a detailed study of geometric permutations of translates. This paper contains some additional information on geometric permutations, which is both interesting in itself and likely to be useful for the study of a small counterexample. It ends with a discussion of how one should continue the proof.

2. Why there is a small counterexample (if any)

Assume F to be a counterexample to the Katchalski-Lewis conjecture. Then F is finite, as Santalo [14] and Hadwiger [6] proved that $T(3) \Rightarrow T - 0$ for infinite families, in the cases int $K = \varnothing$, int $K \neq \varnothing$, respectively. "But how finite?" one may ask. To find out, we consider the *bad* 5-tuples in F, where bad means "not having a transversal."

The bad 5-tuples have the following basic property: Whenever X and Y are two sets from F, some bad 5-tuple is disjoint from $\{X, Y\}$. Or put otherwise: They are not representable by two elements. For if each bad 5-tuple meets $\{X, Y\}$, then $F \backslash \{X, Y\}$ would have no bad 5-tuples and would thus have a transversal by Grünbaum's conjecture.

Now the reader will immediately verify the following fact: If a family of 5-tuples is not representable by two elements, then some subfamily, consisting of $\leq 1 + 5 + 5^2 = 31$ 5-tuples will have the same property. At most 155 translates of K are contained in at least one of these bad 5-tuples, and we conclude: A subfamily of F, with at most 155 sets, form a counterexample.

This crude argument can of course be refined, and a recent result by Tuza [16, Th. 2], who studies the obvious more general problem in finite set theory gives $49 (= \binom{5+2}{5-1} + \binom{5+2-1}{5-1}) - 1)$ instead of 155, so we have the following

THEOREM 1. *If there is a counterexample to the Katchalski-Lewis conjecture, there is one involving ≤ 49 translates.*

In an earlier paper [15], Tuza (and Lehel) formulate an inequality (problem 18(a)) which, if true, would allow us to replace 49 by 25. Here 25 is best possible, as seen by considering the sets $\{i, j, k, x_{i,j,k}, y_{i,j,k}\}$ for $1 \leq i < j < k \leq 5$. But even 25 is a very large number in this connection so it seems necessary to bring the geometry in to improve on Theorem 1. Section 6 will deal with this.

3. On Hadwiger's characterization of families having a transversal, and related matters

Let L induce the geometric permutation $i_1, i_2, \ldots, i_N$ on the disjoint convex sets $1, \ldots, N$ in the plane. Then, of course, for every $k < l < m$, there is a line inducing the geometric permutation i_k, i_l, i_m on $\{i_k, i_l, i_m\}$ (namely L). Hadwiger [7] proved the converse: The existence of the permutation $i_1, \ldots, i_N$ satisfying the condition just stated, implies the existence of a transversal L' for $\{1, \ldots, N\}$. As Hadwiger himself pointed out by an example, one cannot always find what one would have liked to find: a transversal L' inducing the permutation $i_1, i_2, \ldots, i_N$ which was given to start with.

When working on Grünbaum's conjecture, I discovered

THEOREM 2. *Assume that conditions are as above, only that for every $j < k < l < m$ there is now a transversal for $\{i_j, i_k, i_l, i_m\}$, inducing i_j, i_k, i_l, i_m. Then $\{1, \ldots, N\}$ has a transversal inducing $i_1, i_2, \ldots, i_N$.*

Below we present the simple proof, which is different from that of Wenger [19], who discovered Theorem 2 independently. The interesting thing is that we also have:

THEOREM 3. *If the sets are disjoint translates of one convex set, then the conclusion of Theorem 2 holds if one just assumes that for every $j < k < l$ there is a transversal inducing the permutation i_j, i_k, i_l.*

PROOF OF THEOREM 2. Note that the sets may be assumed to be compact. We apply the shrinking process, which in [17] was called the Hadwiger s.p., but which should rather be the Klee s.p., as it first appeared, it seems, in [13]. It consists in performing N homotheties, all with the same factor $\lambda, 0 \le \lambda \le 1$, one to each set, with center in that set, letting λ decrease from 1 to 0. After applying this process, we may assume that there are three sets, A, B, C, say, such that A and C touch a horizontal line L from below, while B touches L from above, with $B \cap L$ between $A \cap L$ and $C \cap L$. The given permutation π has the form $\ldots A \ldots B \ldots C \ldots$. Then L is easily seen to be the only line inducing ABC on $\{A, B, C\}$, and, as Hadwiger shows, L is a transversal for $\{1, \ldots, N\}$. We must check that L induces π. By the condition on 4-tuples L induces the restriction of π to $\{A, B, C, X\}$ for every $X \in \{1, \ldots, N\}\backslash\{A, B, C\}$. What remains to be shown then is that L induces the restriction of π to $\{A, B, C, X, Y\}$ for those pairs $\{X, Y\}$ that are not separated by any of A, B, C in π. It suffices, by symmetry, to consider the cases $\pi = \ldots X \ldots Y \ldots A \ldots B \ldots C \ldots$ and $\pi = \ldots A \ldots X \ldots Y \ldots B \ldots C \ldots$.

In the first case, it must be shown that L does not induce $YXABC$. It suffices to prove that L does not induce $YXAB$ when π is $XYAB$.

Similarly, in the second case it suffices to prove that L does not induce $YXBC$ when π is $XYBC$. Upon reflection in L and replacement of B, C with A, B, we thus see that it suffices to do the first case.

Let L' be a line which induces $XYAB$, while L induces $YXAB$. Then L' has to cross L somewhere in $\mathrm{conv}(A \cap L \cup B \cap L)$, and meet X and Y in the open lower half-plane (the reader is encouraged to draw a figure). Now there arises a bounded domain, the boundary of which consists, in cyclic order, the piece of L connecting X and A, a part of bdry A, the piece of L' between A and Y, a part of bdry Y, the piece of L' between Y and X, and a part of bdry X. That component of $L' \backslash X$, which is disjoint from Y and A (and is contained in the lower half-plane), must escape from that domain somewhere, but this is easily seen to be impossible.

PROOF OF THEOREM 3. In view of Theorem 2, it suffices to prove that if there are lines inducing BCD, ACD, ABD, and ABC, then there is a line inducing $ABCD$. By shrinking we obtain (up to symmetry) either the same situation for A, B, and C as obtained above, or a situation where A, B, and D touch L (A and D from below, B from above, between A and D).

Consider the first case first. As above, L is a transversal for $\{A, B, C, D\}$, and we have to show that neither $DABC$, $ADBC$, nor $ABDC$ is induced by L.

$DABC$. It is clear that the centers of the translates (which may be assumed centrally symmetric, as in [17]), are convexly independent with cyclic order $DBCAD$. The existence of geometric permutations ACD and ABD then show that a segment in the boundary of $\mathrm{conv}(A \cup D)$ meets both C and B. This is the segment which connects A and D but is not in the boundary of $\mathrm{conv}(A \cup B \cup C \cup D)$. Thus we have a geometric permutation $ABCD$ or $ACBD$. $ABCD$ is impossible, as only L induces ABC, so we have $ACBD$. Now we have BCD by assumption, and also BAD (as the present case assumes $DABC$). An argument similar to the one just used then shows that we have either $BACD$ or $BCAD$. In [17] it was shown that with the given cyclic order $BACD$ is impossible, so we have $BCAD$. But, as first shown by Katchalski et al. in [10], $ACBD$ and $BCAD$ cannot coexist for disjoint translates A, B, C, D.

$ADBC$. We assume ABD, so B must meet one of the two segments forming bdry $(\mathrm{conv}\, A \cup D) \backslash (A \cup D)$. B cannot meet the upper of these, as shown by a glance at a drawing. Thus B meets the lower, which forces convex independence of the centers, with cyclic order $DBCAD$. As above, we conclude that $DBCA$ is a g.p. because DBA and DCA are g.p.'s. In the present case, $ADBC$ is also assumed so that incompatible g.p.'s are again obtained.

$ABDC$. Now the g.p. BCD shows that C must meet the upper of the two segments forming bdry $(\mathrm{conv}(B \cup D)) \backslash (B \cup D)$. Likewise, C must meet the lower of those two forming $\text{bdry}\,(\text{conv}(A \cup D))\backslash(A \cup D)$, but this is clearly impossible.

The case when A, B, and D touch L, A, and D from below, B from above, between A and D, is dealt with as follows. We perform a further shrinking, with all four homothety centers on L, until one more permutation becomes critical. But if this is ABC or BCD, we are back to the case above, so we may assume it to be ACD. Thus, C touches one of the two segments forming $\text{bdry}\,\text{conv}(A \cup D)\backslash(A \cup D)$. As C meets L, this segment must be the one contained in L. But then the shrinking can go on unless A, B, C, and D now have become points, in which case L clearly induces $ABCD$.

REMARK. It is not possible to extend Theorem 3 to the case of positive homothets, as can be shown by four appropriately chosen circles.

4. On compatible geometric permutations for translates

In [10] and [11], Katchalski, Lewis, and Liu have obtained some basic information on families of translates having more than one g.p.. Quite a lot of geometric detail is involved, and I found, when doing the cases $N = 5$, $N = 6$, which they omitted in their first paper, that it was better to develop the topic *ab ovo*, basing it on §6 of [17]. The treatment follows, and the first main result will be:

THEOREM 4. *The general form of a compatible pair of distinct geometric permutations for N disjoint convex translates in the plane is* $\{W_1KW_2, W_1K'W_2\}$ *where* $\{K, K'\}$ *is one of the word pairs* $\{AB, BA\}$, $\{ABC, BCA\}$, *and* $\{ABCD, BDAC\}$.

The proof of Theorem 4 starts with the case $N = 4$. One checks immediately that the only types of pairs of g.p.'s which are not mentioned in Theorem 3, are

$$\{ABCD, BADC\} \quad \text{and} \quad \{ABCD, ADCB\}.$$

But the rules I_1 and I_2 in [17] exclude just such pairs. I_1, which is very easy to prove as it does not require the sets to be translates, was first given in [12]. I_2 excludes also pairs $\{ABCD, ADC'B\}$, with $C \neq C$, and the case with $C' = C$ was found independently by Katchalski et al. (cf. [10]).

For $N \geq 5$ it suffices, using induction, to prove that any two distinct g.p.'s have an end in common, i.e., that they can be presented as PW, PW'. For then W and W' are either distinct g.p.'s of $N-1$ sets, or W is the reverse of W'. But the latter case is excluded as $PABC$ is incompatible with $PCBA$, so we may assume that the pair is

$$\{PW_1KW_2, PW_1K'W_2\} \quad \text{or} \quad \{PW_1KW_2, P\overline{W_2}\,\overline{K'}\,\overline{W_1}\}$$

where $\overline{W}$ means W in reverse, and (K, K') is one of the pairs mentioned in the theorem. The first pair is just what we want. If K is formed by three or four sets, some two sets have different orders in K and $\overline{K}'$, and we are finished by the rule I_2, unless W_1 and W_2 are both empty. But $\{PK, P\overline{K}'\}$ has the required form. If $K = AB$ then W_1 and W_2 together have at least two sets, so I_2 is violated again.

It remains to check the assertion about a common end, i.e., that for $N \geq$ 5 a pair of the form $\{A...D, B...C\}$ with $|\{A, B, C, D\}| = 4$ cannot exist. By what we know for $N = 4$ it suffices to show that for $N = 5$ the permutations $A \cdot B \cdot C \cdot D$ and $B \cdot D \cdot A \cdot C$, where in each case a fifth set X replaces one of the dots, are incompatible. Now $AXBCD$ is incompatible with $B \cdot D \cdot A \cdot C$ because, by the rule I_2, $AXBC$ is incompatible. Thus, by symmetry, the only possibility left is

$$ABXCD, \; BDXAC.$$

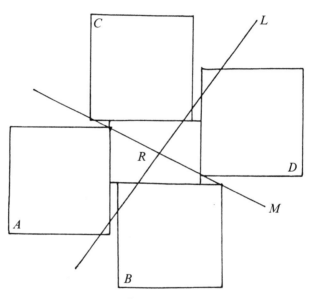

FIGURE 1

In [17] it was proved that if $ABXCD$ and $BDAC$ coexist, then it can be assumed that $A, B, \ldots$ are squares behaving as in Figure 1, where X, not drawn, is contained in the rectangle R. L and M are the transversals inducing $ABXCD$ and $CADB$. It is clear that they will continue to do so if A and D are pushed horizontally and B and C are pushed vertically until they all meet X. One may then also push C to the left, B to the right, A upwards, and D downwards until they all just touch L while they are still met by M. Introducing coordinates so that the SE-corner of C is $(c, 1)$, while the NW-corner of B is $(-b, -1)$ one finds that $A \cap \text{conv}(C \cup D) \neq \varnothing$ is expressed by $c^2 + bc - 2b - 2c + 2 \leq 0$, and that $D \cap \text{conv}(A \cup B) \neq \varnothing$ is expressed by $b^2 + bc - 2b - 2c + 2 \leq 0$. By adding the inequalities, one gets $(b + c - 2)^2 \leq 0$, so that $b + c = 2$ and $b = c = 1$. Thus, L is the line

$y = x$, while M has to be the line $y = -x$. Restoring any of the squares A, B, C, D to its original position, one finds that it misses M.

We will end this section with some remarks on the main result by Katchalski et al., namely that no four g.p.'s can coexist under our conditions. We know already that for $N \geq 5$ any two g.p.'s have an end in common. We shall see that the same holds for any three g.p.'s. Assume the opposite, i.e., that three g.p.'s for the same family have the forms $P \ldots R \ldots Q, P \ldots Q \ldots R,$ $Q \ldots P \ldots R$. Let X be a fourth set. By symmetry we may assume that the first permutation is $P \ldots R \ldots X \ldots Q$. Now, by rule I_2, $PRXQ$ is incompatible with $PQYR$ for any Y (also $Y = X$). Thus, our permutations must be

$$P \ldots R \ldots X \ldots Q, \, P \ldots X \ldots Q \ldots R, \, Q \ldots P \ldots R.$$

By rule I_2 again we specialize them further to

$$P \ldots RX \ldots Q, \, P \ldots X \ldots QR, \, QP \ldots X \ldots R,$$

and then to

$$PRX \ldots Q, \, P \ldots XQR, \, QPX \ldots R.$$

But a fifth set Y would have to behave exactly like X and cannot exist.

It is now easy to reduce the proof of noncoexistence of four g.p.'s for $N \geq 5$, to the corresponding proof for $N = 4$. For by what we have just proved we can assume them to be (where A, B, C are simple letters and $W_1, \ldots, W_4$ words)

$$AW_1, \, AW_2, \, AW_3, \, BW_4C.$$

If $A \notin \{B, C\}$, W_1 and W_2, say, would have to end in, say, B, to have a common end with the last g.p. But then W_3 also has to end in B, to have a common end with the first and the last g.p. Then, changing B to A, we may assume the g.p.'s to be

$$AW_1, \, AW_2, \, AW_3, \, AW_4.$$

But assuming that for smaller values of N no four distinct g.p.'s are coexistent, we must have, say, that W_2 is W_1 in reverse, which is impossible by rule I_2.

The argument just given also deals with the case $N = 4$. For if three permutations have no common end, they must be $PRXQ, PXQR, QPXR,$ say. A fourth one of the form, say, $P \ldots$, must equal $PXRQ$ or $PQRX$ in order to be compatible with $PRXQ$ and $PXQR$, but then it is incompatible with $QPXR$.

If every three of four permutations have a common end, we say that all four have a common end. Thus they have the forms $PW_1, PW_2, PW_3,$ and PW_4. But among the four words $W_1, \ldots, W_4$, one has to be the reverse of another, so the rule I_2 is contradicted again.

It should be remarked that the pigeon-hole-type argument given in [10] also deals nicely with the case $N = 4$.

It is also interesting to find all triples of consistent g.p.'s for every N. We shall now see that for $N \geq 4$ they are of the form $(W_1 K W_2,\ W_1 K' W_2,\ W_1 K'' W_2)$ where K, K', K'' are three distinct g.p.'s of four sets.

We may clearly assume $N \geq 5$, so we know, by induction and because any three consistent g.p.'s have a common end, that such a triple has the form

$$\{PW_1 K W_2,\ PW_1 K' W_2,\ PW_1 K'' W_2\}$$

or the form

$$\{PW_1 K W_2,\ PW_1 K' W_2,\ P\overline{W}_2 \overline{K}'' \overline{W}_1\}.$$

Thus we just have to discuss the latter form further. Now there are clearly at least two sets which occur in opposite order in words K and $\overline{K}''$ since K and K'' are distinct g.p.'s. This means that, by the rule I_2, we can assume that W_1 and W_2 are empty. But K, K', and $\overline{K}''$ are distinct g.p.'s, being the same as K, K', and K''.

In the case $N = 4$, it is easy to convince oneself that there are just four possible triples of g.p.'s that are consistent with rules I_1 and I_2. They are

$$\{AXYZ,\ AYZX,\ AZXY\}, \qquad \{AXYZ,\ AYZX,\ AYXZ\}$$
$$\{ABXC,\ BCXA,\ CAXB\}, \qquad \{ABCD,\ BDAC,\ ACDB\}.$$

It turns out that only the first two triples can actually exist, and they can be extended in the obvious way to triples of g.p.'s for any $N \geq 4$. The necessary example for the first triple is found in [10], and the reader will certainly manage to construct an example, with hexagons which are almost squares, for the second triple. As for the somewhat detailed hexagon study which seems to be necessary for the exclusion of the last two triples, it is best presented as part of a broader study of small families of translates, so we omit it here. We formulate the result as

THEOREM 5. *For $N \geq 3$ there are exactly two types of triples of consistent g.p.'s of convex translates in the plane. They are* $\{W_1 XYZW_2,\ W_1 YZXW_2,\ W_1 ZXYW_2\}$ *and* $\{W_1 XYZW_2,\ W_1 YZXW_2,\ W_1 YXZW_2\}$

5. On Eckhoff's result concerning $T(3)$

In [3] Eckhoff proves a basic result that gives one a good first impression of what a family of translates satisfying $T(3)$ looks like. He introduces a distance function and proves that if 1 and N, say, are the sets with maximal distance between the circumcenters, then the following holds: The lines containing bdry(conv$(1 \cup N))\setminus(1 \cup N)$ jointly meet all the sets $1, \ldots, N$. We can improve this a little bit, to get a yet clearer picture! The *segments* forming bdry(conv$(1\cup N))\setminus(1\cup N)$ meet the sets $2, 3, \ldots, N-1$. The easy argument follows. We have to prove that the set 2, say, meets conv$(1 \cup N)$. We know from above that L, one of the lines containing bdry(conv$(1 \cup N)\setminus(1 \cup N)$, meets 2, so assume now that L induces the g.p. $2, 1, N$. By shrinking

$1, 2$, and N around their circumcenters, one may assume that L also separates 2 from 1 and N. It follows from Eckhoff's definition that the distances between the circumcenters are multiplied by the shrinking factor λ, so we still have that the centers of 1 and N are the furthest apart. If one shrinks a little bit further, the three new sets $1'$, $2'$, N' will still satisfy $T(3)$ because of the transversal inducing $2, 1, N$. Thus, the lines containing $\operatorname{conv}(1' \cup N') \backslash (1' \cup N')$ jointly meets $1'$, $2'$, and N' by the result cited first. But the line L touches 2 and the corresponding line L' then misses 2. We remark that Eckhoff's result does not require the sets to be disjoint, as is the case here.

6. Some final remarks

Even if Tuza's result could be sharpened, the proof of Theorem 1, as given, would only give a reduction to 25 sets for a possible counterexample to $K - L$. We will briefly describe how one can get further down. Consider a counterexample. As shown in the preceding section, one may assume that the two segments forming $\operatorname{bdry}(\operatorname{conv}(1 \cup N)) \backslash (1 \cup N)$ jointly meet the remaining sets $2, 3, \ldots, N-1$. We now move the set 1 in the segment direction, away from N, until some triple including 1 becomes critical, i.e., has just one transversal with this transversal not having the segment direction. (If this never happened, the family would clearly have a transversal in the segment direction.) The critical triple is $\{1, A, B\}$, say. Note that $N \notin \{A, B\}$. In the same way, one can obtain a critical triple $\{N, C, D\}$ with $1 \notin \{C, D\}$.

Now, since we have a counterexample to $K - L$, the unique transversal for $\{1, A, B\}$ misses at least three sets X, Y, and Z. Thus, we have three bad quadruples $\{1, A, B, X\}$, $\{1, A, B, Y\}$, and $\{1, A, B, Z\}$; and, similarly, $\{N, C, D, X'\}$, $\{N, C, D, Y'\}$, and $\{N, C, D, Z'\}$. (Since we are interested in having a small number of sets, it should be observed here that the uniqueness of the two transversals implies that one may assume either $\{X, Y, Z\} \cap \{N, C, D\} \neq \phi \neq \{X', Y', Z'\} \cap \{1, A, B\}$ (if the transversals are distinct) or $\{A, B\} = \{C, D\}$ (if they are the same).)

The bad quadruples obtained show that there must also be a family of bad quintuples for which the only representing pairs miss $\{1, A, B\}$ or $\{N, C, D\}$.

We will not go into details here, but only mention that "49" can be reduced to "x" where x is a little below 20, this way. What was obtained above could also have been obtained by shrinking, but the moving of 1 and N makes for a stronger link between the geometric situation and the bad quadruples and quintuples.

The study of a possible counterexample, specified in terms of bad quadruples and quintuples, will certainly be a complicated matter. The strategy that seems most promising is to use geometry to get maximal information on possible behaviour of small families satisfying $T(3)$, so that larger families can be studied without too many geometric considerations. This section originally contained a suggestion that it might be profitable to try to prove (in an obvious notation)

$$T(3) \Rightarrow T(4) - 1 \quad \text{and} \quad T(4) \Rightarrow T - 1.$$

This statement clearly implies $T(3) \Rightarrow T - 2$, and, if true, it could be proved by considering only families of ≤ 12 sets. Unfortunately, for any N, Bezdek [1] has given a counterexample, with circles, to $T(3) \Rightarrow T - 1$. For at least six circles, it is known from [5] that $T(4) \Rightarrow T$. Thus, $T(3) \Rightarrow T(4) - 1$ is false already for seven circles. The status of $T(4) \Rightarrow T - 1$ is unknown. There are two statements, weaker than $K - L$, which ought to be checked if $K - L$ turns out to be too difficult. They are

$$T(3) \Rightarrow T(4) - 2 \quad \text{and} \quad T(4) \Rightarrow T - 2.$$

Together, they would give $T(3) \Rightarrow T - 4$, and they have the advantage that for the first one, only families of ≤ 12 sets have to be studied, while the second one involves the stronger condition $T(4)$. For more information on $T(4)$ (which ought to be studied in any case, of course) see [3].

7. Acknowledgment

The author is grateful to the DIMACS organization, which supported a very stimulating three-week visit to Rutgers University during its special year in Discrete and Computational Geometry. This paper is based on a talk presented there.

References

1. A. Bezdek, personal communication, cited in [4].
2. L. Danzer, B. Grünbaum, and V. Klee, *Helly's theorem and its relatives*, Convexity, Proc. Sympos. Pure Math. vol 7, Amer. Math. Soc., Providence, RI, 1963, pp. 100–181.
3. J. Eckhoff, *Transversalenprobleme in der Ebene*, Arch. Math. **24** (1973), 195–202.
4. J. Goodman, R. Pollack, and R. Wenger, *Geometric transversal theory*, New Trends in Discrete and Computational Geometry, Springer-Verlag, Berlin and New York.
5. B. Grünbaum, *On common transversals*, Arch. Math. **9** (1958), 465–469.
6. H. Hadwiger, *Über einen Satz Hellyscher Art*, Arch. Math. **7** (1956), 377–379.
7. ____, *Über Eibereiche mit gemeinsamen Treffgeraden*, Portugal. Math. **16** (1957), 23–29.
8. M. Katchalski and T. Lewis, *Cutting families of convex sets*, Proc. Amer. Math. Soc. **79** (1980), 457–461.
9. M. Katchalski, T. Lewis, and A. Liu, *Geometric permutations and common transversals*, Discrete Comput. Geom. **1** (1986), 371–377.
10. ____, *Geometric permutations of disjoint translates of convex sets*, Discrete Math. **65** (1987), 249–259.
11. ____, *The different ways of stabbing disjoint convex sets*, unpublished manuscript.
12. M. Katchalski, T. Lewis, and J. Zaks, *Geometric permutations for convex sets*, Discrete Math. **54** (1985), 271–284.
13. V. Klee, *Common secants for plane convex sets*, Proc. Amer. Math. Soc. **5** (1954), 639–641.
14. L. A. Santaló, *Complemento a la nota: Un teorema sôbre conjuntos de paralelipedos de aristas paralelas*, Publ. Inst. Math. Univ. Nac. Litoral **3** (1942), 202–210.
15. Z. Tuza, *Critical hypergraphs and intersecting set-pair systems*, J. Combin. Theory Ser. B **39** (1985), 134–145.

16. ____, *The minimum number of elements representing a set system of given rank*, J. Combin. Theory Ser. A **52** (1989), 84–89.

17. H. Tverberg, *Proof of Grünbaum's conjecture on common transversals for translates*, Discrete Comput. Geom. **4** (1989), 191–203.

18. P. Vincensini, *Figures convexes et variétés linéaires de l'espace euclidien à n dimensions*, Bull. Sci. Math. (2) **59** (1935), 163–174.

19. R. Wenger, *A generalization of Hadwiger's theorem to intersecting sets*, Discrete Comput. Geom. **5** (1990), 383–388.

DEPARTMENT OF MATHEMATICS, UNIVERSITY OF BERGEN, ALLEGT. 55, N-5007 BERGEN, NORWAY

DIMACS Series in Discrete Mathematics
and Theoretical Computer Science
Volume 6, 1991

Invariant-Theoretic Computation
in Projective Geometry

NEIL L. WHITE

A small group of researchers has recently been investigating the use of the algebra of projective invariants, as opposed to the usual coordinate algebra of elementary analytic geometry, for symbolic computation applied to projective and affine geometry problems. For problems in $(d-1)$-dimensional projective space, these invariants are just the determinants of d-tuples of points, which are called brackets. The bracket algebra has the advantage of being much closer to the underlying geometry than the usual algebra. The purpose of this paper is to give an exposition of the bracket algebra and some of its applications. We begin with an overview of the bracket algebra and the straightening algorithm, which is a normal form algorithm in that algebra. We describe the Bokowski-Sturmfels method of final polynomials, which uses the bracket algebra to give compact proofs of the nonrealizability of certain polytopes and oriented matroids. We then describe the Grassmann algebra or Cayley algebra, which is very useful for deriving bracket algebra expressions which are equivalent to synthetic geometric conditions. The Cayley factorization problem is the problem of going in the reverse direction: given a bracket algebra condition, effectively find the synthetic geometric condition which is equivalent to it. We give an algorithm which solves this problem in the multilinear case.

Bracket algebra and straightening

Let E be a finite set of points $\{e_1, e_2, \ldots, e_n\}$ in $(d-1)$-dimensional projective space over a field F. We will use homogeneous coordinates, so each point is represented by a d-tuple, which we take to be a column of the

---

1991 *Mathematics Subject Classification.* Primary 13A50, 15A75, 51A05, 52C25.

following matrix:

$$
\begin{pmatrix}
x_{1,1} & x_{1,2} & \cdots & x_{1,n} \\
x_{2,1} & x_{2,2} & \cdots & x_{2,n} \\
\vdots & \vdots & \cdots & \vdots \\
x_{d,1} & x_{d,2} & \cdots & x_{d,n}
\end{pmatrix}
$$

Now we assume that the entries of the above matrix are algebraically independent indeterminates over F.

We define the *bracket ring B of E* (*over F in rank d*) to be the subring of $F[x_{1,1}, x_{1,2}, \ldots, x_{d,n}]$ generated by all brackets

$$
[e_{i_1}, e_{i_2}, \ldots, e_{i_d}] = \det
\begin{vmatrix}
x_{1,i_1} & x_{1,i_2} & \cdots & x_{1,i_d} \\
\cdots & \cdots & \cdots & \cdots \\
x_{d,i_1} & x_{d,i_2} & \cdots & x_{d,i_d}
\end{vmatrix}.
$$

These brackets are not algebraically independent, but rather they satisfy the following relations:

(1) $[a_1, a_2, \ldots, a_d] = 0$ if any $a_j = a_k$, $j \neq k$.

(2) $[a_1, a_2, \ldots, a_d] = \text{sign}(\sigma)[a_{\sigma(1)}, a_{\sigma(2)}, \ldots, a_{\sigma(d)}]$ for any permutation σ of $\{1, 2, \ldots, d\}$.

(3) $[a_1, a_2, \ldots, a_d][b_1, b_2, \ldots, b_d]$
$= \sum_{j=1}^{d}[a_1, a_2, \ldots, a_{d-1}, b_j][b_1, b_2, \ldots b_{j-1}, a_d, b_{j+1}, \ldots, b_d]$.

It is a classical theorem (called the Second Fundamental Theorem of Invariant Theory, for projective invariants) that all relations among the brackets are consequences of relations of the above three types. The relations of the third type are called the *Grassmann-Plücker relations* or *syzygies*, and they correspond to generalized Laplace expansions in B.

Now we wish to consider how to tell if two elements of B (written in terms of brackets) are equal (as elements of $F[x_{1,1}, x_{1,2}, \ldots, x_{d,n}]$). This can be extremely difficult to do directly from the definition. Let $<$ denote a linear order on E. Let $M = [a_1, a_2, \ldots, a_d][b_1, b_2, \ldots, b_d] \cdots [c_1, c_2, \ldots, c_d]$ be a monomial in B. We represent M by a *tableau* T,

$$
T = \begin{matrix}
a_1 & a_2 & \cdots & a_d \\
b_1 & b_2 & \cdots & b_d \\
\vdots & \vdots & \ddots & \vdots \\
c_1 & c_2 & \cdots & c_d
\end{matrix}.
$$

We may then write every element of B as a linear combination of tableaux over F. By the antisymmetry of brackets (relation (2) above), we may assume that each row of a tableau T is ordered, $a_1 < a_2 < \cdots < a_d$, etc. If, in addition, each column of T is nonstrictly ordered, $a_j \leq b_j \leq \cdots \leq c_j$, then we say that T is a *standard tableau*.

Example:

$$T = \begin{array}{cccc} a & c & f & g \\ b & d & g & h \\ b & e & f & h \end{array}$$

is nonstandard, assuming that alphabetical order is our linear order on the points, since the third column is not correctly ordered.

THEOREM 1. *The standard tableaux form a basis of B as a vector space over F.*

In fact, there is a well-known algorithm, known as the *straightening algorithm*, for computing the unique linear combination of standard tableaux equal to a given bracket polynomial [7, 12, 17, 29]. We will skip the details of this algorithm, which repeatedly applies certain relations derived from the Grassmann-Plücker relations. These relations may also be viewed as a special case of a Gröbner basis, and hence the straightening algorithm is a normal form computation with respect to the Gröbner basis. This connection is explained in [20].

Since the number of standard tableaux of a given shape is exponential in the number of entries, the straightening algorithm is inherently very inefficient on large tableaux.

Given this inefficiency, why do we bother with this algebra, which even apart from issues of computational efficiency is more complicated to use than ordinary coordinate algebra? There are several reasons.

(1) We have eliminated the actual coordinates, allowing us to work symbolically with the brackets, independently of a choice of basis.

(2) The brackets are projective invariants, and are therefore the most natural language for doing projective geometry.

(3) The bracket algebra is much closer to the geometry, in a sense that we will later make precise via the Cayley algebra.

First, however, we examine a more direct application of bracket algebra.

The Bokowski-Sturmfels method of final polynomials

Final polynomials are a method of proving nonrealizability of various combinatorial configurations in projective space, by producing a bracket polynomial which can be shown to be zero (by the straightening algorithm), but which must be nonzero if the desired geometric realization of the combinatorial configuration is possible.

The types of problems to which this method applies include realizations of (1) simplicial spheres as polytopes, (2) other cell complexes, (3) matroids, (4) oriented matroids, and (5) partial matroids or partial oriented matroids.

EXAMPLE 1. We give as an example the uniform non-Pappus oriented

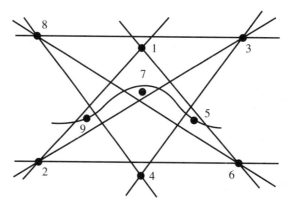

FIGURE 1. The uniform non-Pappus oriented matroid.

matroid of Figure 1, which we wish to show is nonrealizable by straight lines in the plane.

An ordered triple of points is oriented counter-clockwise if and only if its bracket is positive, and the curved lines of the figure indicate which triples are to be regarded as counterclockwise. Thus $2, 4, 6$ is counterclockwise, since 4 is below the line $\overline{26}$, and 2 is to the left of 6. The following final polynomial, after verification by the straightening algorithm that it is zero, proves that the configuration is nonrealizable (i.e., nonstretchable), since every term of the polynomial must be *positive* if the configuration is realizable.

$$
\begin{aligned}
& [246][184][175][437][197] \;+\; [129][184][175][437][467] \\
+\;& [138][194][247][175][467] \;+\; [156][184][247][437][197] \\
+\;& [345][184][247][176][197] \;+\; [489][247][175][176][143] \\
+\;& [597][247][184][176][143] \;+\; [678][247][175][194][143] \\
& \qquad\qquad\quad +\; [237][194][184][175][467] \;\;=\; 0.
\end{aligned}
$$

Final polynomials were first studied by Bokowski and applied to polytopality problems, as well as oriented matroid problems. White [23] proved implicitly that a matroid is nonrealizable if and only if it has a final polynomial with a single nonzero monomial. Dress and Sturmfels independently saw, using a real version of the Hilbert Nullstellensatz, that an oriented matroid is nonrealizable if and only if it has a final polynomial (see [2]). Although a final polynomial gives a fast and easily understandable proof of nonrealizability, finding a final polynomial is quite difficult. Bokowski and Richter [2] have had some success in finding final polynomials by turning realizability problems into linear programming problems. Nonfeasibility of the LP problem implies the existence of a dual LP solution, which is then used to construct the final polynomial.

Cayley algebra

We now describe an algebra that allows the direct encoding of synthetic projective (or affine) geometric statements into the bracket algebra. This algebra is the Cayley algebra, which is essentially the same as the classical Grassmann algebra. There is also the possibility of translation in the reverse direction, from the bracket algebra back to the Cayley algebra, and thence to synthetic statements. This process is called Cayley factorization, and will be considered in the next section. A more thorough exposition of the Cayley algebra may be found in [1], [7], or [18].

Let V be a vector space of dimension d over a field F of characteristic zero. Let $\Lambda(V)$ denote the exterior algebra of V (many algebra books, including [4, 8, 9, 14, 15], contain the definition and details of exterior algebra, though Marcus refers to the exterior algebra as the Grassmann algebra). We will write the exterior product in $\Lambda(V)$ as $\vee$ rather than the usual $\wedge$, and refer to it as the *join* operation. As is well known, this product is associative, distributive over addition, and antisymmetric. Now,

$$\Lambda(V) = \bigoplus_{k=0}^{d} \Lambda^k(V),$$

where

$$\dim_F \Lambda^k(V) = \binom{d}{k}.$$

In particular, if we choose a basis $\{e_1, \ldots, e_d\}$ of V over F, then a basis for $\Lambda^k(V)$ over F is

$$\{e_{i_1} \vee e_{i_2} \vee \cdots \vee e_{i_k} \mid 1 \leq i_1 < i_2 < \cdots < i_k \leq d\}.$$

Let $a_1, \ldots, a_k \in V$, and write $A = a_1 \vee a_2 \vee \cdots \vee a_k$ or $A = a_1 a_2 \cdots a_k$ for the join of the k vectors. Then A is called an *extensor of step* k, or a *decomposable k-vector*. It is not always possible to write a sum of two or more extensors of step k as another extensor of step k, hence we also have indecomposable k-vectors in $\Lambda^k(V)$. For example, if a, b, c, and d are linearly independent in V, then $ab + cd$ is an indecomposable 2-vector in $\Lambda^2(V)$.

Let $B = b_1 b_2 \cdots b_j$ be an extensor of step j. Then

$$A \vee B = a_1 \vee a_2 \vee \cdots \vee a_k \vee b_1 \vee \cdots \vee b_j = a_1 a_2 \cdots a_k b_1 \cdots b_j$$

is an extensor of step $j + k$. In fact, $A \vee B$ is nonzero if and only if $a_1, a_2, \ldots, a_k, b_1, \ldots, b_j$ are distinct and linearly independent. If $\overline{A}$ denotes the span of $\{a_1, a_2, \ldots, a_k\}$ when $A \neq 0$, then $\overline{A}$ is called the *support* of A. Assuming $A \neq 0$, then $\overline{A}$ is well defined. We note that A is determined up to scalar multiple by $\overline{A}$, a well-known fact from exterior algebra. If $a_1, a_2, \ldots, a_k, b_1, \ldots, b_j$ are distinct and linearly independent, then $\overline{A \vee B} = \overline{A} + \overline{B}$, and thus the join operation on extensors corresponds to the lattice join of subspaces of V in the independent case, that is, in the case that $\overline{A} \cap \overline{B} = 0$. It is for this reason that we have chosen to call the exterior product the "join" and we will shortly define a second operation which corresponds similarly to the lattice meet.

An extensor A of step d is, up to a fixed nonzero multiple, the determinant of its d constituent vectors. We normalize so that this multiple is unity, thus such an extensor is just the bracket considered earlier, $A = a_1 a_2 \cdots a_d = [a_1, a_2, \dots, a_d]$. We note that $\Lambda^d(V)$ may be identified with the scalar field F.

We now define the *meet* operation. If $A = a_1 a_2 \cdots a_j$ and $B = b_1 b_2 \cdots b_k$, with $j + k \geq d$, then

$$A \wedge B = \sum_\sigma \operatorname{sign}(\sigma)[a_{\sigma(1)}, \dots, a_{\sigma(d-k)}, b_1, \dots, b_k] a_{\sigma(d-k+1)} \cdots a_{\sigma(j)}.$$

The sum is taken over all permutations σ of $\{1, 2, \dots, j\}$ such that $\sigma(1) < \sigma(2) < \cdots < \sigma(d-k)$ and $\sigma(d-k+1) < \sigma(d-k+2) < \cdots < \sigma(j)$. Such permutations are called *shuffles* of the $(d-k, j-(d-k))$ split of A.

An alternate notation for such signed sums over shuffles is the Scottish (named for Turnbull [22]) or *dotted* notation, which we will frequently employ. We simply place dots over the shuffled vectors, with the summation and $\operatorname{sign}(\sigma)$ implicit. Similarly, shuffles may be defined over splits into any number of parts, and denoted by dots. In this paper, the parts into which dotted sets of vectors are split are determined by the brackets, including perhaps one part determined by those vectors which are outside all brackets. If we wish to sum over several shuffles of disjoint sets, we will use separate symbols (triangle, square) over the vectors of each shuffled set. Thus

$$A \wedge B = [\overset{\bullet}{a}_1, \dots, \overset{\bullet}{a}_{d-k}, b_1, \dots, b_k] a_{d-k+1} \cdots a_j.$$

If $j + k = d$ then $A \wedge B = [a_1, \dots, a_j, b_1, \dots, b_k]$. This is a scalar of step 0, and we must be careful to distinguish it from $A \vee B$, a scalar with the same numerical value but of step d. Thus $\Lambda^0(V)$ is a second copy of the scalar field F in $\Lambda(V)$.

The following facts about the meet are perhaps not obvious from the definition. A meet of two extensors is again an extensor. The meet is associative and anticommutative in the following sense:

$$A \wedge B = (-1)^{(d-k)(d-j)} B \wedge A.$$

We have given the definition for both join and meet in terms of two extensors, for the sake of simplicity, but the definitions are extended to arbitrary elements of $\Lambda(V)$ by distributivity. The extended operations remain well defined and associative. The meet is dual to the join, where duality exchanges vectors with covectors (extensors of step $d-1$). The meet corresponds to lattice meet of subspaces, $\overline{A \wedge B} = \overline{A} \cap \overline{B}$, provided $\overline{A} \cup \overline{B}$ spans V. Thus the meet operation corresponds to our geometric intuition in the case that A and B themselves are nondegenerate and that $\overline{A} \cap \overline{B}$ has as small dimension as possible. The *Cayley algebra* is the vector space $\Lambda(V)$ together with the operations $\vee$ and $\wedge$. We will say that C is a *Cayley algebra expression* if C consists of a finite numbers of symbols for points (vectors) of V joined by $\vee$, $\wedge$, $+$, and scalar multiplication. A *Cayley algebra statement* is an equation involving Cayley algebra expressions, typically just $C = 0$.

We now illustrate the translation of geometric incidences into Cayley algebra statements. Consider the affine plane over F. By its usual embedding into a projective plane, we have a vector space V of dimension 3 over F, with its subspaces of dimensions 1 and 2 corresponding to points and lines, respectively. Thus we can represent both projective and affine geometries in a symbolic, coordinate-free way in the Cayley algebra.

EXAMPLE 2. Consider 3 coincident lines, $\overline{ab}$, $\overline{cd}$, $\overline{ef}$, as shown in Figure 2.

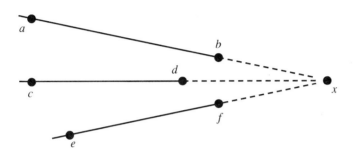

FIGURE 2. Three coincident lines.

Then $ab \wedge cd = \alpha x$, where α is a scalar, and $x \wedge ef = 0$. Thus

$$0 = (ab \wedge cd) \wedge ef = ([acd]b - [bcd]a) \wedge ef = ([a\overset{\bullet}{c}d]\overset{\bullet}{b}) \wedge ef = [a\overset{\bullet}{c}d][\overset{\bullet}{b}ef].$$

In similar fashion, any incidence theorem or incidence relation in projective geometry may be translated into a conjunction of Cayley algebra statements, and, conversely, Cayley algebra statements may be translated back to projective geometry just as easily, provided they involve only join and meet, not addition. We will refer to Cayley algebra expressions (resp. statements) involving only join and meet as *simple* Cayley algebra expressions (resp. statements). Now, simple Cayley algebra statements may be expanded into bracket statements by the definitions and properties of join and meet. Most Cayley algebra statements resulting from geometric incidence relations have step 0 or d. If a simple Cayley algebra statement $C(a, b, \dots) = 0$ has $C(a, b, \dots)$ of step k, $k \neq 0$ or d, then this is the equivalent of

$$C(a, b, \dots) \vee x_1 \vee \cdots \vee x_{d-k} = [C(a, b, \dots), x_1, \dots, x_{d-k}] = 0$$

for all $x_1, \dots, x_{d-k} \in V$. This in turn is equivalent to $[C(a, b, \dots), y_1, \dots, y_{d-k}] = 0$ for all $y_1, \dots, y_{d-k} \in Y$ where Y is a basis of V. Thus every simple Cayley algebra statement is equivalent to a finite conjunction of bracket statements. However, the converse problem, that of writing a bracket statement as a simple Cayley algebra statement, when possible, is not easy.

This is the problem we refer to as *Cayley factorization*.

Cayley factorization

To explain the importance of this problem, let us introduce one more step away from the geometry, namely, the introduction of coordinates. Until now, we have been dealing with invariant languages with respect to the projective general linear group. Our Cayley algebra and bracket algebra statements may be expanded in terms of the coordinates of the vectors, but then we have statements in a larger algebra which includes noninvariant expressions. We may represent the situation in the following diagram:

(1)   Projective geometry
            $\updownarrow$
(2)   Cayley algebra
            $\downarrow$ $\uparrow$ Cayley factorization
(3)   Bracket algebra
            $\downarrow$
(4)   Coordinate algebra

Consider now the situation in computer-aided geometric reasoning. Considerable success has already been attained by automated geometry-theorem-proving programs in proving theorems of projective and Euclidean geometry, by going directly to the coordinate algebra [5, 13]. Suppose, however, that we wish to use the computer in a more interactive fashion, where the computer reduces the problem to an algebraic one, does some symbolic manipulation, and then we wish to interpret these algebraic results geometrically in order to decide how to proceed. From ordinary coordinate algebra, it is computationally much more difficult to find such geometric interpretation in general than it is from the bracket algebra. Indeed, any such interpretation program would at some point have to ascend to invariant statements, and therefore contain in some disguised form a Cayley factorization algorithm. However, in projective synthetic geometry, we can do our symbolic computation directly in the bracket algebra, since brackets are projective invariants, and if we have a Cayley factorization algorithm, then we can more easily find such geometric interpretation when it exists. The main result of this paper is such an algorithm for an important special case, that of *multilinear* bracket expressions, wherein each point (or vector) occurs only once in each bracket monomial. A preliminary version of the algorithm appeared in [25].

Some more immediate applications of Cayley factorization are in structural rigidity of bar-and-joint and bar-and-body frameworks [26, 27, 6], convex polytopes [3], scene analysis [6], and splines [28]. The infinitesimal rigidity of certain kinds of frameworks, the realizability of certain convex polytopes, the realizability of "scenes" as projections of higher-dimensional objects, and the dimensions of spline spaces have been characterized in terms of bracket statements. Cayley factorization of these statements would provide direct geometric interpretation of these conditions. In the case of bar-and-body frameworks, the bracket conditions are always multilinear, hence our algorithm applies.

We mention parenthetically that the work of Havel [11] on distance geometry is similar in spirit to our approach. He translates statements in Euclidean geometry into equations involving Cayley-Menger determinants. Again the reverse translation is fairly easy, and the corresponding geometric statements may be regarded as occurring in an arbitrary quadratic space rather than Euclidean geometry specifically. Cayley-Menger determinants may be expanded into expressions in the inter-point distances, which are the invariants of the Euclidean group. One may ask whether there exists a Cayley-Menger factorization algorithm which would allow a translation of statements in the distance algebra back to geometric statements.

EXAMPLE 3. We now give an example of a bar-and-joint framework to illustrate some of the above ideas. A bar-and-joint framework is a structure built out of rigid bars, or line segments, attached at flexible joints. An infinitesimal motion of such a framework is an assignment of velocity vectors to the joints so that the lengths of the bars are preserved instantaneously (at time 0). For example, a triangular framework of three bars is infinitesimally rigid, meaning that the only infinitesimal motions are the Euclidean motions of the entire framework. However, this framework has a special position in which it is not infinitesimally rigid, namely when all three bars are collinear. Then we can assign a velocity vector of 0 to two of the joints and a vector perpendicular to the line of the bars to the third joint.

It is known that the infinitesimal rigidity of a framework is a projective invariant, that is, applying a projective transformation to the framework does not change its rigidity or lack thereof. It follows that the special positions in which a given framework is nonrigid should be expressible in terms of projective invariants, namely brackets. White and Whiteley [26] give an explicit combinatorial algorithm to find a single bracket polynomial which does this in the case of a framework which is minimally rigid in general position.

We consider the particular framework obtained by realizing the graph $K_{3,3}$ in the Euclidean plane. We think in terms of homogeneous coordinates (although this is a coordinate-free approach!), hence the appropriate vector space is R^3. The White-Whiteley algorithm now adds appropriate "tie-down" bars, in this case at vertices a and b, directed outward, and then directs all the remaining edges so that each vertex has out-degree two. The pure condition then has one bracket monomial for each such choice of directions, with an appropriate sign attached, and each such monomial having one bracket for each non-tied-down vertex. This is illustrated in Figure 3; for details consult White and Whiteley [26]. Thus this framework has an infinitesimal motion if and only if

$$[cdb][dae][ebf][fac] - [cbf][dac][ebd][fae] = 0.$$

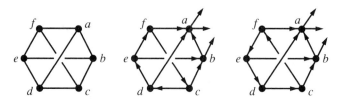

FIGURE 3. $K_{3,3}$ as a bar-and-joint framework.

This is an algebraic statement at level 3 in the above diagram, and we could, of course, expand out each of these 3-by-3 determinants to get a level 4 expression in terms of the coordinates of the 6 vertices. This would be a big mess of 2592 terms, before cancellation. But we want to go in the opposite direction, namely to Cayley factor this bracket expression. Although this bracket polynomial is not multilinear, it does have a Cayley factorization; namely, the above bracket condition is equivalent to

$$(ab \wedge de) \vee (bc \wedge ef) \vee (cd \wedge af) = 0.$$

This is immediately translated into a synthetic statement, which we recognize as the condition of Pascal's Theorem, hence the result is that the framework has an infinitesimal motion if and only if the 6 joints lie on a conic (possibly degenerate).

EXAMPLE 4. To illustrate a multilinear condition, we give an example of a bar-and-body framework, as shown in Figure 4. This framework consists of three rigid planar bodies, free to move in the Euclidean plane, and connected by six rigid bars, with the connections at the ends of each bar allowing free rotation of the bar relative to the rigid body.

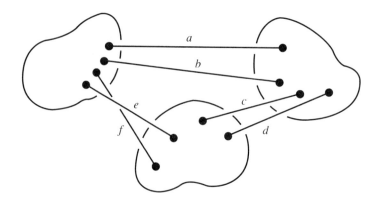

FIGURE 4. A bar-and-body framework.

Let $V = R^3$, and $W = \Lambda^2(V) \cong R^3$. We think of the endpoints of the bars as elements of V, and hence the lines determined by the bars are two-extensors of these points, or elements of W. In White and Whiteley [27], a direct combinatorial algorithm similar to that for bar-and-joint frameworks is given which provides the bracket condition for such a framework to have an infinitesimal motion. In our example, we find that the framework has an infinitesimal motion if and only if $[abc][def] - [abd][cef] = 0$. Now, this bracket polynomial may be Cayley factored as $ab \wedge cd \wedge ef = 0$, as in Example 2. Now we switch to thinking of $a, b, \dots, f$ as 2-extensors in V rather than elements of W, and recall that we have duality between V and W, hence between $\Lambda(V)$ and $\Lambda(W)$. Thus the above becomes $(a \wedge b) \vee (c \wedge d) \vee (e \wedge f) = 0$. This is the desired geometric condition for the existence of an infinitesimal motion, and is in this case just an application of the classical theorem of Arnhold-Kempe that the centers of relative motion of three pairs of bodies must be collinear.

Returning to the general consideration, let $P(a, b, c, \dots)$ be a bracket polynomial which is the expansion of a simple Cayley algebra expression. Then P is homogeneous, that is, for each (variable) vector x which is an argument of P, there is a positive integer n_x such that each monomial of P has precisely n_x occurrences of x among the vectors in its brackets. In fact, n_x is also the total number of occurrences of x in the simple Cayley algebra expression. If $n_x = 1$ for all x, then P is *multilinear*. Now suppose that $ab \cdots e$ occurs explicitly as a join of vectors in the simple Cayley expression. We will refer to $ab \cdots e$ as an *atomic extensor* in that case. Each atomic extensor appears in P as a dotted set of symbols, where undotted symbols all in the same bracket may be considered a special case of being dotted. EXAMPLE 5. Let $d = 5$. Then

$$(abcd) \wedge (efgh) \wedge (amn) \wedge (bekl) = ([a\overset{\bullet}{e}fgh]b\overset{\bullet\bullet\bullet}{c}d) \wedge (amn) \wedge (bekl)$$

$$= ([a\overset{\bullet}{e}fgh][b\overset{\bullet\bullet}{c}amn]\overset{\bullet}{d}) \wedge (bekl)$$

$$= [a\overset{\bullet}{e}fgh][b\overset{\bullet\bullet}{c}amn][\overset{\bullet}{d}bekl],$$

where the dotting in the bracket polynomial indicates the signed sum over all permutations $\sigma(a), \sigma(b), \sigma(c), \sigma(d)$ such that $\sigma(b) < \sigma(c)$ with "$<$" refering to alphabetical order. Note that in this nonmultilinear example, some of these permutations cause repeated letters in a bracket, making those terms zero.

The multilinear Cayley factorization algorithm

Now we are ready to describe the Cayley factorization algorithm for multilinear bracket polynomials. Let the input be $P(a, b, \dots, z)$, a bracket polynomial of rank d which is multilinear in the N points $a, b, \dots, z$.

Step 1. We first find the atomic extensors. We define $a \sim b$ if $a = b$ or if $P(a, b, \dots, z) = -P(b, a, \dots, z)$. The latter is checked by applying the straightening algorithm to $P(a, b, \ldots, z) + P(b, a, \ldots, z)$, or, equivalently (in the multilinear case), to $P(a, a, \ldots, z)$, to see if zero is obtained. The transitivity of $\sim$ cuts down on the number of pairs of points we have to check, although it is still $O(N^2)$ calls of the straightening algorithm in the worst case. The equivalence classes are precisely the maximal atomic extensors, if a factorization exists, by a result in [**24**]. If there do not exist two atomic extensors whose sizes sum to at least d, then return **no factorization possible**.

Step 2. If there exists an atomic extensor E of step d, then apply straightening with the d elements of E first in the linear order. The result must have E as the first row of every resulting tableau, that is, $[E]$ is now an explicit factor, $P = [E] \cdot Q$. Remove E, store $P = E \wedge Q$, and proceed with the bracket polynomial Q replacing P. Repeat step 2 as appropriate. If $P = 1$, then we are DONE, and the required Cayley factorization may be reconstructed.

Step 3. Find two atomic extensors $E = \{e_1, e_2, \ldots, e_k\}$ and $F = \{f_1, f_2, \ldots, f_\ell\}$ such that $E \wedge F$ could be a primitive factor, if they exist. We need only check pairs of extensors E, F such that $k + \ell \geq d$. By a primitive factor, we mean that $(e_1 e_2 \cdots e_k \wedge f_1 f_2 \cdots f_\ell)$ occurs explicitly in some Cayley factorization of P. Note that any simple Cayley expression must have such a primitive factor. A necessary condition for $E \wedge F$ to be a primitive factor is that if the straightening algorithm is applied to P, using an ordering in which $e_1 < e_2 < \cdots < e_k < f_1 < f_2 < \cdots < f_\ell < x$ for all $x \in A - (E \cup F)$, then the result is in the form

$$(*) \quad \sum_{x_1, \ldots, y_1 \ldots} \begin{matrix} e_1 & \cdots & \cdots & \cdots & \cdots & e_k & \overset{\bullet}{f_1} & \cdots & \overset{\bullet}{f_{d-k}} \\ \overset{\bullet}{f_{d-k+1}} & \cdots & \overset{\bullet}{f_\ell} & x_1 & \cdots & \cdots & \cdots & \cdots & x_{2d-k-\ell} \\ y_1 & \cdots & \cdots & \cdots & \cdots & \cdots & \cdots & \cdots & y_d \\ \vdots & & & \vdots & & & & & \vdots \end{matrix}$$

That is, every tableau has E in the first row, part of F filling up the rest of the first row, and the rest of F in the second row. The sum is over various terms with different choices for the x's, y's, etc.

Actually, we are ignoring two global signs, since

$$E \wedge F = \pm F \wedge E = \pm([\overset{\bullet}{f_1}, \ldots, \overset{\bullet}{f_{d-k}}, e_1, \ldots, e_k] \overset{\bullet}{f_{d-k+1}} \cdots \overset{\bullet}{f_\ell})$$

$$= \pm(\pm[e_1, \ldots, e_k, \overset{\bullet}{f_1}, \ldots, \overset{\bullet}{f_{d-k}}] \overset{\bullet}{f_{d-k+1}} \cdots \overset{\bullet}{f_\ell}),$$

but this sign has no bearing on the existence of a factorization, and can be reconstructed if needed.

Step 4. If such E and F do not exist, then return **no factorization possible**. If they do exist, then choose new symbols $g_1, g_2, \ldots, g_p$, where $p = k +$

$\ell - d$, and store $G = E \wedge F$. Let G replace E and F in the collection of atomic extensors. Proceed with the new polynomial

$$P = \sum_{x_1, \ldots, y_1, \ldots} \begin{matrix} g_1 & \cdots & g_p & x_1 & \cdots & x_{2d-k-\ell} \\ y_1 & \cdots & \cdots & \cdots & \cdots & y_d \\ \vdots & & & \vdots & & \vdots \end{matrix}$$

where $x_1, \ldots, y_1, \ldots$ are the same as above. Now P has one less bracket per term than previously.

Step 5. Recompute the atomic extensors by trying to extend the current ones.

Step 6. Go to Step 2. This completes the algorithm.

THEOREM 2 [24]. *If P is a multilinear homogeneous bracket polynomial, then the above algorithm finds a Cayley factorization of P if one exists.*

We now illustrate how the algorithm works.

EXAMPLE 6. Let

$$P(a, b, \ldots, k) = 2[a, b, c][d, e, \overset{\blacktriangle}{f}][\overset{\blacktriangle}{g}, h, k]$$
$$- [\overset{\bullet}{a}, c, e][\overset{\bullet}{b}, d, \overset{\blacktriangle}{f}][\overset{\blacktriangle}{g}, h, k]$$
$$+ [\overset{\bullet}{a}, c, \overset{\blacktriangle}{f}][\overset{\bullet}{b}, d, e][\overset{\blacktriangle}{g}, h, k].$$

We have chosen an example of a multilinear bracket polynomial which is already in dotted form for the extensors ab, fg, and hk. Completing Step 1, we find that each of these is maximal, but that cd is also an atomic extensor.

Letting $E = ab$, $F = cd$, and applying straightening with respect to alphabetical order, we find

$$P = [a, b, \overset{\blacksquare}{c}][\overset{\blacksquare}{d}, e, \overset{\blacktriangle}{f}][\overset{\blacktriangle}{g}, h, k].$$

Thus we have a single dotted tableau of the form required for Step 3, as in $(*)$,

$$P = \begin{matrix} a & b & \overset{\blacksquare}{c} \\ \overset{\blacksquare}{d} & e & \overset{\blacktriangle}{f} \\ \overset{\blacktriangle}{g} & h & k \end{matrix} \quad .$$

Letting $W = E \wedge F$, $W = w$ is an extensor of Step 1, and

$$P = \begin{matrix} w & e & \overset{\blacktriangle}{f} \\ \overset{\blacktriangle}{g} & h & k \end{matrix} \quad .$$

Now we recompute atomic extensors, obtaining we, fg, and hk, and we see that $we \wedge fg$ is already a factor. Letting $U = u = we \wedge fg$, $P = uhk$, and by Step 2, we are done, as uhk is now an atomic extensor. Reconstructing,

$$P = uhk = u \wedge hk = (we \wedge fg) \wedge hk = (((ab \wedge cd) \vee e) \wedge fg) \wedge hk,$$

up to a possible minus sign.

We note that P is zero when the incidences of Figure 5 hold.

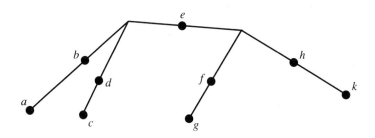

FIGURE 5. The geometric condition equivalent to P.

Clearly, several other simple Cayley expressions are equal to $\pm P$, for example, $(cd \wedge ((fg \wedge hk) \vee e)) \wedge ab$.

Implementation

The Cayley factorization algorithm has been implemented in FORTRAN on an Apollo 3000. It was successfully tested on a number of inputs, both factorable and nonfactorable. For fewer than 15 points it runs very quickly. However, 15 or 16 points is approximately the limit that this implementation can handle. The program is limited to 1500 tableaux during each call of the straightening algorithm, because of memory limitations of the Apollo, and this stack size has been exceeded for some inputs of 15 and of 16 points, after running for several hours, with most of the time spent in Step 3. FORTRAN source code may be obtained from the author.

Questions

Some questions remain. The first is to find an algorithm which can handle nonmultilinear polynomials. This becomes much more complicated and a straightforward modification of the multilinear algorithm will not suffice.

The second concerns the fact that examples are known of non-Cayley-factorable bracket polynomials, which become Cayley factorable when multiplied by a single bracket. Sturmfels and Whiteley [21] show that this always happens if we multiply by an appropriate product of brackets of high degree. Such extra bracket factors may correspond to the nondegeneracy factors that occur in Wu's method (see [5]). How can such factors of small degree be found practically?

REFERENCES

1. M. Barnabei, A. Brini, and G.-C. Rota, *On the exterior calculus of invariant theory*, J. Algebra **96** (1985), 120–160.

2. J. Bokowski and J. Richter, *On the finding of final polynomials*, European J. Combin. **11** (1990), 21–34.

3. J. Bokowski and B. Sturmfels, *Polytopal and nonpolytopal spheres: an algorithmic approach*, Israel J. Math. **57** (1987), 257–271.

4. N. Bourbaki, *Eléments de mathématique, algèbra*, Chapter 3, Diffusion, C.C.L.S., Paris, 1970.

5. S. Chou, W. F. Schelter, and J. Yang, *Characteristic sets and Gröbner bases in geometry theorem proving*, Computer-Aided Geometric Reasoning (H. Crapo, ed.), INRIA Rocquencourt, France, 1987, pp. 29–56.

6. H. Crapo, *Invariant-theoretic methods in scene analysis and structural mechanics*, J. Symbolic Comput. **11** (1991), pp. 523–548.

7. P. Doubilet, G.-C. Rota, and J. Stein, *On the foundations of combinatorial theory*. IX, *Combinatorial methods in invariant theory*, Stud. Appl. Math. **57** (1974), 185–216.

8. W. H. Greub, *Linear algebra*, Grundlehren Math. Wiss., vol. 97, 2nd ed., Springer-Verlag, Berlin and Academic Press, New York, 1963.

9. ____, *Multilinear algebra*, Grundlehren Math. Wiss., vol. 136, Springer-Verlag, Berlin, 1967.

10. F. Grosshans, G.-C. Rota, and J. Stein, *Invariant theory and superalgebras*, CBMS Regional Conf. Ser. in Math., no. 69, Amer. Math. Soc., Providence, RI, 1987.

11. T. Havel, *Some examples of the use of distances as coordinates for Euclidean geometry*, J. Symbolic Comput. **11** (1991), pp. 579–594.

12. W. V. D. Hodge and D. Pedoe, *Methods of algebraic geometry*, vols. 1 and 2, Cambridge Univ. Press, London, 1946.

13. B. Kutzler and S. Stifter, *On the application of Buchberger's algorithm to automated geometry theorem proving*, J. Symbolic Comput. **2** (1986), 389–398.

14. M. Marcus, *Finite dimensional multilinear algebra*. I, Dekker, New York, 1973.

15. ____, *Finite dimensional multilinear algebra*. II, Dekker, New York, 1975.

16. T. McMillan, *Invariants of antisymmetric tensors*, Ph.D. thesis, University of Florida, Gainesville, FL, 1990.

17. C. Procesi, *A primer in invariant theory*, Brandeis Lecture Notes **1** (1982).

18. G.-C. Rota and J. Stein, *Applications of Cayley algebras*, Accademia Nazionale dei Lincei atti dei Convegni Lincei 17, Colloquio Internazionale sulle Teorie Combinatoire, Tomo 2, Roma, 1976.

19. ____, *Symbolic method in invariant theory*, Proc. Nat. Acad. Sci. U.S.A., Mathematics **83** (1986), 844–847.

20. B. Sturmfels and N. White, *Gröbner bases and invariant theory*, Adv. in Math. **76** (1989), 245–259.

21. B. Sturmfels and W. Whiteley, *On the synthetic factorization of projectively invariant polynomials*, J. Symbolic Comput. **11** (1991), pp. 439–454.

22. H. W. Turnbull, *The theory of determinants, matrices, and invariants*, Blackie and Son, London, 1928.

23. N. White, *The bracket ring of a combinatorial geometry*. I, Trans. Amer. Math. Soc. **202** (1975), 79–95.

24. ____, *Multilinear Cayley factorization*, J. Symbolic Comput. **11** (1991), 421–438.

25. N. White and T. McMillan, *Cayley factorization*, Preprint no. 371, Institute for Mathematics and Its Applications, Minneapolis, Minnesota, 1987.

26. N. White and W. Whiteley, *The algebraic geometry of stresses in frameworks*, SIAM J. Algebraic Discrete Methods **4** (1983), 481–511.

27. N. White and W. Whiteley, *The algebraic geometry of motions of bar-and-body frameworks*, SIAM J. Algebraic Discrete Methods **8** (1987), 1–32.

28. W. Whiteley, *The geometry of bivariate S_2^1-splines*, preprint, 1991.

29. A. Young, *On quantitative substitutional analysis* (3rd paper), Proc. London Math Soc. (2) **28** (1928), 255–292.

DEPARTMENT OF MATHEMATICS, UNIVERSITY OF FLORIDA, GAINESVILLE, FLORIDA 32611